Executing Design for Reliability within the Product Life Cycle

Executing Design for Reliability within the Product Life Cycle

Ali Jamnia and Khaled Atua

CRC Press
Taylor & Francis Group
Boca Raton London New York

CRC Press is an imprint of the
Taylor & Francis Group, an **informa** business

First published in paperback 2024

CRC Press
2385 NW Executive Center Drive, Suite 320, Boca Raton FL 33431

and by CRC Press
4 Park Square, Milton Park, Abingdon, Oxon, OX14 4RN

CRC Press is an imprint of Taylor & Francis Group, LLC

© 2020, 2024 by Taylor & Francis Group, LLC

ISBN: 978-0-8153-4897-9 (hbk)
ISBN: 978-1-03-283883-0 (pbk)
ISBN: 978-1-351-16572-3 (ebk)

DOI: 10.1201/9781351165723

Visit the Taylor & Francis Web site at
http://www.taylorandfrancis.com

and the CRC Press Web site at
http://www.crcpress.com

I would like to dedicate this book to Norah, my daughter and my best friend, and the memory of my late father, the teacher.

Khaled Atua

In turn, I too would like to dedicate this work to Naseem, my daughter, the apple of my eye, and to the memory of my late father.

Ali Jamnia

Contents

List of Figures xv
Preface...xxi
Authors xxiii

1. **Reliability, Risk, and Their Significance**..1
   Introduction..1
   A History of Reliability and Background on Design for Reliability (DfR) 2
   Reliability Metrics...4
   Cost and Return of Reliability 5
   Quality, Reliability, and Robustness 7
   Organizational DfR Maturity and Reliability Competency..................................8
       A Cross-Functional Process..9
   Product Risk 13
   Notes..14

2. **Design for Reliability Process** 17
   Introduction..17
   Reliability Requirements, Planning, and Execution in the Design Process.................18
   Elements of Design for Reliability..18
       Product Use Profile 19
       Life Expectancy, Design Life, and Failure Rate 20
       Product Reliability Prediction and Modeling...20
           Reliability Modeling 20
           Reliability Testing 21
           Reliability Data Analysis...22
           Reliability Allocation 23
   Other Aspects of Reliability 23
       Reliability and Its Association with Product Risk 24
   Note..25

3. **The Design Process and the V-Model** 27
   Introduction..27
   Product Development Models 27
       The V-Model for Life Cycle Management...29
       Engineering Activities...31
   Systems and Requirements 32
       Voice of the Customer and Voice of Stakeholders 34
       Design Documents..35
           Subsystem Design Document...38
           Subsystem Architecture Document 38
           Assembly Design Document 38
   Developing Product Requirements 39
       Defining the Product to Be Developed 39

Measuring What Customers Need 40
Development of Product Requirements 41
     Translation from Needs to Requirements 42
     Product Requirements Document 42
Quality Function Deployment 43
     Using QFD in Product Development 44
     Requirements Decomposition 45
The V-Model in a Nutshell 47
Reliability Planning and Execution in the V-Model 47
Notes 50

4. Reliability Requirements, Modeling, and Allocation 51
Introduction 51
Developing Reliability Requirements 52
     Reliability Metrics 52
          Mean Time between Failures 52
          Mean Time to Failure (MTTF) 53
          Failure Rate 53
          Reliability and Life 54
          Availability 54
          Mean Cumulative Function 54
          Probability of Success Rate and Out-of-Box Failures 55
     Required Inputs for Developing Requirements 55
          Reliability Requirements Based on Voice of Business 56
          Reliability Requirements Based on Voice of Customer 57
          Reliability Requirements Based on Use Profile 57
               Product Mission 57
               Product Use Distribution 58
               Use and Operating Conditions 58
     Reliability Modeling and Allocation 58
          Reliability Modeling 59
               Example 61
          Reliability Requirement Allocation 63
               Example 64
               Developing the Apportionment Table 65
               Reliability Requirements Feasibility and Gap Analysis 66
               Case Study 66
     Other Considerations in Reliability Requirements Development 71
     A Summary of Criteria for Reliability Requirements 72
Notes 72

5. Reliability Planning 73
Introduction 73
Design for Reliability Plan Process 76
     Business Plan 76
     Design and Development Plan 76
     Needed Inputs 78
          Functional Requirements Document and System Architecture 78
          Use Profile Document 80

Detailed Design Documents..81
Failure Modes and Effects Analysis Documents..................................82
Test Articles...82
Design for Assembly (DfA) and Design for Serviceability (DfS).........82
Design for Reliability Plan...83
Baseline Reliability Model 83
Design Documents...83
Reliability Growth, Evaluation, and Demonstration Testing 83
Final Reliability Model 84
Highly Accelerated Stress Screening 84
Design for Reliability Deliverables..84
Notes...85

6. **Reliability Statistics**...87
Introduction...87
Basic Definitions 88
Continuous Distributions and Reliability Analysis 90
Normal Distribution Model...91
Example 92
Exponential Distribution Model..92
Example 93
Example 94
Gamma Distribution 94
Application to Cumulative Damage..95
Case Study: Pressure Sensor First Failure after Partial Physical Damages........95
Chi-Square Distribution..97
Failure-Truncated Test 98
Time-Truncated Test...98
Example 98
Lognormal Distribution...99
Weibull Distribution...100
A Modified Weibull Distribution...101
Example 102
Selecting the Right Distribution for Continuous Variables 102
The Exponential Distribution..103
The Lognormal Distribution 103
The Weibull Distribution 104
Discrete Distributions and Their Applications to Reliability 104
Binomial Distribution...104
Example 105
Geometric Distribution 106
Example 106
Hypergeometric Distribution..106
Example 107
Poisson Distribution 108
Example 108
Case Study: New Product Launch...109
Case Study: Expected Service Calls 111
A Discussion on Reliability Metrics and the Bathtub Curve 112
Notes...114

7. Predictive and Analytical Tools in Design 117
   Introduction.. 117
   Stress versus Strength 117
      Cascading the Use Profile into Component Specifications...................... 118
         Case Study... 118
      Uncertainty in Strength and Stress: Single-Point Solutions versus Distributions...... 120
         Example 121
         Example 121
      Interaction of Strength and Stress Distributions....................................... 123
   Safety Factor or Design Margin.. 126
      A Statistical Approach.. 126
         Example 126
      Engineering Analysis and Numerical Simulation 129
         What-If Scenarios 129
      Stress Derating 130
   Tolerance Analysis... 132
      Worst-Case Analysis.. 133
         Electrical Circuits 134
      Root Sum of Squares Method... 135
         Sensitivity Analysis 136
         Example 136
      Monte Carlo Analysis.. 137
      Unintended Consequences.. 139
      Functional Tolerance Concerns.. 141
   Design of Experiments... 141
      Case Study... 142
         DoE Test Design... 143
         Test Runs and Output Signal Strength 144
   Physics of Failure 147
      Failure Classifications... 147
         Reversible Failures 148
         Irreversible Failures 148
         Sudden Failures 149
         Progressive Failures 149
         Chemical Failures.. 149
         Failure Modes and Mechanisms... 150
   Failure Modes and Effects Analysis.. 151
      Example 152
   Life-Expectancy Calculations 156
      Life Expectancy for Pure Fatigue Conditions.. 156
         Example 156
      Life Expectancy for Random Vibration Conditions............................... 158
         Example 158
      Life Expectancy for Pure Creep Conditions 159
      Life Expectancy for Creep–Fatigue Interactions.................................... 159
         Example 160
         Example 161
      Design Life, Reliability, and Failure Rate 162
   Device Failure Rate Prediction 162

Failure Prediction Using Databases 163
   Temperature Effects 163
   Electrical Stress Effects... 163
   Environmental Factors 164
   Calculating System Failure Rate 164
   Case Study.. 165
Failure Rates Based on Physics of Failure 168
   Case Study.. 169
Notes 172

8. Component and Subsystem Reliability Testing 175
Introduction.. 175
Robustness versus Reliability Testing 175
The What and How of Reliability Testing.. 177
   Types of Reliability Testing 178
      Reliability Design Margin Development and Characterization 179
      Reliability Demonstration Test.. 179
      Production Reliability Stress Screening... 179
Highly Accelerated Limit Testing.. 180
   HALT Chamber and Test Setup.. 182
      Examples... 186
      HALT and Realistic Failures.. 186
Reliability Demonstration Testing 188
   Accelerated Life Testing 189
      Known Stress–Life Relationships 190
      Unknown Stress–Life Relationships 194
   Reliability Test Duration and Sample Size 198
      Single-Use or Nonrepairable Components...................................... 199
      Success-Run Reliability Test Duration and Sample Size 201
      Limited Sample Availability.. 209
      Cumulative Damage Reliability Testing... 210
   Test Duration with Anticipated Failures 218
      Example: Test Design with Anticipated Failures........................... 218
Degradation Testing 219
   Case Study: Lead–Acid Battery Remaining Capacity 220
Notes 222

9. System Reliability Testing.. 223
Introduction.. 223
The What and How of System-Level Reliability Testing 224
   Parameter Diagram... 224
      Keep Records 225
      Develop Measurement Baselines 227
      Test Parameter Settings 227
      Test Data Generation and Outputs 227
      Automation of Functions in the Test 227
   Failure Modes and Effects Analysis... 228
   Duane Model 229
      Duane Model Example 231

Crow–AMSAA NHPP Model 233
  Case Study: Crow–AMSAA NHPP 234
Grouped Data Reliability Growth Model 236
  Case Study: Grouped Data Reliability Growth Model 237
Hardware and Software Reliability Growth of an Integrated System 244
  Example 244
System Reliability Demonstration Testing 245
Probability Ratio Sequential Testing 247
  Case Study: PRST 248
  A χ^2 Approach 250
Time-Truncated MTBF Demonstration Testing 252
  Example 252
General Considerations in System Reliability and Life Testing 255
Accounting for Manufacturing Defects 255
Sample Size and Test Duration 255
System Demonstration Test Plan: "Zero" versus "r" Failures 256
  Example 257
Sample Size Calculations for Service Life System Demonstration Testing 259
Repairable Systems Reliability over Service Life 259
  Example 260
System Accelerated Life Testing 260
Notes 263

10. Reliability Outputs 265
Introduction 265
Reliability Outputs 267
Baseline Model and Its Application 267
Reliability Model Case Study: Baseline Model 267
Design-Related Analyses 269
Reliability Testing Output 269
  Reliability Model Case Study: Test Data 269
Final Reliability Model 270
  Bayes' Theorem 270
  Reliability Model Case Study: Final Reliability Model 271
Production Screening Testing 273
Preventive Maintenance 275
  Case Study 276
Predictive Maintenance 280
Notes 280

11. Sustaining Product Reliability 281
Introduction 281
Field-Related Sources of Noise 282
Reliability Model and Field Data Misalignment 283
Mismatch of Field Data and Selected Reliability Model 285
  Example 287
  Other Considerations When Fitting a Distribution to Field Data 288
Use Profile Discrepancies 288
Manufacturing Quality Variations 290
Mixed-Failure Mechanisms 293

Field Failure Investigation Process..295
  Define..296
    Failure Observation and Determination...296
    Failure Isolation and Scope Determination..297
  Measure...298
    Failure Verification..298
  Analyze..299
    Failure Analysis and Root Cause Determination......................................299
    Root Causes...300
  Improve..302
    Corrective Actions and Verification 302
  Control...303
    Effectiveness and Control 303
  Keys to Successful Investigations 303
  Corrective Action and Preventive Action (CAPA) Core Team 304
Warranty and Service Plan Data Review 305
Notes...306

12. **A Primer on Product Risk Management**..307
Introduction...307
A Risk Management Tool 307
  Failure Modes and Effects Analysis..308
  Hazard, Hazardous Situation, and Harm 311
    Risk Assessment Code Table 313
  Controlling Critical or Safety-Related Items 317
Other Risk Management Tools 317
  Fault Tree Analysis 318
    Case Study..319
    Fault Tree Diagram Symbols 320
    FTA Process..322
Notes...323

13. **Relating Product Reliability to Risk** 325
Introduction...325
Reliability, Risk, and Safety 325
Relating Reliability to Risk...326
  Risk Analysis 327
  Reliability Concerns Leading to Hazards 329
  FMECA and Criticality Index 330
    Case Study...330
Fault Tree Analysis 333
  Boolean Operation 333
  FTA P_1 Calculations 334
Notes...336

Appendix 339

References 341

Index 347

List of Figures

Figure 1.1 Return on investment and cost of unreliability in the product life cycle 6

Figure 1.2 Reliability–cost trade-off 7

Figure 1.3 Reliability–quality–robustness relationship 8

Figure 1.4 Design for reliability organization maturity 9

Figure 3.1 Typical stages of a product's life cycle 28

Figure 3.2 The steps and stages of the V-model 29

Figure 3.3 A requirements-based electromechanical engineering design process flow 30

Figure 3.4 A pictorial representation of system of systems 33

Figure 3.5 A pictorial representation of system of systems and its lower-level elements 34

Figure 3.6 Requirements and architecture hierarchy 36

Figure 3.7 The primary elements of the quality function deployment 43

Figure 3.8 The secondary elements of the quality function deployment 44

Figure 3.9 Inputs and outputs of mechanical/electromechanical requirements decomposition 45

Figure 3.10 Inputs and outputs of electronics requirements decomposition 46

Figure 3.11 A typical V-model for product development along with the roadmap. The highlighted area indicates the segments that have already been discussed 48

Figure 3.12 Design for reliability and the V-model 49

Figure 4.1 Elements of reliability block diagrams 59

Figure 4.2 An example of developing a system reliability model: a tank filling system with no redundancy 61

Figure 4.3 An example of developing a system reliability model: a tank filling system with redundancy 62

Figure 4.4 An example of developing a system reliability model reducing component reliability into a single system value 62

Figure 4.5 Composition of failure rate over product service life for a new product in the field 71

Figure 5.1 Steps in developing a model including its production build and verification 74

Figure 5.2 The interaction of the cross-functional team and its influence on design. The impact and influence of design for reliability (DfR) is highlighted...75

Figure 5.3 A visual overview of the design for reliability process 77

Figure 5.4 An example of a system view of an electromechanical product 79

Figure 5.5 An example of a system architecture of an electromechanical product.........80

Figure 5.6 An example of a use profile distribution...81

Figure 6.1 Probability density function distribution of time to failure.............................88

Figure 6.2 Cumulative distribution function of time to failure.............................88

Figure 6.3 Survival distribution function of time, t 89

Figure 6.4 The progression from a histogram of times-to-failure data to a mathematical representation model...90

Figure 6.5 Pressure sensor structure diagram 96

Figure 6.6 Hazard rate versus operating time. The influence of preventive maintenance has been shown 113

Figure 6.7 Average failure rate for population mix and MTBF for a product 114

Figure 6.8 Average failure rate of fielded products experienced by a manufacturer during various stages of the product life.................................114

Figure 7.1 Impact of input variations and distributions on expected load-carrying capacity of a column 122

Figure 7.2 An injection-molded enclosure with snap-fit features 122

Figure 7.3 Impact of dimensional variation and distributions on part stresses 124

Figure 7.4 The relationship between stress and strength and their potential overlap...125

Figure 7.5 Depiction of the free body diagram for a cantilever beam.............................127

Figure 7.6 Numerical simulations are used to set design and/or configuration decisions 130

Figure 7.7 Numerical simulation of a magnetic field to develop the transfer function...130

Figure 7.8 Comparison of outcome under two different environmental conditions 131

Figure 7.9 An assembly block for tolerance calculation demonstration.........................133

Figure 7.10 An RCL circuit for tolerance calculation demonstration.............................134

Figure 7.11 Assembly block example for tolerance calculations.......................................136

Figure 7.12 An example to illustrate various tolerance analysis techniques..................139

Figure 7.13 Monte Carlo analysis results for the gap size in Figure 7.11. This analysis indicates an interference of 0.04 to a maximum gap of 0.14 may exist..140

Figure 7.14 An often-overlooked issue: deformation of housing due to tightening of the nut on the right-hand side 141

Figure 7.15 A simplified configuration of an air-in-line sensor........................142

Figure 7.16 Air-in-line sensor design critical parameters and DOE factors 143

Figure 7.17 An example of PCBA chatter and its response 148

Figure 7.18 Fatigue S–N curve for 300M alloy 157

Figure 7.19 A distribution of Minor's index (*R*) values. Failure occurs when *R* reaches 1. Therefore, a calculated R value greater than 1.0 indicates that the component is already in a failed state 172

Figure 7.20 Failure rate of the supercapacitor and its growth in the 10-year life...........172

Figure 8.1 Steps of the DfR within the V-model 176

Figure 8.2 Relationship of design capability, robustness, and reliability in terms of stress, strength, and time 177

Figure 8.3 A general overview of a parameter diagram.....................................178

Figure 8.4 Operating and destruct limits defined from highly accelerated limit test 180

Figure 8.5 (A) Using HALT in the design process or (B) in comparing product configurations from, say, different manufacturers...........................182

Figure 8.6 A typical setup of a HALT chamber 183

Figure 8.7 An example of printed circuit board assembly setup in HALT chamber along with the data logging equipment for input and output signals 183

Figure 8.8 HALT typical temperature (hot/cold) and vibration stress profiles 184

Figure 8.9 Missing components at thermal cycling or vibration test during HALT 186

Figure 8.10 Crossover effect of instantaneous failure rate observed in HALT and in the field..187

Figure 8.11 Relationship of design specification limits, operating limits, and selected elevated stress for ALT..189

Figure 8.12 Selection of elevated stress levels for multilevel ALT...................194

Figure 8.13 Graphical data analysis and extrapolation of probability of failure at normal-use stress level using Weibull plot..................................195

Figure 8.14 A linear pneumatic pump with a reciprocating piston expected to wear out the Teflon lining...196

Figure 8.15 Weibull plot of the time to failure and the probability of failure at three elevated temperature levels.. 197

Figure 8.16 Graphical analysis at three different temperature levels and the projection of use temperature 198

Figure 8.17 Ratio of test duration to design life (L_D) for different failure trends to demonstrate reliability... 202

Figure 8.18 Probability of failure plot for the current design prior to supplier improvements. The current design has a 2% probability of failure with 95% confidence at 50 cycles 206

Figure 8.19 A depiction of an engine control module.. 212

Figure 8.20 Battery remaining capacity (RC) degradation and projection over testing cycles... 221

Figure 9.1 A depiction of the main elements of a parameter (P-) diagram 225

Figure 9.2 An example P-diagram for the operation and errors of an electromechanical system... 226

Figure 9.3 Reliability (MTBF) growth and projection of goal achievement using reliability growth analysis.. 241

Figure 9.4 Difference in projection of the time to achieve reliability target after each testing phase 243

Figure 9.5 Repairable device reliability growth across different builds and test case coverage 247

Figure 9.6 PRST test plan, accept and reject boundaries, and actual test results and decision.. 250

Figure 9.7 PRST test plan with truncation boundaries for testing................................... 251

Figure 9.8 Different decisions for different discrimination ratios 252

Figure 10.1 DfR process with typical outputs deliverables including HASS................. 266

Figure 10.2 System baseline reliability model apportionment 268

Figure 10.3 Prior gamma distribution process with test data using Bayes' theorem... 271

Figure 10.4 Final reliability model using prior distribute (baseline model) and test data... 273

Figure 10.5 HASS test setup as an output of HALT results and findings....................... 274

Figure 10.6 Reliability–cost trade-off analysis for different PM intervals...................... 277

Figure 10.7 Optimum PM interval for the water pump based on reliability– cost-of-service trade-off... 278

Figure 11.1 Fielded product reliability monitoring: MTBF and top 10 failures Pareto 284

Figure 11.2 Electronic PCBA as SRUs in an avionics product 285

Figure 11.3 Fitting time to failure histogram to a continuous distribution model (for field data of 120 failures)..286

Figure 11.4 Critical values of correlation coefficient for reliability data. (Adopted from 104C.D. Tarum, Determination of the critical correlation coefficient to establish a good fit for Weibull and log-normal failure distribution, SAE Paper 1999-01–057, Detroit, March 1999.) 287

Figure 11.5 Typical field data for motor failures with different duty cycles/usage 289

Figure 11.6 Typical distribution of different duty cycles/usage for fielded product 289

Figure 11.7 Probability of failure analysis for motor failures with different duty cycles/usage..290

Figure 11.8 Typical field data for motor bearing failures with bad batch......................291

Figure 11.9 Probability of failure plots after separating the bad batch data set............292

Figure 11.10 Probability of failure of the compressor using three-parameter Weibull..292

Figure 11.11 Typical field data for mixed failure mechanisms of a compressor............293

Figure 11.12 Outer wall damage due to material nonhomogeneities in manufacturing..294

Figure 11.13 Separate analysis for field data for mixed-failure mechanisms................294

Figure 11.14 Early life failure probability distribution analysis using lognormal distribution..295

Figure 11.15 Reliability-based investigation process flowchart..297

Figure 11.16 Electronic subsystem MTBF and threshold triggers based on field data 298

Figure 11.17 Good and returned sample power cord from the field 299

Figure 11.18 Failure reporting and the role of the core team..305

Figure 12.1 A block diagram for a fault tree analysis example 319

Figure 12.2 The top events in a fault tree analysis example 319

Figure 12.3 The top two event layers in a fault tree analysis example...........................320

Figure 12.4 The fault tree analysis of car accident example. Note that the details of the engine and break system have been ignored in this case study.......321

Figure 12.5 Symbols in a fault tree analysis 321

Figure 13.1 A basic flow diagram for risk analysis...327

Figure 13.2 A flow diagram relating reliability analysis to risk...329

Figure 13.3 An example of AND operation calculation 334

Figure 13.4 An example of OR operation calculation 335

Figure 13.5 An example of P_1 calculation within a fault tree...335

Figure 13.6 An example of P_1 calculation for an entire fault tree......................................336

Preface

The Need for This Book

At an early stage of development, design teams should ask questions, such as: *How reliable will my product be? How reliable should my product be?* and *How frequently does the product need to be repaired/maintained?* To answer these questions, design teams need to develop an understanding of how and why their products fail. They should develop a means and understanding to predict these failures and their future frequencies. The set of activities used to accomplish this task is called *design for reliability (DfR)*. It involves mathematical modeling, laboratory testing, and even possibly analysis of field failures.

Basic reliability calculations are simple but they can be quite sophisticated if statistical theories are employed. A rudimentary review of the literature reveals a relatively large number of good resources for the detailed study of this topic. *Reliability Theory and Practice* (Bazovsky 2013); *Statistical Reliability Engineering* (Gnedenko et al. 1999); *System Reliability Theory, Models, Statistical Methods and Applications*, 2nd ed. (Rausand and Hoyland 2004); *Probability Statistics and Reliability for Engineers and Scientists*, 3rd ed. (Ayyub and McCuen 2011); and *Reliability Engineering* (Elsayed 2012) may be enumerated among others. These series of books explore the "theory" of reliability and its associated calculations. For the uninitiated, the task of managing these equations—outside of an academic environment—can become quite daunting and quite impractical within the fast pace of the product development arena.

Others have tried to provide a solution to this dilemma by offering practical approaches. O'Connor and Kleyner's *Practical Reliability Engineering*, 5th ed. (2012) may be considered the pre-eminent resource. Others include *AT&T Reliability Manual* (Klinger et al. 1990), *Practical Reliability of Electronic Equipment and Products* (Hnatek 2003), *Reliability Improvement with Design of Experiments* (Condra 2001), and MIL-HDBK-338B (1988). These resources enable a practicing engineer to develop an appreciation for conducting reliability analyses of test or field data—provided the data are well behaved.

There is still a third category of books that are not replete with theories or equations. The authors of these books are more concerned with establishing and managing reliability activities within an organization as opposed to applying reliability principles to a given product or concept. Among these, the out-of-print *Reliability Assurance for Medical Devices, Equipment and Software* by Richard S. Fries (1990) and *Medical Device Reliability and Associated Areas* by B.S. Dhillon (2000) may be named.

The problem still remains that when design engineers are faced with designing for reliability, they are often at a loss on where to begin. What is missing in the reliability literature is a set of practical steps for designing for reliability without the need to turn to someone with heavy statistics background—typically within a quality or statistics department. This book addresses the question of reliability from a design process point of view and this what we believe every engineer should know. In fact, we are presenting design for reliability within a product development life cycle model (PDLM).

Within the context of PDLM, we first begin with the DfR process, including establishing a link between reliability requirements and risk associated with failure consequence. We treat the methodology of this topic in detail along with needed calculations. In fact, to make sure that we are all on the same page, we provide a rather detailed treatment of the V-model to base the foundations of product development and the placement of reliability within the model.

In the second step, we discuss—in some detail—reliability requirements and their allocations to subsystems and components. This may not be done without conducting reliability feasibility and analysis methods to ensure the suitability of the requirements and their allocation. Reliability feasibility analysis helps the design team identify gaps in reliability, define the strategy, and outline the reliability plan. We have given a chapter to review the elements of a design for a reliability plan, and its inputs and expected outputs.

With a new product development (NPD) process, once requirements are defined to some extent, the detailed design activities begin in earnest. In line with this set of activities, we present and emphasize design activities or calculations to ensure a reliable design. These analytical and often predictive methods are great tools that enable design engineers to examine the corner cases of their design and not just be content with the nominal design. In the current market-driven fast NPD cycles, we can no longer depend on the design–test–fail–redesign mindset of the past. Design engineers need predictive tools to examine not only the nominal design but also the design when components and conditions are outside expectations.

After a brief digression to provide a background in reliability statistics, we explain in detail various accelerated life testing calculations and results analyses, especially with real-life situations when data are either incomplete or unavailable. We also explain how to optimize design of reliability demonstration testing over the service life or design life of systems. Test design optimization includes calculating sample size and defining proper accelerants and stresses to be used to achieve life testing in a short period of time. It also includes the impact of these accelerants on different modules and system architecture elements such as electronics, pneumatics, and moving parts.

The book also covers and presents a simple technique to blend theoretical and predicted reliability with the actual test or field data to establish a realistic reliability model using the Bayesian method. Some special topics related to reliability are addressed in this book. This includes simple approaches to project reliability, measuring system reliability growth, and optimizing reliability testing prelaunch. Other special topics addressed in this book are a rationale and simple calculations of the impact on reliability when designing optimum preventive maintenance, its impact on reliability, and the reliability–cost trade-off of preventive maintenance.

Finally, we cannot speak of reliability and probability of failure without considering the risk and impact of these failures on either end users or the mission of the product. Risk and reliability are interconnected and we have provided a primer on this interconnectedness.

Authors

Dr. Ali Jamnia is a systems level, seasoned mechanical engineer with over 25 years of innovative product development from concept to market as well as product support experience. He has a working knowledge of Design for Six Sigma (DFSS) and DMAIC methods of problem solving. Additionally, he has a deep understanding of reliability issues and implementing operational mechanisms to monitor and track reliability metrics related to fielded products. He has expertise in both analytical product design as well as failure analysis. His experience in the innovation domain includes not only "inventing" but also, and possibly more important, applying innovative thinking combined with state-of-the-art manufacturing methods (such as MIM) to replace less efficient traditional methods to reduce piece part cost and eliminate product rework. He holds 9 issued patents and 12 patent publications. He has authored more than 20 papers and 3 books including *Introduction to Product Design and Development for Engineers.*

Dr. Khaled Atua is a reliability expert with experience in electromechanical devices and in ship design and construction. His experience covers areas such as medical devices, ship structural design, and mechanical engineering. He received his PhD from the University of Maryland, College Park, in 1998, where his primary area of research was reliability-based structural design of ships. Dr. Atua has published 22 technical papers and presentations, and his work has been cited and referenced in *Ship Design and Construction* by the Society of Naval Architects and Marine Engineers (SNAME). He has been a lecturer at Alexandria University (Egypt), The American University in Cairo, and the Arab Maritime & Transport Academy in Alexandria. Dr. Atua has worked as the reliability manager and head of reliability for multiple US companies and corporations, including Hospira Inc., Baxter International, and Beckman Coulter Inc.

1

Reliability, Risk, and Their Significance

Introduction

The classical textbook definition of reliability as defined in many statistics and reliability engineering books is as follows: the probability that a product will perform its intended function for a specified time under specified conditions. However, given this definition, the interpretation of reliability varies based on the interest of the stakeholder. For instance, if you ask the end user or the buyer what the reliability characteristics of a purchased product should be, the answer may be one or all of the following:

- It never fails.
- No surprises, no unscheduled downtime.
- Get me up and running quickly after failures occur.
- Highest productivity and throughputs.
- Available whenever needed.
- Low cost of ownership.

For a manufacturer or a business, reliability is usually understood in terms of the following elements:

- Low cost of product surveillance, failure investigation, and field corrective action
- Low cost of service, including warranty, inventory for repair, and spare parts
- Low cost of liability, including regulatory or noncompliance, and low legal cost
- Maximized profit, including expedited product launch with low risk, increased customer loyalty, and elimination or rescue of lost sales

On the one hand, it is the responsibility of the reliability engineering team to examine the reliability of the product and provide feedback to the design team. In this effort, both the end user's and the business' concerns are addressed. On the other hand, it is the responsibility of the design team to define reliability requirements for the product. These requirements should fulfill both user and business concerns, i.e., end user expectation of failure-free life and longevity, and business needs in terms of low cost.

A History of Reliability and Background on Design for Reliability (DfR)

We find it interesting to consider the etymology of the word *reliability* as we delve into developing an understanding of how to develop and execute a plan to make products more reliable. The word *reliable* is a combination of *rely* and *able*, meaning the capacity on which one may depend on. Interestingly, reliable is based on the 1560s Scottish word *reliabill*, which was not commonly used prior to the 1800s.[1] The word *reliability* was (is) synonymous with dependability and consistency, meanings that were utilized in the fields of psychology and statistics when references were made to psychological tests and measurements (McLinn 2010). The use of the word *reliability* is quite widespread and is not limited to technical fields. As Saleh and Marais (2006) pointed out, there are over 3000 books in the Library of Congress on this subject or with this title, and a Google search on "reliability" returned over 12 million entries.

Reliability engineering owes its beginnings to Walter Shewhart (in the 1920s and 1930s) who founded statistical process control by applying statistical methods to identify and solve manufacturing problems. It is interesting to note that his initial paper was not well received by the engineering community because "laws of chance have no proper place among scientific production methods" (Freeman as quoted by Saleh and Marais, 2006). Another impetus for the field of reliability engineering was the American system of manufacturing. This system was based on the concept of mass production using interchangeable components—a process that was popularized by Henry Ford.

In the 1940s, electronic equipment played a significant role in World War II. Use of electronic devices such as radars, radios, and other electronic sensors became possible because of the vacuum tube. However, these tubes frequently failed and needed to be replaced repeatedly, almost as much as five times more than any other component (Coppola 1984).

Eventually, in 1952, the US Department of Defense created the Agree Group on Reliability of Electronic Equipment (AGREE) in collaboration with the American electronics industry. The goal of AGREE was, first, to identify and recommend improvements that would lead to more reliable products; second, to create a mechanism to influence increased reliability in both government and civilian programs; and, last, to provide training material on reliability (Coppola 1984).

In the same year, ARINC (Aeronautical Radio Inc.) published a report titled "Terms of Interest in the Study of Reliability." Recall that in the 1920s and 1930s, the engineering community believed that the concept of chance would not play a role in scientific production methods. This report turned that opinion and mindset on its head by recognizing the probabilistic nature of reliability, and hence, a departure from a definition based on a deterministic dependability and consistency. During the same period, ARINC was tasked by the Navy to study the vacuum tube field failures to identify root causes. This information was then relayed back to manufacturers of these tubes for design improvements and ultimately better reliability (Saleh and Marias 2006).

The realization of the stochastic nature of reliability was not fully based on a study of electronic equipment. In part, the works of Weibull as well as Epstein and Sobel on fatigue and cumulative damage, which would ultimately lead to defining a general distribution model as well as life testing (in 1951 and 1953, respectively), were crucial to this realization (Saleh and Marias 2006).

In a way, one may say that the birth of reliability engineering was celebrated when formal societies of reliability were created by the IEEE and the US Department of

Defense in the 1950s. *IEEE Transactions on Reliability*, founded during this period, continues to present the latest theories and discoveries in this field. The premise of reliability engineering, established in the 1950s, was based on three objectives (Saleh and Marias 2006):

1. Collect field failure data followed by root cause identification to improve reliability
2. Understand and develop specific and measurable reliability requirements
3. Develop means of predicting product reliability prior to building and testing units

Two reports were issued by the end of the 1950s . The first was the aforementioned AGREE report and the second was "Reliability Stress Analysis for Electronic Equipment," also known as TR-1100, from the Radio Corporation of America. TR-1100 was arguably the first document that provided a model for predicting the reliability of a product purely based on failure rates of its components. Thus, this document satisfied the need for developing means for predicting reliability. It was the predecessor for the military handbook MIL-HDBK-217, which was published in 1961 (Saleh and Marias 2006). The AGREE report provided evidence that developing and demonstrating specific and measurable reliability requirements is quite possible (Saleh and Marias 2006).

It is important that we recognize that one of the most important assumptions of reliability, namely, constant failure rate (or the exponential distribution), used to this date, was established in the 1950s. It was this model that enabled reliability predictions based on known component failure rates. This approach and the known component failure rates of military devices were collected in MIL-HDBK-217. Today, most engineers consider any calculated values based on this military handbook irrelevant and erroneous. Even though this may be a correct assessment, they do not realize that the handbook was primarily developed to predict reliability of military equipment. Later, Bellcore and Telecordia updated the same data for the components and products used within their own specific industries.

The 1960s brought a new degree of sophistication to the field of reliability engineering by witnessing higher degrees of specialization. One area of pursuit was to examine new statistical distribution functions to model field failures and their behaviors. At the same time, new reliability models departed from a straight part-count by developing deeper understandings of redundancies in a system as well as using prior knowledge of product behavior to predict future field failures. Alongside these theoretical pursuits, a great deal of attention was given to testing and test methodologies. As a result, different time-to-failure distributions were identified. In this era, focus began to shift from components to systems (Saleh and Marias 2006, Azarkhail and Modarres 2012).

In the following two decades (the 1970s and 1980s), two areas of reliability flourished. The first area was in accelerated life tests methods, namely, HALT and ALT. HALT stood for highly accelerated life test, and ALT stands for accelerated life test.[2] These tests moved the engineer's focus from components to assemblies and systems. The second area of reliability that became prominent was *software reliability*. It has become obvious that as systems become more sophisticated, software plays an increasingly more significant role in their reliability (Saleh and Marias 2006).

The decade that followed (the 1990s) showed a growth and interest in understanding why and how components and systems fail. Although theories associated with physics of failure had been around for a couple of decades, with the advent and availability of personal computers, solving complicated equations became accessible. On the one hand, the reliability engineer could deploy Monte Carlo techniques to evaluate variations associated with material properties, design features and operating conditions based on the product's strengths and stresses, and estimate probability of failure. On the other hand, using failure modes and effects analyses as well as HALT and ALT data along with any available field data, the reliability engineer could develop an understanding of how the product may fail and its distribution in time (Azarkhail and Modarres 2012).

Ultimately, the goal of reliability is to provide predictions about the behavior of a product in the field. However, looking back in the past and evaluating historical failures—in and of themselves—does not provide any benefit to predicting future failures. If we were to look at any of the approaches that had been proposed by this time, no one method could provide an accurate means for future reliability predictions in a general sense. The two decades of the 2000s seem to be witnessing a degree of maturation in this field as practicing engineers and researchers alike have come to this realization. The solution may be found in hybrid methods that combine data from a variety of sources including generic handbook data to information from physics of failure simulations as well as accelerated test data and/or predicate and field information (Azarkhail and Modarres 2012).

Reliability Metrics

Reliability can be measured using different metrics; depending on an organization's role in the marketplace, a different measure may be needed. The most commonly used metric is the mean time between failures (MTBF), or sometimes mean time to failure (MTTF), for components or non-repairable systems. These two metrics, which are often misunderstood, are typically cited in marketing materials as a measure of the durability of a device. For instance, if device A has a larger MTBF (or MTTF) than device B, then device A must last longer. In the course of this book, we will discuss these metrics in more detail.

Organizations—typically manufacturing—may be interested in a *failure rate per month* metric. From a service cost and warranty point of view, a manufacturing organization may be interested in the number of failures in a given month and the reasons behind those failures. Still, as consumers of products, we may have no interest in a product's failure rate; rather, we are interested in the *availability* of a product for use. For example, every time that we get into our cars, our expectation is that we can start the car and drive to our intended destination.

Other types of metrics include *mean time between swaps* (MTBS) and *mean time between service* (MTBS). Here is a situation in which an acronym may be used by two different organizations with two distinct meanings. The first is typically used by a leasing organization to measure the length of time that customers are willing to hold on to a product by renewing their lease. The second term may be indicative of an average time when a product would need to be returned to a service organization for calibration or adjustments without a need to be repaired. In the context of repairs, another metric is *mean time to repair* (MTTR), which is an indication of how long repairing a product may take.

The point that we are trying to make here is that before settling on a particular metric, it is prudent to fully explore what we intend to gain or represent by that metric.

Cost and Return of Reliability

DfR and ensuring reliability in a new design are not a matter of simply calculating a number at a certain statistical confidence level. Rather, it is an elaborate effort that covers a number of areas. It includes defining reliability requirements, developing a reliability model, planning reliability activities, and ensuring reliability and margins when selecting parts. Furthermore, it involves executing reliability testing, developing and executing plans for monitoring and tracking reliability in the field. Finally, it covers performing failure analysis, proposing design fixes, and rolling out field corrective actions as needed. DfR is a costly process, but the return of investment in reliability sometimes is justified by the savings to the manufacturer, especially in the early phases of the design process.

Significant savings can be achieved by implementing comprehensive DfR best practices by investing in the following:

1. Activities in the *concept phase*
   a. Defining user needs in terms of reliability, operating conditions, and stresses
   b. Gathering voice of customer and defining the business case in terms of maximum service cost up front
2. Activities in the *design phase*
   a. Creating reliability models, identifying critical components
   b. Allocating reliability to components based on user needs, duty cycle, and operating conditions; defining reliability thresholds for subsystems
   c. Conducting reliability analysis and feasibility, defining top failures in failure mode and effects analysis (FMEA)
   d. Defining technical specifications for parts per reliability allocation and requirements, and selecting reliable parts
3. Activities in the *verification phase*
   a. Developing and following a reliability growth plan
   b. Simulating realistic usage stress level in testing, and properly testing for reliability demonstration
   c. Reviewing supplier reliability
4. Activities in the *design transfer and production* phase
   a. Monitoring and tracking reliability by setting failure thresholds on components and subsystems
   b. Setting proper screening stresses in manufacturing processes and testing
   c. Proactively updating the service model

The sooner design flaws are found and failure mechanisms removed, the lower the overall cost would be to the business. We can associate approximate cost-saving return on investment (ROI) with each of these four stages, as illustrated in Figure 1.1.

Investing in DfR up front during the design and development process could save a business millions of dollars. For example, we have familiarity with an original equipment manufacturer (OEM) that spent in excess of $12 million on investigating and the corrective

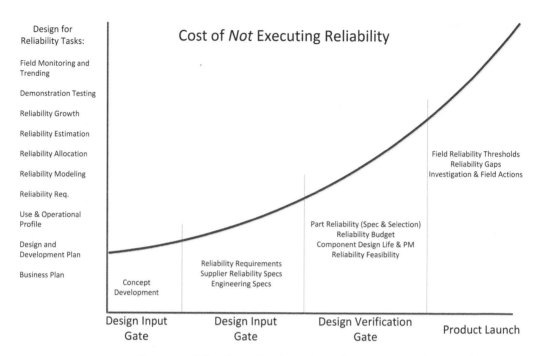

FIGURE 1.1
Return on investment and cost of unreliability in the product life cycle.

actions of a failure of a fielded device. This was due to the OEM's termination of preventive maintenance (PM) of a component to save money. The rationale was that the PM interval was too short based on the supplier's recommendations. No one realized that the supplier's data was based on immature reliability test results and a poorly designed reliability test. This decision was made without any reliability assessment of the component's failure rate trend in the field and what the failure rate would be if the PM was eliminated. Eventually, the OEM executed proper life testing of the component and the analysis of its results at a cost of $100,000. Had this activity taken place earlier, a savings of nearly $12 million would have been realized. Can we say it was an expensive but worthy lesson learned?

This story is not unique. Often, businesses make decisions to ignore reliability to save money in the short run only to pay severalfold more at a later date. There is no shortage of available stories in the news. One example is that of Philips North America that in 2019 recalled its wearable patient monitors due to their defective lithium-ion rechargeable batteries. Philips indicated that this defect may impact the operation of the monitor in its report to the US Food and Drug Administration (FDA) (www.accessdata.fda.gov/scripts/cdrh/cfd ocs/cfRES/res.cfm?id=171236). Although we are not aware of the cost of this issue borne by Philips, we can estimate it to be in several millions of dollars if not in tens of millions.

We need to keep in mind that overspending on design for reliability may create an expensive product that would not be competitive in the marketplace. An optimization of investment in reliability in the design should be the goal of the design team and aligned with the business goals for profitability and customer satisfaction trade-off, as shown in Figure 1.2.

FIGURE 1.2
Reliability–cost trade-off.

Quality, Reliability, and Robustness

There is often confusion on the differences between the reliability, robustness, and quality of a product. Earlier, we defined reliability as the probability that a product will perform its intended function for a specified time under specified conditions. Robustness is often defined as insensitivity to variation of the ability of a system or product to perform its intended functions dependably in the presence of noise factors (Alippi 2014). The definition of quality is a bit more elusive. For the students of Deming, it means to conform to requirements. More recently, Juran has defined it as fitness of purpose (Juran and De Feo 2010).

In the context of product development, there are two elements. The first is the applied loads (or the application/use of the product), and the second element is the inherent strength or capabilities of the product. Both of these elements are described as distributions because of the variations and variabilities in both the use of the product and its manufacturing. Figure 1.3 depicts a graphical relationship between capabilities and loading in light of quality, reliability, and robustness. If a design does not exhibit a proper margin between the capabilities and the loading, we expect failures to take place. Design capabilities or strengths—a subset of the design quality—are expressed in design specifications and component selections. If the distribution of design capabilities and stresses due to loading conditions are far apart, a product can support a wider range of permutation variations that exists in its components. As a result a design may be said to be robust with no expected failures. Over time, the inherent strength (or design capabilities) of a product decreases and its distribution approaches that of the applied loads. Remember that unlike robustness, reliability is defined as the ability of the design to perform the

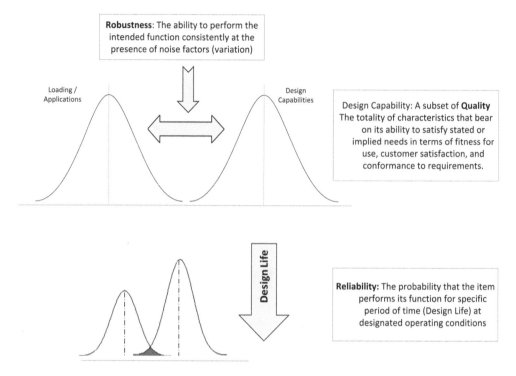

FIGURE 1.3
Reliability–quality–robustness relationship.

same function over time. Should the two overlap, we can say that a reliability failure has taken place.

In our experience, 70% to 80% of fielded products' failures are not due to reliability issues that manifest themselves as component wear out or degradations. Rather failures are often contributed by marginal designs or a lack of robustness.

Organizational DfR Maturity and Reliability Competency

Design for reliability is a systematic, streamlined, concurrent engineering approach in which reliability engineering methodologies and tools are fully integrated into the overall product development process. This effort will satisfy customer requirements for reliability and robustness throughout the design life of a product. The execution of DfR in any organization requires a maturity of both the organization's reliability process and the individual team members' engagement and interest. Needless to say, DfR maturity varies between different organizations and industries. Aerospace and automotive industries, for example, are advanced and disciplined in executing and ensuring the reliability of their products. Other industries such as consumer products may be lagging behind.

To provide the reader with a basis for comparison, we conducted an analysis of the DfR elements and the maturity levels of a typical consumer OEM to an automotive

FIGURE 1.4
Design for reliability organization maturity.

manufacturer based on an adaptation of IEEE1624-2008: Standard for Organizational Reliability Capability. This analysis is presented in Figure 1.4. As we can see in this figure, for the consumer OEM, a lack of reliability planning up front in the design and development process affects the maturity of executing DfR later in the testing and verification phase. The details of the adaptation is provided in Table 1.1.

A Cross-Functional Process

Executing DfR is a cross-functional team responsibility led by the reliability engineering department. It starts with reliability planning and model development. The members of the cross-functional teams executing the DfR activities and tasks should have the appropriate technical knowledge, training, and experience in product use, technologies, or tools as appropriate for their role. In order for an organization to grow its DfR competency, a certain skill set among reliability leaders, owners, and executers should be available in the organization. Table 1.2 lists this required reliability body of knowledge and skill sets. As shown in this table, it ranges from reliability planning and analysis to testing. These skill sets are not typically possessed by a single person or persons in an organization, rather it is the collective expertise of a number of individuals who collaborate as needed on projects.

A cross-functional team should include reliability engineering, mechanical engineering, electrical engineering, systems engineering, quality engineering, manufacturing and service engineering, software engineering, verification and validation, and failure analysis and reporting personnel.

TABLE 1.1

Organization Design for Reliability Maturity Matrix Adopted from IEEE Std.1624-2008: Standard for Organizational Reliability Capability

DfR Critical Elements	DfR Criteria	Assessment Criteria		
		Level 0: Tool or process not used or used improperly. Foundational elements not in place.	Level 1: Tool or process used properly. Foundational elements in place. Standard work in place.	Level 2: Some advanced elements in place. All foundational elements in place and improvement measurable.
		Level 0	Level 1	Level 2
Understand and quantify customer reliability requirements and operating/usage environment	Usage environment understood	The typical product usage environment is understood.	The product use and misuse environment is understood and documented. VoC standard work captures specific reliability requirements.	Customer needs and wishes are translated into product design requirements using QFD
	Reliability goals well defined	Reliability goal not defined or stated as "no worse than previous design."	Reliability goals are defined quantitatively, e.g., reliability 90% for 10,000 hours with 65% confidence.	Reliability goals are derived from customer-specified or implied customer requirements, and can be quantified relative to the competition.
Ensure reliability goals are met prior to launch	Reliability deliverables defined at toll gates	Reliability deliverables are not defined or required at every toll gate in development process.	Reliability Plan or reliability APD deliverables defined and reviewed at every Tollgate	Reliability deliverables have standard templates, work instructions, and pass criteria
	Reliability is a milestone at toll gates	Reliability is sometimes discussed at monthly project review or TG reviews.	Quantitative reliability metrics reviewed at monthly project review or PPG.	Development risk and reliability growth progress discussed at monthly project review or PPG, and followed up with actions.
Identify, quantify, and manage risk to both customer and producer	Risk utilized effectively To drive reliability	Risk assessment efforts do not drive actions that have meaningful impact or are not living documents.	System level FMEAs are conducted with cross functional participation. FMEA facilitators have been properly trained.	System, design, and process FMEAs are linked, flow down, and updated throughout development.
	Reliability testing vs. analysis and prediction	Organization typically relies on testing and not on modeling or simulation.	Modeling and simulation utilized on critical areas of design.	Modeling and simulation highly leveraged, which has led to reduced testing.

(Continued)

TABLE 1.1 (CONTINUED)

Organization Design for Reliability Maturity Matrix Adopted from IEEE Std.1624-2008: Standard for Organizational Reliability Capability

DfR Critical Elements	DfR Criteria	Assessment Criteria		
		Level 0: Tool or process not used or used improperly. Foundational elements not in place.	Level 1: Tool or process used properly. Foundational elements in place. Standard work in place.	Level 2: Some advanced elements in place. All foundational elements in place and improvement measurable.
		Level 0	Level 1	Level 2
	Design reviews focus on eliminating risk	Design reviews are not conducted or scheduled as needed.	Design reviews are conducted with a cross-functional group. Standard work requires specific reliability deliverables reviewed.	Design reviews are conduct throughout development process and actions followed up.
	DfX implemented in design	Transition from engineering to manufacturing is considered "over the wall" or with limited collaboration.	Manufacturing, service, and others involved in upfront design. Design for manufacturing/assembly standard work in place.	All individuals responsible for development and production are involved at the earliest stages of product design.
Verify design meets customer expectations prior to launch	Reliability test plans	Reliability test plans are not evident.	Reliability test plans are defined and utilized. Testing needed to achieve reliability goals are defined.	Design margin testing and user environment life testing conducted.
	Reliability testing strategy	Unknown if limited test samples used in development are sufficient. Organization believes more testing is the answer to better reliability.	Standard reliability tests verify known failures and identify unknown failures. Standard reliability testing is performed on all products.	Testing includes output from tools such as FMEA as well as warranty history.
Continuously build a knowledge database	Lessons learned implemented	A small percentage of parts and processes reused.	Engineering standards have been developed and utilized. Standard work for releasing engineering standards has been documented.	Modular designs, platforms, and preferred part reuse is standard. Engineering standards and guidelines developed and updated regularly.
	Reliability field data	Previous product history not captured or difficult to locate. Multiple storage locations.	Product failures are documented and accessible by development teams. Development process requires failure history to be reviewed prior to developing new products.	FMEA and reliability test plan templates capture learnings.

(Continued)

TABLE 1.1 (CONTINUED)

Organization Design for Reliability Maturity Matrix Adopted from IEEE Std.1624-2008: Standard for Organizational Reliability Capability

DfR Critical Elements	DfR Criteria	Assessment Criteria		
		Level 0: Tool or process not used or used improperly. Foundational elements not in place.	Level 1: Tool or process used properly. Foundational elements in place. Standard work in place.	Level 2: Some advanced elements in place. All foundational elements in place and improvement measurable.
		Level 0	Level 1	Level 2
Training and competency development	Reliability training	Informal quality or reliability training may be provided to some employees, but no Sustainable program is evident.	A list of reliability training opportunities is provided.	Reliability practitioners, design engineers, and manufacturing engineers are trained in root cause analysis and corrective action.
	Reliability training for RE	Informal quality or reliability training may be provided to some employees, but no Sustainable program is evident.	Reliability practitioners are trained in reliability concepts and statistical methods for reliability prediction and data analysis.	Reliability management is trained on how specific reliability activities can impact reliability throughout the product lifecycle.
	Reliability training For managers	Informal quality or reliability training may be provided to some employees, but no Sustainable program is evident.	Reliability training is made available to the managers and leaders.	Business managers appreciate how reliability impacts business.

TABLE 1.2

A Body of Knowledge and Skill Set for Executing and
Growing Design For Reliability and Robustness

System architecture and cascading

Functional and reliability block diagrams

Statistical probability models in reliability and maintainability

Sample size determination for test

Reliability test design and data analysis

Accelerated life testing and reliability predictions

HALT, HAST, and HASS

Hypotheses testing

Measurement system analysis (MSA)

Statistical process control (SPC)

Parameter diagram (P-diagram)

Fault tree analysis (FTA)

Design and process failure modes and effects analysis

Weibull and warranty analysis

Physics of failure

Failure analysis, and corrective and preventive actions; DMAIC
  (define, measure, analyze, improve, control)

Change control management process

Risk management models

Product Risk

Risk is the product of the probability of occurrence of an undesired event multiplied by the severity of the event's effect. Undertaking any new activity involves facing and dealing with elements and factors that are unknown at the start. These unknowns may have varying effects—some positive and some potentially negative. Risk analysis and risk management refer to a set of activities that would anticipate unexpected events and estimates their relative ability to derail the program along with their probability of occurrence. Within a product development mindset, there are two different categories of risk. One category focuses on project risks. Issues such as completing the design within the needed timelines or making sure that suppliers are aligned and can deliver ordered parts are within the scope of project risk. The second risk category focuses on the product itself. Issues such as product failure and the mechanisms of failure or product misuse belong to this category. In this work, we are primarily concerned with the second category of risk in which a product failure may have a detrimental effect on the end user. Typically, as the design team progresses in its development efforts, there comes a time when the team should ask some fundamental questions. These are:

1. What can potentially go wrong with the product?
2. How serious will it be when it happens?
3. How often will it happen?
4. And, finally, will the product benefits outweigh its risks?

These questions may be addressed by a technique called failure mode and effects analysis (FMEA). Should this analysis be conducted at the design level, it is referred to as a *design* FMEA. This analysis may also be conducted for manufacturing or service procedures. In this case, it would be called *process* FMEA. There are similar approaches for the misuse of a product and even one for cybersecurity concerns.

In all FMEAs, the team starts by entertaining what may go wrong and what would fail. Then, the effects of these failures are entertained. While one failure may have an insignificant impact on its user or environment, another failure may have a detrimental effect. The former is quantified by its *occurrence* rate and the latter by its *severity*. These two fields are often expressed numerically and their product, as mentioned earlier, is a measure of a product's Risk.

When the question of what can potentially go wrong is asked, the real but unexpressed concern is this: What kind of harm is the end-user exposed to?[3] But, just because something can go wrong, on the one hand, it does not always go wrong; and on the other hand, if it goes wrong, it does not necessarily cause harm. By way of example, consider a car having a flat tire. Depending on whether the flat tire is the front or back tire, or if it happens when the car is parked in a driveway or going down the highway at 65 mph, the outcomes may be very different. Another example is having a slippery walkway after a rain shower. It is obvious that not every walkway is slippery when it rains, nor does everyone who walks on a slippery walkway slip and fall. So, how should we look at what can go wrong and the potential of harm?

Let us consider this: *Harm* (or *mishap* as used in MIL-STD-882E) is defined as "an event or series of events resulting in unintentional death, injury, occupational illness, damage to or loss of equipment or property, or damage to the environment." However, before harm comes to anyone or anything, there has to be a potential source of harm defined as *hazard* (MIL-STD-882E). Following the same logic, for a harm or mishap to take place, either people or property have to be exposed to one or more hazards. This exposure is called a *hazardous situation*. For instance, lightening is a hazard; being in a storm with lightening is a hazardous situation; harm is getting hit by lightning. Finally, *severity* is the indication of the consequences of harm if it were to befall.

Now that four terms associated with risk have been defined, we would like to note that there are two other factors to be considered in determining risk. The first, as was mentioned earlier, is the frequency of *occurrence*. This metric provides a measure of the likelihood of occurrence of the specific hazardous situation (or the failure effect). The second factor called *detection* is a measure of how easily the hazardous situation (or failure effect) can be detected. In other words, if a shark is present in the area of the beach with people swimming, how easily can it be seen and pointed out? Now, with the idea of detection introduced, the definition of *risk* may be updated as the product of severity, occurrence, and detection.

Notes

1. See www.etymonline.com/word/reliable. The use of the word reliability is first attributed to Samuel T. Coleridge, who in 1816 referred to his friend Robert Southey as "…inspires all that ease of mind on those around him or connected with him, with perfect consistency, and (if such a word might be framed) absolute reliability." As quoted in Saleh and Marais 2006, p. 249.

2. It should be mentioned that nowadays HALT stands for highly accelerated limit test. This particular methodology—as we will see later—stresses an assembly or system to its operational and destructive limits, but it does not provide any data on the life of the unit under test. ALT, however, is designed to provide a measure of assembly or a system's life.

3. Regulated industries such as avionics and medical devices are very focused on risk. Recent events such as toys that are harmful to children or hoverboards that catch fire are reminders that even consumer product developers should be mindful of the impact of their products on the public.

2

Design for Reliability Process

Introduction

For many engineers, design for reliability (abbreviated as DfR) may be an unfamiliar concept. For this reason, we decided to provide an overview of the rest of the book in this chapter. This should provide a bird's-eye view of the DfR process for the reader so that the details provided in the following chapters may easily be placed in a bigger picture.

The term *design for reliability* represents a concurrent engineering mindset where the development team has recognized that the product is greater than the sum of its design elements. Reliability concerns include design, manufacturing, assembly, and service; and at times, other concerns such as cost and environment.

DfR should not be an afterthought. It is important that the reliability team (as one of the major stakeholders) be invited as principal stakeholders to the table at the early stages of design. Design for reliability activities should:

1. Define the processes and DfR activities and tools to ensure and demonstrate that the design and selected components are reliable and suitable for functions of the design
2. Identify the most appropriate materials to deliver a reliable design
3. Identify the actual reliability and life span of components/assemblies, over which they are expected to perform functions safely and as intended

Elements of manufacturability and assembly along with serviceability and reliability have been addressed elsewhere (see Jamnia 2017). However, in this volume, the goal is to focus on a simple and easy approach to the execution of the design for reliability process. We aim to provide guidelines on

1. Implementing the appropriate DfR tools at each phase of the product design and development
2. Explaining easier and simpler methods for interpreting reliability data from testing and the field
3. Calculating reliability of the design based on available data

Additionally, we will provide a treatment of reliability data when they are not textbook perfect, and one noisy and incomplete. In this regard, our goal is to be practical minded and provide examples of situations that we (as reliability subject matter experts) have experienced in the course of our careers.

Reliability Requirements, Planning, and Execution in the Design Process

As will be explained in more detail in Chapter 3, design for reliability begins at the product requirement level, when operational and use profiles, as well as customer expectations, are reviewed, discussed, and approved. Next, reliability requirements based on customer expectations along with past performances or competitive products are established, typically by the reliability owner. Depending on the industry and product life cycle durations, it is possible that reliability requirements are achieved through reliability growth programs and several products released post the initial launch.

When components have been selected or designed, theoretical reliability models (calculations) may be developed based on the bill of materials and industry standards, such as MIL-HDBH-217 or Telcordia, or on the basis of predictive and analytical models and approaches. At this step of reliability activities, a deeper understanding of system and subsystem behavior is developed and system-level reliability requirements may be allocated to various subsystems. On the bases of information collected or calculated, a redesign of a part or an entire assembly may be required.

Meanwhile, the reliability owner conducts a reliability feasibility analysis to identify any gaps between the target reliability and the inherent reliability of the selected components and parts. Based on this analysis, she or he defines the DfR tasks required to close the gap or to ensure that the target reliability can be attained. The DfR tasks are not limited to reliability testing, rather these tasks include up-front activities and predictive techniques such as stress derating, design margin analysis, and simulation tolerance analysis, even conducting life calculations.

The next steps are to design reliability tests and determine the number of needed samples. This data is typically communicated for planning the first (or subsequent) engineering builds. By conducting qualitative reliability testing such as highly accelerated limit testing (HALT), the design team will learn whether the prediction of failure modes is accurate; and at the same time what additional failure modes may exist. As a result of these HALTs, components or assemblies may need to be redesigned, and documents such as design failure modes and effects analysis (DFMEA) files may need to be updated.

Other reliability qualification testing may be conducted to ensure the product can withstand and endure storage, operating, and misuse stresses. These test are well defined and explained in published standards and can be altered or scaled down to suit the design application.

Once these iterative steps are completed, reliability demonstration tests (RDTs) are conducted to provide objective evidence that reliability requirements can be met. These tests are typically conducted in an accelerated fashion to shorten the test duration to meet project deadlines. The outcome can determine product life expectancy and evaluate service maintenance schedules.

Elements of Design for Reliability

At an early stage of the development, design teams should ask questions, such as: *How reliable will my product be? How reliable should my product be? How will my product fail? What tools can be used to ensure and improve design reliability?* and *How frequently does the product need to be repaired or maintained?* To answer these questions, design team members need to develop an understanding of how and why their products fail. They should develop a means to predict these failures and frequencies in future.

The proper answers to questions such as how reliable will my product be and how reliable should my product be are defined as the probability of a product or device performing adequately for the period of time intended under the operating conditions encountered. Note that this definition is based on the assumption that the product operates properly at the beginning. It should be noted that reliability concerns those characteristics that are dependent on time: stability, failure, mean time to failure/repair, etc. If harm or injury comes as a result of poor reliability, the manufacturer or distributor may have a responsibility to compensate for these losses and/or injuries. This is a matter for the legal system to settle.

Putting the law and litigation aside, another, albeit unofficial, definition of reliability is the reflection of customers' opinion of what constitutes a good product and maintaining what would be considered as repeat business. So, it may be suggested that if a product has low reliability, the distributor and the manufacturer still suffer—even if no harm or injury has occurred. As Taguchi and Clausing (1990) have explained:

> When a product fails, you must replace it or fix it. In either case, you must track it, transport it, and apologize for it. Losses will be much greater than the costs of manufacture, and none of this expense will necessarily recoup the loss to your reputation.

Design for reliability is an extensive field concerned with understanding and managing failures in equipment and systems (MIL-HDBK-338B, IEEE Std. 1624-2008). There are, in general, three aspects of reliability that concern electromechanical design engineers: mechanical, electrical, and chemical issues. Additionally, shelf life, chemical reactions, solderability, moisture, and aging have important influences on reliability as well. Furthermore, the reliability of the software controlling the device should not be ignored.

In the next sections of this chapter, we will briefly explain the main elements of and the activities that constitute the DfR process. These tasks should be completed in a timely manner beginning at the start of the design and development process, and extending into the product service life.

Product Use Profile

The use or mission profile describes how a product is expected to be used. Typically, use profile and operation environment are among the very first items discussed when a new concept product is under development. For instance, if a new minivan is under development and the brake design is under discussion, the conversation shall focus on the statistics that say a 50th percentile minivan driver applies their brakes just over 320,000 times in a 10-year period, whereas a 90th percentile driver applies their brakes about 530,000 times in the same period (Dodson and Schwab 2006). The implication of this comparison is to contemplate which of these two usages the design should support. Clearly, there are lower levels of stress associated with the lower use. A product developed for the brake at the 50th percentile use will probably demonstrate higher failure rates at higher stresses.

Typically, power consumption, environmental conditions, customer usage, and even storage and transportation conditions are included in the description of a product's usage.

Life Expectancy, Design Life, and Failure Rate

Earlier, it was suggested that reliability is, in a way, a study of failure. Although this statement is true, it is not complete. Reliability is also about understanding product usefulness or effectiveness.

Often, *life expectancy* is equated with *design life*. Design life is defined as the length of time during which a product/system/component is expected to operate within its specifications. However, this definition appears to suggest that failures do not happen at all, though this is not always the case. It should be clarified that reliability is among the specifications within which the design is expected to operate. Reliability, as a specification of the design, can be defined in terms of mean time between failure (MTBF), or failure rate. So for a given design, the reliability specification assumes that failure will occur during the design life or life expectancy at a certain acceptable limit. For this reason, we will define the design life as the length of time that a product/system/component operates within its expected failure rate and/or expected reliability target. For example, a repairable product may have the reliability specification set as no greater than 1% monthly failure rate over a design life or life expectancy of 10 years. This implies that for a period of 10 years, on average 1% of the product population fails every month. If consumers or end users operate this product beyond 10 years, the OEM should expect to see an increase in the monthly frequency of the reported failures of the product.

Product Reliability Prediction and Modeling

Design activities may begin in earnest once reliability requirements or goals such as an acceptable failure rate and life expectancy are set, and operational and environmental profiles are understood and defined. One of the first DfR activities within the design process is to model the reliability of the design and predict its behavior.

There are three aspects of modeling the reliability of a product. First, the reliability of a product may be modeled using theoretical values based on feedback and published data from various industries and databases. The second is through predicate data and information obtained from test data and field data of fielded devices, legacy products, and similar applications. Finally, the third method may be through modeling using predictive means, an approach that is often called physics of failure.

Reliability Modeling

To begin reliability modeling of a product, a system needs to be segmented into logical blocks and the relationship between these block be identified. The resulting *reliability block diagram* (RBD) provides a pictorial relationship between various logical (and often functional) sections of a system. The reliability block diagram can be arranged in such a way that the blocks are either in series or in parallel. Hence, should the reliability of each block be known, the reliability of the completed system may be calculated. The reliability of each block and subsequently for the entire system may be evaluated either theoretically or through testing. For instance, consider a system that is used to control and drive a pump. The logical blocks may be the control mechanism module, the pumping mechanism module, and the housing. All three module (or blocks) are required for the function of the system to be performed, which is producing the required water flow rate.

There are a number of ways to model the reliability behavior of a block or module. Three of the most common models are the *exponential distribution*, the *lognormal distribution*, and the *Weibull distribution* models. Unfortunately, there is no simple answer to the question of what is the right distribution to choose. The chosen distribution model should represent the failure trend behaviors of the module of interest. This includes the time to failure distribution of the components and failure mode and mechanism. This information is not typically available at the onset of the project; however, it should be obtained systematically and collected from testing as the design is developed and progresses.

There are reference documents and software that may provide expected (or typical) values for active and/or passive electronic components, similar to MIL-HDBK-217. There are also databases that provide failure rates for printed circuit board assemblies (PCBAs) based on industry-reported data. Based on this information, it is possible to develop a theoretical reliability model for the device under development. In many instances, an assembly's failure rate is an algebraic sum of its components failure rates (including certain stress factors). This approach provides the strictest sense of reliability calculations in which the failure of any component will flag the failure of the entire assembly.[1] The advantage of this approach is that it will very quickly provide information and data on whether the design will be able to meet the required reliability requirements or if a redesign will be needed.

Once the reliability of an assembly or module is predicted and statistically represented by a distribution, the overall system reliability may be created by assembling the individual models belonging to each module into a model based on the system architecture, the RBD, and module interrelationships. Reliability prediction and modeling provide the designer with an overall evaluation of any reliability gaps.

Reliability Testing

It is possible to develop a best-guess estimate for both the expected reliability of a product and its preventive maintenance schedule. These estimates will be based on a reliability allocation and analysis of the product and whether there are redundancies; assumption of the exponential failure distribution; and the use of generic and published failure rates or predicate devices. However, as mentioned, this would only be an estimate. It is through systematic reliability testing that we can verify whether these estimates are acceptable and assumptions appropriate. We need to remember that this estimate of reliability does not account for the actual use of the parts in the design and its applications. Rather, it depends on failure rates obtained from other discoveries, applications, and different environments. Also, published reliability data do not account for the interactions between subsystems, nor for the shipping, manufacturing, or service-induced failures in the new design. Categorically, a disproportionate number of failures is rooted in one or a combination of the following areas: poor design or design elements, defective materials, weak manufacturing techniques (i.e., excessive variations), and/or poor service procedures. Proper reliability testing exposes these deficiencies. In general, there are four types of reliability testing:

1. Tests that seek to induce failures
2. Tests that aim to grow reliability during the design and development process
3. Tests that are designed to demonstrate the actual reliability of the final design
4. Tests that seek to identify problematic products

Typically, in the design stages of a product's development, the reliability engineers try to gather as much information from test-to-failure activities as possible. HALT is used to push components, assemblies, and eventually the finished product well beyond their stated limits and usage. HALT is designed such that certain stresses, mainly step temperature, thermal cycling, vibration, and a combination of them, are applied in a very short period of time to induce failures. These stresses are not meant to simulate real-life conditions; rather, they are applied to stimulate failures of the weakest elements of the design. HALT seeks to quickly expose design flaws. Its goal is either that design improvements are implemented as early as possible prior to product launch or that considerations are given to a service maintenance plan to replace less reliable components on a regular basis.

Once a product is launched, manufacturing becomes the primary activity in the product's life cycle. In order to identify manufacturing defects or flawed components batches of finished products are subjected to *highly accelerated stress screening* (HASS). The purpose of this test is to separate flawed products from robust units by inducing failures of manufacturing defects and flaws without damaging those units which do not possess imperfections. The applied stresses in HASS are similar to HALT, but their levels and duration are much shorter. HASS enables engineers to understand whether excessive manufacturing or process variations exist. Another major difference between HALT and HASS is that in HALT, the product is tested in small quantities; whereas in HASS, as many as 100% of production quantities are tested.

It is important to mention and emphasize that although both HALT and HASS are tools within the reliability tool kit, they do not provide the information needed to make life expectancy and failure rate predictions. The primary reason is that induced failures are the result of accumulated stresses at various levels. They do not induce failure over the normal rate of wear over time or usage or duty cycle of the design. For this reason, life and/or reliability predictions should be based on other accelerated life tests (ALTs). If the relationship between stress and life is known, an ALT is performed at a single elevated stress and the design life is prorated to estimate normal use. If the stress-life model is unknown, an ALT is conducted at multiple levels of operating stresses (e.g., temperature) as follows. One stress is applied at a constant level until failure is induced. The time of failure and the level of stress are recorded. Then, the same stress at a different level is applied until failures are obtained. Again, the time of failure and this different level of stress are recorded. By fitting the data into an appropriate model, life expectancy at nominal conditions may be calculated. Accelerated life testing is a useful technique in that it shortens test duration and helps the design team meet its deadline for product launch.

Accelerated life testing at the system level often takes the form of reliability growth testing (RGT) or reliability demonstration testing (RDT). RGT is used to project the final reliability of a system using early prototypes and design data through applying the test–find–fix–test technique. Reliability growth testing can be applied on both hardware and software reliability growth. RDT is different that RGT in that it is not used to project future reliability; rather it is used to provide objective evidence that a design's reliability requirements have been met. Various accelerate reliability tests will be discussed later in more detail in Chapters 8 and 9.

Reliability Data Analysis

Reliability and life testing generate a fair amount of data. This data should be analyzed and then compared to and/or blended with the theoretical predictions. Testing does not always result in failures, especially at component levels. Reliability test time, especially at system level, is often cut short due to market pressures to launch the new product. The sample size and testing budget are hardly adequate to explore and identify all issues and failures that are experienced by the end user in real-life use. Yet, in light of all these obstacles, we need to be able to make sense of the reliability data that has been collected and to extract as much information as possible.

Reliability test data analysis can be misleading at times, especially when there is no knowledge of the failure modes, or trends of the failure mechanisms, or when more than one failure mechanism is produced in the test. Incomplete and mixed failure modes reliability data analysis, along with accelerated life testing data will be discussed in detail in Chapters 8, 9, and 11.

Reliability Allocation

Once the theoretical model prediction and product testing are completed, it may be possible to identify the contribution of component failures to assembly failures, and the contribution of assembly failures to subsystem failures. Hence, a table—called apportionment—is developed that would contain the percent contribution of component/subassembly failures to the total system failures. It would then be a straightforward calculation to cascade the system-level reliability requirements into subsystems using the apportionment table. This data may then be used by the service organization to make predictions on the number of service components required to be kept on inventory or by the commercial unit to establish warranty repair schedules and pricing.

At this stage of product development, the design is complete, samples have been tested, and requirements have been verified. The product is ready to be launched. The reliability model is then revised and updated using testing data, and then field data after product launch. Reliability allocation is explained in detail in Chapter 4.

Other Aspects of Reliability

MIL-HDBK-338B (Section 7.1) defines reliability engineering as "the technical discipline of estimating, controlling, and managing the probability of failure in devices, equipment and systems. In a sense, it is engineering in its most practical form, since it consists of two fundamental aspects:

1. Paying attention to detail
2. Handling uncertainties

However, merely to specify, allocate, and predict reliability is not enough."

This handbook goes on to recommend that each organization should develop a series of design guidelines that design engineers may access and use to ensure product reliability at the design stage because once the design is frozen, making changes is difficult and costly; and, once the product is launched and marketed, making any meaningful changes is nearly impossible due to associated high costs.

Design engineering teams typically develop the initial reliability assessment inputs such as operational and use profiles, followed by failure modes and effects (FMEA), fault-tree analysis (FTA), and other similar tools. As the design and development progresses, the team focuses on analyzing the details of design by utilizing finite element analysis for component mechanical stress analysis, or circuit design and part derating analysis. Eventually, the team's focus shifts toward the initial build and verification testing.

The reliability aspects that may be missed by the team are component management and obsolesce. Unfortunately, both of these areas are often treated haphazardly by the sustaining team and within a change management process—if one is in place. Thus, it is highly recommended that as the detailed design progresses, three documents be developed. These are the preferred components list, critical and/or safety components list, and the preferred supplier list. These documents need to be developed collaboratively among components, reliability, design, and manufacturing engineering alongside the purchasing and supplier quality teams (MIL-HDBK-338B 1998).

A world-class organization generally strives to maintain an up-to-date parts database. An impetus for maintaining such as database is cost and management. Another motivation for establishing a parts database is reliability concerns. A parts database enables the design team to identify component reliability measures along with functional and physical requirements. Furthermore, part equivalency maybe established early on that would ease the burden of proving part replacements should a component become obsolete.

The critical and/or safety components list is a reflection of the DFMEA file where component failures may have determinable impact on the function or safety of the product. In regulated industries where listings with agencies such as Underwriters Laboratories (UL) or TUV are required, any critical component change requires notification and possibly technical file updates. Typically, an organization's compliance team owns this list; however, by having a clear understanding of this list, the design and reliability teams select these critical components not only based on a part's useful life failure rate buts also on its infant mortality as well as wear-out behaviors.

Finally, suppliers should provide quality and reliable components. In larger corporations, supplier quality teams audit and approve vendors prior to order placement and purchasing. In smaller corporations, purchasing or quality teams may conduct this audit. An aspect of these audits should include reliability concerns.

Reliability and Its Association with Product Risk

As reliability activities for a new product begin to wind down with product launch, another set of reliability workflow steps begin to wind up. As the product is launched into the market, it would be as if a gigantic reliability test has been designed with numerous noise factors. An aspect of sustaining reliability activities is to collect field data along with customer complaints and feedback. This enables the organization to monitor the behavior of the product in the field and the impact of failures on its users.

In a world-class organization, the failure modes of a design and their associated effects, as well as causes, are subjects of scrutiny in a FMEA study. Associated with each failure mode, a *risk priority number* (RPN) is calculated, which is typically the product of the rate of occurrence of the failure mode and the severity of its effect. At the design stage and during FMEA studies, the probabilities associated with each of these factors are often estimated and RPNs calculated based on these estimates. Then, the design team focuses on the failure modes with the highest RPN values and attempts to either eliminate these risks or mitigate their effect.

If we note that the value used for the occurrence of the failure mode is in reality its associated failure rate and the value used for the severity of its effect is the probability of harm, we can then see how risk is the product of reliability failure rate and the probability of harm associated with a specific failure mode.

As postlaunch data is analyzed, it is possible to calculate the failure rate associated with each component. Furthermore, based on customer complaints and the frequency of specific complaints, the severity of each failure may be quantified. Hence, it may be possible to calculate and monitor product risk based on actual field data and to ensure that risk thresholds remain within safe thresholds.

Note

1. The reason we call this a strict reliability is because in many instances a system or assembly may continue to function even if one or several minor components fail. For instance, should a car radio fail, the car continues to drive. Or, should an indicator LED fail on a radio, it still continues to broadcast.

3

The Design Process and the V-Model

Introduction

In Chapter 2, we quoted MIL-HDBK-338b (section 7.1) as saying that "[reliability] is engineering in its most practical form." In this sense, design for reliability (DfR) is inherently an integral aspect of the design and development process. Since engineering is at the core of product design, we like to first review elements of the new product development process to better understand the role and contribution of DfR with this process. The topic of this chapter may appear as a digression, but it is necessary to present a complete picture.

A product begins with an idea, and through engineering efforts this mental concept turns into a physical reality. Should this embodiment of the concept fulfill a market need—be it real or imagined—it may become a financial success. Throughout the last six or seven decades, a number of approaches have emerged in order to organize the way products are developed: on the one hand to bring a certain degree of efficiency and on the other hand to ensure that the finished product is what the customer had asked for and wanted. The majority of these methodologies start with an effort to understand the market and/or customer requirements. The last step in them ends with a litany of best practices to launch the product. Some authors place a greater emphasis on the front end, i.e., the marketing and business needs, while others concentrate on the back end, i.e., product launch. Typically, sources that focus on engineering design practices tend to neglect nonengineering aspects of product development.

Product Development Models

Stages of product development and life cycle are defined differently by different people. For instance, Crawford and Di Bennedetto (2003) define the life cycle of a product to have five stages, namely, *prelaunch, introduction, growth, maturity,* and *decline.* This view reflects someone who has a bird's-eye view. Cooper (2001, 2005) provides a five-stage, five-gate Stage-Gate® model for developing new products.[1] The five stages are *scoping, build business case, development, testing and validation,* and *launch.* In Cooper's model, the product development activity would not proceed unless it passes through management review gates. Cooper focuses on management activities leading up to and through launching a product. De Feo (2010) proposes a six-step process defined as follows: *project and design goals, customer identification, customer needs, product* (or *service) features, process features,* and *controls and transfer to operations.*

If we were to examine these authors' points of view, their emphasis is on developing a new product and launching it into the market. They do not consider any postlaunch activities such as support and/or service, nor do they talk about engineering undertakings.

Hooks and Farry (2001) consider that product life ends with its "disposal" as the last stage in their life cycle model: *manufacturing, verification, storage, shipping, installation, upgrading,* and *disposal*. Similarly, Pancake (2005) extends the steps of the product life cycle beyond launch by including support as the last step of the seven steps of product development: *discovery process and product strategy, product requirements, design and development, verification and validation, manufacturing, quality assurance and regulatory,* and finally, *product launch and support*. Stockhoff (2010) cites several different models: in some, the first step is *idea generation* and in others *establishing market concepts* or *project initiation*. The last step is *commercialization, launch,* or *post-implementation*.

Clearly, there is no standard definition or a one-size-fits-all process. However, in a simplistic way, the product life cycle may follow these generalized steps to a greater or lesser degree: As shown in Figure 3.1, first, the idea for a product is conceived (*market opportunity*). On the basis of this idea, some analysis is conducted to provide evidence that the idea has commercial merits (*scoping/business case*). When investors are willing to support the idea financially, engineers are engaged to first develop concepts (*concept generation*) and then provide detailed design (*design*).

The next step is to fabricate and test prototypes (*verification and validation*). When all parties are satisfied with test results, the product is introduced in one or two small markets (*limited launch*) and customer response along with behavior of the product—particularly any failures—is closely studied. Eventually, manufacturing ramps up to full production (*full launch and production*). Shortly after, the service organization begins to service failed units. As time goes on and manufacturing continues, issues associated with product failures and reliability, process and/or design improvements, component changes, and part obsolescence will require design changes (*engineering notices of change*). Eventually, the business recognizes that the product is no longer profitable and decides to first end manufacturing (*end of manufacture*) and then stop any service activities (*end of service*). In some instances, the manufacturer may even actively remove the product from the field (*retire product*).

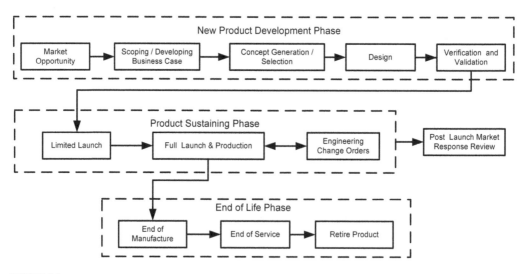

FIGURE 3.1
Typical stages of a product's life cycle.

The V-Model for Life Cycle Management

A product development model that has gained some currency in the United States is called the V-model (Forsberg and Mooz 1998). A pictorial view of this model is presented in Figure 3.2. This model has also been referred to as the V&V model to represent that verification and validation of the outcome is an integral aspect of this model.

Briefly, the left hand branch of the "V" corresponds to a series of activities to define the tasks to be completed, starting from a very high level overview of what should be done broken down to smaller tasks to eventually down to manageable pieces. Once all tasks are properly defined, implementation may begin at or near the apex of the "V." Once the smaller tasks are completed, they are integrated into larger pieces and tested to ensure that they work as expected. This aspect of the work is represented by the right-hand branch of the "V." Integration and testing continue with larger assemblies until the entire system is put together and tested. Finally, the end product is validated to ensure that it is the right response to the initial question (or need). Even though Figure 3.2 outlines 12 steps from the beginning to end, the number of these stages is rather arbitrary. The intention is to show the progression from the big picture to development of the details and then expand back to a confirmed full picture.

The versatility of this approach is that it is a task-oriented model. It should be readily obvious that the V-model is not specific to developing only engineering products. Rather, it has been used extensively for project management and software development.

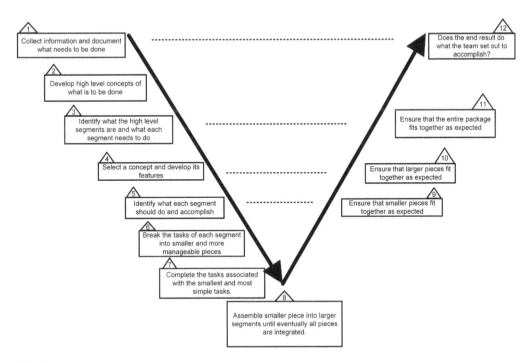

FIGURE 3.2
The steps and stages of the V-model.

So, what do these 12 stages of the V-model mean to a design engineer? To answer this question, note the first five stages:

1. Collect information and document what needs to be done.
2. Develop high-level concepts of what is to be done.
3. Identify what the high-level segments are and what each segment needs to do.
4. Select a concept and develop its features.
5. Identify what each segment should do and accomplish.

From an electromechanical product development point of view, the first item is clearly a marketing, business, and ultimately financial question and concern. This is a question and concern for the decision-makers.

Stages 2 and 3 involve engineering input and work. This process begins with the development of system-level requirements (stage 3) cascaded from product requirement definitions (stage 2). From the system-level requirements, further decomposition occurs to produce subsystem and possibly sub-subsystem requirements, and the functions (i.e., mechanical, electrical, and software) allocated to them. This is akin to stages 4 and 5 of Figure 3.2 and is presented in more detail in Figure 3.3, which will be discussed shortly in the next section.

At stages 6 and 7 of the V-model, details of each segment (i.e., components and subassemblies) are developed. In engineering language this means that engineering drawings are developed. Finally, at stage 8 of the V-model, the smaller components are fabricated and assembled into larger subassemblies. These subassemblies may be tested to ensure that they meet their intended functions (stage 9). In stage 10, larger assemblies are made and verified, and finally, at stage 11, the entire product is assembled and tested to ensure that it was designed and built per design specification. At the last stage of the V-model (i.e., stage 12), the product is tested to the customers' specifications and needs.

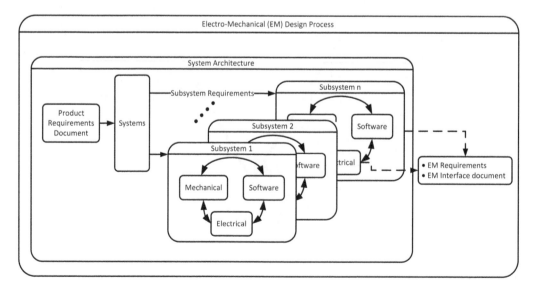

FIGURE 3.3
A requirements-based electromechanical engineering design process flow.

Incidentally, there is an element that is implied but not explicitly called out. A lower step cannot be attempted unless the upper level is completed and approved. In other words, a subsystem requirement may not be developed if system-level requirements (and architecture) are not clearly defined. This mindset ensures that design inputs are properly defined before engineering attempts to develop design outputs (i.e., solutions).

This approach and model has given birth to an engineering discipline called systems engineering, which is the disciple of understanding what market needs are and translating these needs into technical terms not only at the product level but also translating how various lower-level assemblies and components should function and interface with one another. This is done to ensure that at the top level, everything works as intended.

Engineering Activities

Earlier we mentioned that stages 4 and 5 of the V-model (Figure 3.2) focus on concept development and further decomposition to produce subsystem and assembly requirements. Their associated functions (i.e., mechanical, electrical, and software) are also allocated to them. The details of this cascading process for an electromechanical system are shown in Figure 3.3. The outcomes of this cascading process are *electromechanical requirements* and the *electromechanical interface document*.

When many engineers are first introduced to this concept, i.e., cascading product requirements to lower levels, their reaction may be "What does this mean and what does that have anything to do with my design!?!" Unfortunately, the explanation of many product development models (here the V-model) may not provide a clear set of instructions on how to easily cascade *requirements* into assembly design specifications. This lack of clarity, as so properly pointed out by Ogrodnik (2013), is because the process has not provided guidelines and details on the required set of activities. In a way, this shortcoming may be due to the fact that systems engineering has not been actively and properly integrated with other functional (i.e., traditional) design engineering fields. In other words, systems and design engineers do not speak the same technical language.

The reader may note that we have not mentioned any related "engineering" processes or activities as integral parts of the proposed product development models. Explanations of many product development models tend to gloss over engineering activities as a given set of workflows. Admittedly, there are references that attempt to explain product development methodology from an engineering point of view. Among these are Ullman, *The Mechanical Design Process* (2010); Gerhard Pahl et al., *Engineering Design: A Systematic Approach* (2007); and Stuart Pugh, *Total Design: Integrated Methods for Successful Product Engineering* (1991). Ogrodnik (2013) provides a critical review of the models proposed by Pugh as well as Pahl et al. He eventually concludes that these models do not give a full picture of actual engineering activities. His point is that the presented models visualize the processes but not the activities. In response to available but incomplete models, he presents a didactic "divergent–convergent" model of his own.

We are not really sure if there is a great deal of difference between these various models, per se. As mentioned earlier, almost all begin with a reference to market need and end with the embodiment of a product. A few have even mentioned the need for sustaining the product and eventually removing it from the market. An often forgotten (or ignored) factor is that even though most of the models present their various steps and stages in a linear fashion, in reality, product development is a highly iterative process, either within each step or in between two or several stages. Another overlooked factor is that product development is never a purely engineering, a purely sales and marketing, or a purely business activity. It is truly an interdisciplinary and interwoven set of activities whose success depend on each other.

The shortcoming of many models may be due to the fact that the activities required in the product development process are not "visualized." Or, if they are referenced, they are considered within a specific discipline such as engineering, finance, or marketing.

We tend to agree with Ogrodnik (2013) that any proposed model should present both the process as well as the activities. This was accomplished in *Design of Electromechanical Products: A Systems Approach* (Jamnia 2017). The set of activities that are required—from an engineering point of view—are outlined in a roadmap and presented in that work.

Because DfR begins at a product's requirements and then their cascades to various component specifications, it is important to review some basic elements of systems, along with the voice of customers (VoC) and its translation into product requirements.

Systems and Requirements

In the last section we made a number of references to systems engineering and requirements. Additionally, the first steps in the product development roadmap (Jamnia 2017) begin with the product requirements document as well as the systems requirements. The goal of this section is to provide a high-level review of the role requirements and requirements management. Associated with these artifacts, there are architecture and design documents. We will define these and provide a definition of their role and place in designing a product.

Requirements are considered to be one of the seven key elements of systems engineering, the others being (systems engineering) management plan, master schedule, detailed schedule, work breakdown structure, technical performance measurement, and technical reviews and audits (Martin 1997).

This sounds impressive but what does "system" mean? And, how does systems engineering differ from other functional engineering disciplines such as mechanical, electrical, or software engineering.

Let's explore this issue: In response to a market need, corporations—large and small—work on providing a "product." This product may be a service, a family of devices, or a combination of the two. The novice, often, thinks of the device (or the service) as the system because it is the "product" that the end user interacts with. Strictly speaking, this is not a correct definition of a system.

For example, consider a blood sugar monitor (the "product"). This product is a corporation's response to a customer's need to monitor his or her blood sugar daily at a time and place that is convenient. However, this product did not just happen: it was designed, manufactured, tested, marketed, sold, and, if damaged, serviced; and finally when the product is no longer used or useful, disposed. One may consider these various organizations and their interactions as the "system." Now consider this: In addition, suppose that this unit uses special strips that capture the end user's drop of blood for analysis in the unit. These strips must be available to the user on a daily basis. Wouldn't the customer support organization and the delivery company (e.g., FedEx, UPS, US Postal Service) be considered as a part of the system as well? As Hatley et al. (2000) point out, "Every system below the level of the whole universe is a component of one or more larger systems." However, to be practical, the "system" and its boundaries must be defined by the design team.

Industry standards define systems in different ways. For instance, IEEE 1220-1998 defines a system as "a set or arrangement of elements and processes that are related and whose behavior satisfies customer/operational needs and provides for life cycle sustainment of the products." ISO/IEC 15288:2008 meanwhile defines a system as "a combination of interacting elements organized to achieve one or more stated purposes." The *NASA Systems Engineering Handbook* (2007, p. 275) gives a more comprehensive definition:

1. The combination of elements that function together to produce the capability to meet a need. The elements include all hardware, software, equipment, facilities, personnel, processes, and procedures needed for this purpose.
2. The end product (which performs operational functions) and enabling products (which provide life-cycle support services to the operational end products) that make up a system.

In general, a system is a combination of the "end" product(s) as well as product "enablers." Product enablers are all the ancillary organizations that make it possible that the end user and/or customer make use of the device or service.

From a more practical point of view, for relatively simple applications, a system is generally defined as the end product or a combination of end products. In more complicated situations, a "system of systems" is defined in an effort to partition complexity into simpler segments. Thus, each system may be considered independently. And the interactions of various systems are considered and accounted for at a higher system of systems level.

In the previous blood monitor example, the *system of systems* (SoS) may be defined as the business of providing blood sugar monitoring to the market. Depending on the decisions made at the SoS level, one of the constituent systems may be the electromechanical monitor. Another may be the disposable items (such as the strips and the sterile scalpel used to draw blood). Figure 3.4 provides a pictorial example of the blood sugar measurement business and its various segments, which may be considered as systems in and by themselves. In this example, the end products are the handheld monitor and the disposables (i.e., strips and lancets); however, these end products can only be realized if they are supported by the enabling systems such as manufacturing, service, or other organizations.

Earlier we mentioned that industry standards indicated that a system is a combination of elements. Let's call these elements building blocks. For example, the elements or building blocks of an automobile consist of an engine, chassis, wheels, and its body among others. By way of another example, we could envision that a display, along with a keyboard, hard drive, and a motherboard are elements or building blocks of a computer. For the

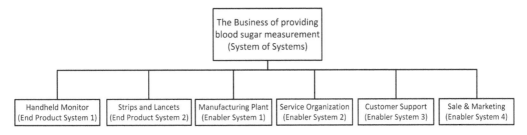

FIGURE 3.4
A pictorial representation of system of systems.

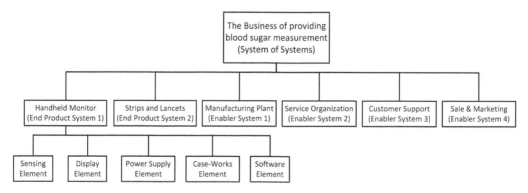

FIGURE 3.5
A pictorial representation of system of systems and its lower-level elements.

blood sugar monitor example, these elements may be sensing, display, power supply, case-works (enclosure), and software. Hence, the systems of systems diagram in Figure 3.4 may be updated as shown in Figure 3.5 to reflect these various building blocks.

Within this context, the role of the systems engineer is to define various building blocks of a system and to ensure that these elements are seamlessly integrated, and that they all function with coherence.

Voice of the Customer and Voice of Stakeholders

Typically, functional (e.g., mechanical, electrical, and software) engineers are trained to solve problems. Often the desire to provide a solution is so strong that it overshadows the problem statement itself. Systems engineers balance this tendency by first focusing on the problem statement and dwelling on it so that it is properly defined. They call this first description of the problem statement the VoC.

VoC is a reflection of a market demand, and hence, an opportunity for financial gain. By suggesting that the VoC is a reflection of market needs and demands, we accept the assumption that the customer knows what she or he wants. While there is debate among entrepreneurs and product innovators on the veracity of this statement,[2] nevertheless, a product that is launched into a market has to meet the recognized or unrecognized/unmet needs of customers to be financially viable. As such, VoC is considered to be the highest-level requirement.

In reality, VoC is not the only voice that is heard by systems engineers. Business has a voice of its own, as does manufacturing and service. For regulated industries such as avionics and medical, compliance with regulatory agencies brings unique voices to the table as well. Collectively, systems engineering brings the voice of stakeholders (VoS) together and develops the highest-level design document typically called a product requirements document (PRD). This document reflects the integration of various voices that go into shaping the initial requirements of a product and its final outcome(s). It is the first product development document.

Product development documents form part of a collection of files (or documents) that profile the development of a product from conception to retirement. Some organizations do not have a specific name for this collection of files, whereas others may refer to it as design evolution collection. In the medical field, this collection is called the design history file (DHF) that is retained under revision control. Because of the simplicity of this terminology and its meaning, in this book we will adopt this term. Often, if there is a legal dispute, it is the documents in the DHF that are provided as evidence. In addition to protecting the business in the case of litigation, there are very practical reasons why these collections should be properly maintained. On the one hand, in regulated industries such as medical or avionics, these files are subject to regulatory body inspections. On the other hand, engineering change notices can be done much more efficiently if the history file is properly maintained.

Design Documents

Design documents for a product contain the requirements and specifications for that product, i.e., they identify what function the product performs and, depending on the hierarchy of the document, how the function is performed. There are typically four levels of documents in a hierarchy: system, subsystem, assembly, and component documents. The number of levels could change depending on the complexity of the product (in simple designs, remove a layer; in more complicated designs, add additional layers).

At the highest-level document, the what of the product function is described. In a way, the voice of stakeholders in a semitechnical language is expressed at this level. Each lower level provides more detail in describing the how of the design. Eventually, the component-level documents provide detailed design descriptions. In the blood sugar monitor example mentioned at the beginning of this chapter, a high-level requirement may be stated as follows:

> Product design shall be reliable.

The next level document may translate the reliability requirement as such:

> Product design shall have a life expectancy of 5 years during which the average failure rate less than 1.0% per month shall be maintained for the population.

This requirement at the lowest level may be satisfied by the following specifications:

> Printed circuit boards shall have an expected life of 10 years and a failure rate of less than 0.25%.
> Selected motor shall be rated for 35 million revolutions.
> Selected connectors shall be rated for 5400 connect/disconnect cycles.

Typically, requirements flow down from *what* to *how* by *which* (subsystem, assembly, or component), where how is the requirements for the lower level and which becomes the description of the structure. The whats to hows flow down forms the requirements documents and the whats to whichs develop the systems architecture.

Figure 3.6 provides a pictorial view of the requirements and their cascade. The top-level requirement document is the product requirements definition (PRD), which reflects the voice of stakeholders as well as performance characteristics of the product or product family.

The architecture documented at the second and third levels describes how various requirements are distributed to multiple subsystems or assemblies. In traditional approaches, there may be a tendency to create the architecture along each engineering

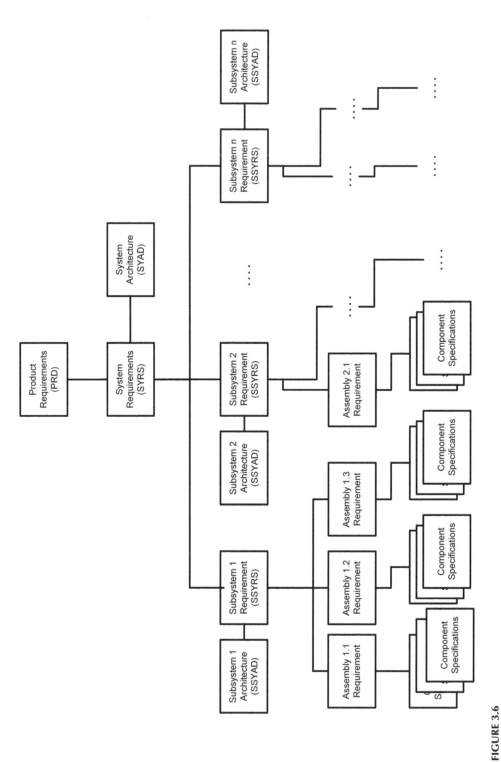

FIGURE 3.6
Requirements and architecture hierarchy.

discipline (i.e., mechanical, electrical, or software). In a systems approach, a cross-functional product development team determines the number of subsystems or assemblies, typically along product functionality, physical, or operational boundaries, and develops the architecture documents.

The system requirements document constitutes the top-level technical definition of the product. This document translates marketing needs into a technical language that sets the basis of the product to be developed. At this stage, the voice of stakeholders is expressed in concrete concepts that are quantifiable and verifiable. This document is usually developed by the technical leads collaborating with the lead systems engineer.

Along with system requirements, the system architecture is also developed dictating various subsystems and their relationship to one another. Subsystems, generally, are defined in such a way to segregate various specific operations or functions of a product, or along physical boundaries. For instance, in an automobile, the powertrain belongs to one subsystem and the chassis to another; in a medical insulin pump, the pumping mechanism would belong to one subsystem and power supply to another. Subsystem design documents contain the requirements and functions allocated to them by the system architecture.

The next layer of the requirements cascade is the assembly design document. This level of the requirements tree identifies various functions within a subsystem. For example, within a pumping mechanism subsystem, there is a tube-load assembly, a pumping assembly, an occlusion sensing assembly, and so on. The term "subassembly" may be used to describe lower levels of an assembly. It should be noted that there may be any number of assemblies under each subsystem; and similar to subsystems, assemblies should be created in operational, functional, or physical blocks. Subsystem architecture dictates the distribution of subassemblies and their relationships.

Those with a strong manufacturing background may equate the assemblies that we are speaking of to those defined by manufacturing processes. The two are not the same. From a design point of view, an assembly document is the lowest-level document containing requirements. However, from a manufacturing point of view, an assembly may be one of several intermediate steps to physically build a portion of the product. Even though the design team may define the architecture with its particular subsystem and assemblies, this structure may not reflect the steps required by manufacturing procedures. As a result, the manufacturing organization may define assemblies differently than what is presented in the design architecture diagram. This does not imply that the manufacturing process does not align with design requirements. On the contrary, each manufacturing step does require its own documentation with requirements decomposed from design specifications.

In Figure 3.6, component specifications are shown under the assembly documents. Component specifications (by definitions) may be any document that describes the component to be used. It may be specifications from vendors or engineering drawings and models.

A facet of component specifications is that it should identify significant (or key) characteristics and/or critical dimensions, and provide justifications and traceability to the requirements tree and ultimately to the PRD. This tracing provides a logical flow of original design intent into product realization. In the medical field, this traceability is a Food and Drug Administration (FDA) requirement.

Typically, high-level documents, i.e., system, subsystem, and assembly documents, are reviewed and approved by members of associated cross-functional teams. Conversely, component specifications are reviewed and approved by specific discipline subject matter experts. In this context, high-level documents include reliability goals, whereas component specifications include elements of how the high-level goals may be achieved. We will speak of this in more detail in later chapters.

Subsystem Design Document

The subsystem design document consists of the requirements allocated by the system architecture to the specific subsystem. So, it is simply a list of the requirements that apply to the same function. The purpose of this document is to group requirements and to provide operating parameters.

Subsystem Architecture Document

The subsystem architecture document divides a subsystem into assemblies and assigns requirements to each assembly. This document describes the design intent and reasons behind design decisions made during the design process. It has four major sections:

1. Theory of operations
2. A description of integrated assemblies
3. Requirements allocation including reliability
4. Design rationale

The theory of operations describes at a high level the subsystem functions. It is a general overview of what the subsystem does. It briefly describes how the subsystem executes its functions.

Another section of the subsystem architecture model describes the assemblies that are integrated within the subsystem. Subassemblies should be grouped by function, which at this level (the lowest level) may be strongly correlated with the physical grouping. Generally, block diagrams describing subassemblies and their interaction with each other are provided in this document.

In the requirements allocation section, requirements are assigned to each subsystem. These requirements, which may span over several assemblies, collectively fulfill the subsystem requirements.

The design rationale is a document where justifications and explanations for the design approach and intent is provided. A proper design rationale document should provide an argument on how the subsystem is expected to meet its reliability goals.

Assembly Design Document

The assembly design document contains the requirements that are assigned to the subassembly. The document also contains a description of how the assembly works. The major sections of the assembly design documents are

1. Theory of operations
2. Design rationale
3. Requirements allocation
4. Relevant drawings or design specifications

The theory of operations provides an overview of how the subassembly works. This is a higher-level view of how the black box of the subassembly works.

The design rationale is where justifications and explanations for the design approach and intent are provided. Considering that this is the lowest-level design document, rationale for component-level choices as well as critical dimensions should be provided in this segment. As an example, the material choice of a critical component or the dimension scheme of a critical interface should be justified here. Similar to the subsystem design rationale, this document should provide a justification of how reliability may be addressed at this level. For instance, should the assembly involve moving parts, a justification of why wear may not be a concern should be provided.

The requirements section is where the requirements assigned to the subassembly are listed.

The relevant drawings section is where drawings or specifications of critical components are referenced and/or discussed.

Developing Product Requirements

Earlier we suggested that the initial steps in the product development process required that we start with developing the product requirements document. We briefly reviewed voice of customer, requirements, requirements cascade, and design documents. Now, we will provide more detail to lay a foundation of a system-level understanding for design and development of electromechanical products.

Since product development begins with creating the product requirements document (PRD) and that PRD depends on understanding customer needs, the questions to be entertained are as follows: *How do we get voice of customer? How do we develop product-level requirements from them?* and *How can we flow these requirements to lower levels?*

The answer to the first question is in the realm of those with expertise in marketing; any detailed treatment of the subject is beyond the scope of this work. However, a simplistic answer is that the business needs to first know who its customers are. Are the intended customers other businesses or individuals? Are customers the end users? This question concerns whether the customer would personally use the purchased item or the customer buys it on the behalf of a third party. The tools for gaining this information are user studies, surveys, and other similar mechanisms.

The answers to the second set of questions on developing the PRD and requirements flow down to subsystems and assemblies will be explored next.

Defining the Product to Be Developed

Based on a design for Six Sigma tool for product development called DMADV,[3] the first step is to define the desired product. Typically, this means that the problem statement is created in this step.[4] This requires a knowledge and understanding of the market and potential financial gains—either through sales and profits or through savings to the bottom line. Writing a clear and concise problem statement may require some practice; however, a complete problem statement is often created through a repetitive process. As we learn more, this statement may be updated and refined further with additional relevant information.

A problem statement should identify intended customers, added value or benefit to the customer, and other relevant known information. As a part of defining the problem statement, project scope and the current versus desired states are also developed.

Once the cross-functional team has agreed on the current versus desired states, the project scope is properly defined. Project scope—as the title implies—describes the boundaries of the project, where it begins and ends. Similarly, it is just as important that the out-of-scope activities are also defined.

Often, products are launched into the market and they fail; not because they are bad products, but because they were not properly scoped. Time and resources were dedicated to inconsequential issues. Another factor that potentially could lead to either costly product launches and/or failures is a lack of understanding of the strengths and weaknesses of the business. To develop an understanding of a business position and standing, a strengths, weaknesses, opportunities, and threats (SWOT) analysis is done.

Cooper et al. (2001) consider the bulk of SWOT analysis to be conducted as a part of the marketing activities, though an aspect of this analysis is to make a realistic self-assessment of the business as well. This implies that for the opportunity at hand, marketing and sales channels along with engineering, manufacturing, and operations need to be evaluated. The results of this activity are generally reported as a summary in the define stage.

The define phase is arguably the most important phase of any project and should not be taken lightly. It is highly recommended to stay in this define phase as long as possible to ensure that the project is as fully explored as possible. Reliability goals should be expressed in the define phase to provide a complete picture.

Measuring What Customers Need

Once the project is properly defined,[5] we move to the measure phase. During this phase, the team will collect the VoC and the voice of the business. This is the realm of systems engineering. In our experience, functional engineers are not typically very involved in this stage.

To collect VoC, systems engineers first identify their customer with the help of their marketing colleagues, and then develop an understanding of them and how they operate a given task or function.

Griffin (2005) provides three techniques for obtaining customer needs: be a user, critically observe users, and interview users for their needs. Each approach has its own benefits and drawbacks; this is a task that requires a high degree of diligence. Cooper (2001) provides a checklist of tips and hints. These tips and hints ensure that the customer needs study considers what its objectives are, uses a structured questionnaire, the right information is solicited, a representative sample of the population is interviewed, the technical people are involved, and the use environment is considered.

Another noteworthy factor is that customers are not always the same as end users. In the case of a medical device, the end users could be clinicians and the customers could be hospital administrators. The needs and wants of both groups should be considered and studied.

Once sufficient data is collected, it should be organized and analyzed. Rarely would any VoC surveys return pure and true customer and/or user needs. Often, this information reflects what the customer may see as a solution to the problem or a different design based on an existing solution.

De Feo (2010) recommends the use of spreadsheets in order to analyze and organize customer statements. The first step is to identify categories for this organization. Typically, *needs* and *requirements* are among this classification. Others may include "preconceived notions" of what the new product should do, look like, and so forth. The second step is to develop the true needs.

As the statements from customers are collected and analyzed, there may be a realization that there may be gaps or issues that have not been communicated clearly. With this additional information at hand, we can complete the customer statement to customer needs spreadsheet.

The third step in analyzing and organizing categorizing customer needs is to place them into categories. This activity may be more difficult and in general will require the participation of the development team. This participation is often in the form of brainstorming, and an effective tool for brainstorming is called the affinity diagram (See Jamnia 2017).

Development of Product Requirements

Admittedly, many young engineers may not have much patience with all this work. Why can't we get to the design already? It may take years to develop a true appreciation for spending time on developing good requirements. Having said this, we would like to share a comment and a personal experience. In our careers, we have come across two typical groups of engineers: one group considers design to be the work that leads to the completion of requirements. Any work beyond this level is often not valued and possibly considered as "details" that can be farmed out. The second group—which we belonged to—considers "design activity" what our colleagues and ourselves did. We considered design to be the actual computer-aided design (CAD) modeling and the creation of the physical embodiment of the product. After all, it is because of engineers like us who know the details of what features on a product may or may not be realized. However, personal experiences led us to change our minds and value having a clear and inconspicuous set of instructions on what needed to be designed. In a way, design activities include both the development of requirements as well as the design specifications that define the physical embodiment of the product.

We read too many stories about products that fail despite extensive testing. Even some very large corporations miss the mark. Hooks and Farry (2001) retell the story of how IBM missed the mark on developing a product to provide a sports news feed to newspapers during the Atlanta Olympics games. IBM's computer system was supposed to channel information about the games and their outcomes to a second computer that would then feed the data to various national and international newspapers. At the time the games began, the system did not work. "Many of the [World News Press Agency] problems were a case of programming a computer to format a certain sport's information one way, while newspapers were expecting it another way" (Hooks and Farry 2001, pp. 83–84).

At the time, IBM personnel felt that had they had a chance to test their system more, they would have been able to figure things out and deliver what the customer wanted. On the contrary, they missed the mark because they did not understand what their customers wanted, expected, and eventually required (Hooks and Farry 2001).

On a more personal note, would more testing of the product that we designed have identified the issues with the design? The sad answer is "Probably not!" In reality, there is not enough time or resources to test every aspect of a product in the hope to ensure that the design team has developed the right product. The right product is and should be identified through its requirement. This is why requirements need to be developed, written, and agreed upon. On the basis of these requirements, design engineers know what to develop and test engineers know what to test. Hence, through a limited number of tests, the design team can identify whether the right product has been developed. Therefore, the question is how to use customer needs to develop product requirements.

Translation from Needs to Requirements

One method to transition from customer needs to product requirements is for the development team to have brainstorming sessions to consider each need separately and agree on what product feature(s) is (are) a response(s) to a specific need. For simple products and line extensions, this is rather easy: just design the extra feature into the device. For instance, there was this instance when the marketing team developed a campaign to support breast cancer research and prevention efforts. Our task was to find a way to apply a pink logo onto a series of our products. The technical challenge was that the product being used by clinicians was subjected to some severe cleaning procedures after each use. We needed to identify a paint that would endure these cleaning cycles. Once the paint was identified, the verification activity was to complete an established set of test regimens that was common and customary to the product line.

Testing to an established set of test procedures did not work for IBM's newsfeed product. Why? Because a new set of tests needed to be developed and no one had thought about it. It was not that the participants did not have the experience or the expertise to develop the test. There were other so-called fires burning. The challenge was that the expectations for the product features were evolving as the launch date was drawing near, and there was not clear communication among the development team. As the expectation for the product function evolves during the detailed design phase of development, the possibility that the finished good meets its intended expectations may be a hit or a miss. Why? Because this approach does not follow a robust process. A robust process, if properly followed, tends to flush out all the factors that need to be considered.

Earlier we asked the question of how does one transition customer needs to product requirements. In other words, our question is how to deliver what the customer needs. We should like to note that the terms *how* and *what* are a part of the systems engineer's vocabulary and we are deliberately using them here.

The first step in translating these whats into hows is to identify the performance characteristics associated with these whats. For instance, a customer need in developing dental hand instruments such as a dental pick is that they have to be lightweight. Also, for ergonomic reasons, the instrument's body should not be too narrow (i.e., small diameter). By conducting user studies or based on prior experience, the development team learns that if a dental instrument weighs between 9 and 12 grams and has a diameter approximately between 8 and 11 millimeters, it has an ideal size. Anything else is too light, too heavy, too small, or too bulky. This is what we mean by translating whats into hows. In this instance, lightweight translates into 9 to 12 grams, and ergonomic means 8 to 11 millimeters. Or, if the customer need for a dental scaler is to be durable and require infrequent sharpening, this translates to a surface hardness of, say, 55 to 60 HRC.

Product Requirements Document

At this point, it is important to initiate what is called a product requirements document and capture the outcome of the product related decisions including the team's understanding of the performance characteristics. By the time this document is completed the following are captured:

- Intended use of the product
- Customer and/or end user performance needs

- Business needs including voice of manufacturing, service, etc.
- Specific markets for the product and any special needs of these markets, such as language variations

As we had alluded to earlier, it is important to resist the urge to develop a solution at an early stage of product development. It is important that the product requirements are developed and defined in a "solution-free" context. When this task is developed, various concepts need to be explored. Again, it is important to remain at concept level and not drive down to details prematurely. Once the system is properly defined, then further details of subsystems and assemblies may be developed.

Quality Function Deployment

In the process of translating customer needs into product requirements and eventually design specifications, we should develop a matrix to relate the what of a customer need to the how that need is satisfied in a product. Developing the matrix of whats against hows is the basis of developing and using a tool that is called *quality function deployment* (QFD). This technique has gained a wide degree of acceptance as a robust tool in the initial stages of requirements and concept development (see, for instance, Jamnia 2017; Juran and De Feo 2010; Weiss 2013; or Ullman 2010).

Although different authors may develop the QFD matrix differently, the construct essentially contains a set of primary elements and a set of secondary elements. As shown in Figure 3.7, the primary elements are the whats and the hows. In addition, there is a list of importance associated with each what, along with the completeness list and the technical weights. These three metrics measure the relative importance of whats, the technical weight of each how as it relates to all the whats that may be impacted by it, and the completeness of design response to each what through the hows.

The secondary elements are shown in Figure 3.8 and include the interaction matrix, which appears as the roof of a house; target values, target direction, and technical benchmark rows that give the impression of the lower floors and the basement of a house; and columns associated with requirement planning and competitive benchmark. Because of

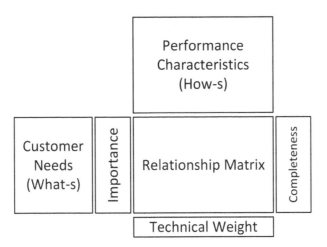

FIGURE 3.7
The primary elements of the quality function deployment.

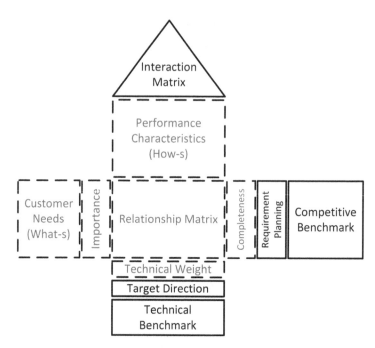

FIGURE 3.8
The secondary elements of the quality function deployment.

this semblance, this construct has been called a house of quality (HoQ) as well (ReVelle 2004; Ogrodnik 2013).

The competitive benchmark column and technical benchmark row provide a means of measuring the proposed hows against competition and their degree of success in responding to customer needs. The target value and direction rows encourage the product development team to focus its attention on quantifying the measures of how to achieve the whats and whether a higher or lower value is better. For instance, the need for a dental probe to measure pocket depth around teeth may be stated as "cause less pain." The "how" to achieve this may be to have a small tip diameter and a smooth surface finish. The target value and direction fields encourage the development team to assign a value, say, 0.1 mm, to the value field and a down arrow ($\downarrow$; "lower is better") to the direction field. Similarly, smooth may get a target value of 0.1 μm and again a down arrow. This mindset enables the team to think in concrete terms and objectively, as opposed to in more subjective terms. The interaction matrix shown as the roof of the house explores if two different hows complement one another, or if they are contradictory and compromises may be needed.

Using QFD in Product Development

The initial steps to build the primary elements of a house of quality are

1. Develop "whats" using VoC surveys and other tools.
2. Using tools such as the analytical hierarchy process, identify the rank and importance of each "what."

3. Identify the "hows" through brainstorming session(s).

4. Identify the "what"-to-"how" relationship.

Now, we are ready to take the steps required to finish the first house of quality. These are

5. Develop the "what"-to-"how" relationship matrix (as shown in Figure 3.7).

6. Calculate the technical weight row and the completeness column.

Once these steps are done, the following two steps may optionally also be taken:

7. Develop and include the secondary elements.

8. Flow down the house of quality to its second generation.

Once the first generation of the house of quality is completed, a second generation should be developed. In this second generation, the hows of the first are used as the whats of the second. Then the design team works on developing high-level design concepts and begin to answer the hows of the second HoQ. Once this is done, the hows are once again flowed down to a third HoQ and the process repeats itself. Through this methodology, the requirements are cascaded down from system to subsystems and eventually to design specifications.

Requirements Decomposition

Requirements decomposition is simply translating input requirements into usable specifications. It is simple, though the challenge is to ensure that functional lines are not crossed. For instance, the mechanical decomposition should not involve elements of electronics or software functions. For instance, suppose that a particular electrical power consumption requirement is 10 watts. It may be tempting to decompose the mechanical function requirements as the "the trace shall be able to carry a power load of 10 watts." It is more appropriate to specify "power traces shall have a width of x mm and a copper weight of y ounces."

Figures 3.9 and 3.10 depict the inputs and outputs of the decomposition process and the information that needs to be specified (or flushed out) for a mechanical requirements decomposition as well as an electronics requirements decomposition. A few points to consider: First, on the input side, one assembly may be impacted by a number

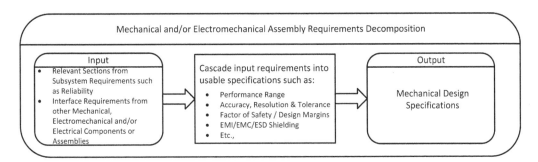

FIGURE 3.9
Inputs and outputs of mechanical/electromechanical requirements decomposition.

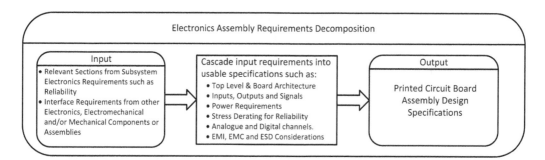

FIGURE 3.10
Inputs and outputs of electronics requirements decomposition.

of cascaded requirements. For instance, there may not only be mechanical requirements such as weight and volume but also concerns such as vibration and shock as well as heat dissipation. Also, on the electronics side, the requirements cascade may include power consumption as well as electrostatic discharge (ESD) or electromagnetic interference (EMI)/electromagnetic compatibility (EMC) issues. In a way, various system and sub-system requirements converge on the assembly and components. Thus, their designs should be traceable back to all their various and potentially conflicting system-level requirements.

Design specifications are the outcome of decompositions. In developing these specifications every effort should be made to phrase them in such a way that they are clear, viable, and measurable. For instance, suppose that a leaf spring exerts pressure on a shaft, and hence, it keeps the shaft from moving through friction forces. The specification for the spring may be developed as follows: "The leaf spring shall be capable of preventing shaft movement at a minimum of 1 N force applied laterally to the shaft at a rate of 0.2 N/sec, in dry conditions at 38°C."

Once design specifications are developed, several concepts are proposed and discussed among the design team for feasibility. The combination of specifications and the chosen concept(s) are from a portion of a set of inputs required to develop the design details and the three-dimensional mechanical CAD models and schematics files for electronics. Other inputs into developing detailed designs may be enumerated as follows:

1. Interface issues and functional tolerance concerns

2. Manufacturing approaches, techniques, and rationale for selection

3. Make-versus-buy decisions

4. Reuse and naming convention

5. Design for manufacturability, assembly, service, reliability, etc.

6. Prototyping and human factors studies

7. A list of critical design components and/or dimensions as well as design margins

8. A traceability matrix that links critical components and/or dimensions to requirements or risk

9. An understanding of service and/or failure conditions experienced by the end user

10. An understanding of reliability concerns and preventative maintenance concerns

Most subsystem and subassembly requirements may be cascaded to design specifications by considering the intended function of the subassembly and the means by which associated components are integrated to deliver the required function.

The V-Model in a Nutshell

For argument's sake, let us review the V-model that we introduced at the beginning of this chapter. Figure 3.11 depicts a typical V-model showing the stages of product development. In this chapter, requirements and their cascade (steps 1 through 3) were discussed. Once step 3 is well understood, design engineering work begins in earnest by developing one or several high-level designs. On the basis of a high-level design selection (step 4), focus will be placed on developing assembly requirements and specifications (step 5). With the specifications at hand, the actual three-dimensional modeling begins (step 6) leading to detailed component design (step 7). Although Figure 3.9 gives the impression that these steps are taken sequentially, in practice there is generally a high degree of interaction between steps 6 and 7 with occasional looping back to step 5 to update assembly requirements and specifications. Needless to say, if the assembly requirements are modified, the higher-level requirements need to be checked to ensure uniformity and traceability. Steps 8 through 12 focus on the right-hand side of the V-model by building and testing/verifying various assemblies, subsystems, and eventually the system and the final product. A more detailed treatment of the V-model in engineering product development may be found in Jamnia (2017).

Reliability Planning and Execution in the V-Model

Figure 3.12 provides a pictorial view of the reliability planning process and its relationship to the V-model. As depicted here, design for reliability begins at the product requirements level when the operational and use profile as well as customer expectations are reviewed, discussed, and approved. Next, reliability requirements based on customer expectations as well as past performances or competitive products are established. It is common that, initially, reliability goals are set. Once the design is established and these goals demonstrated, then documents are updated and goals converted into requirements. Furthermore, reliability growth is anticipated to happen in a period of 24 or 36 months post product launch.

Once components have been selected or designed, theoretical reliability models may be developed based on the bill of materials and industry standards such as MIL-HDBH-217 or Telecordia. At this step, a deeper understanding of system and subsystem behavior is developed by allocating the system-level reliability requirements to each subsystem. This information may dictate a redesign of a part or an entire assembly.

The next steps are associated with stages 8 and 9 of the V-model. The task here is to develop reliability test designs and include the number of needed samples into the first (or subsequent) builds. By conducting highly accelerated limit testing (HALT) and accelerated life testing, expected failure modes may be verified and any potential new ones be uncovered. Typically, at this stage of product development high-level design failure modes and

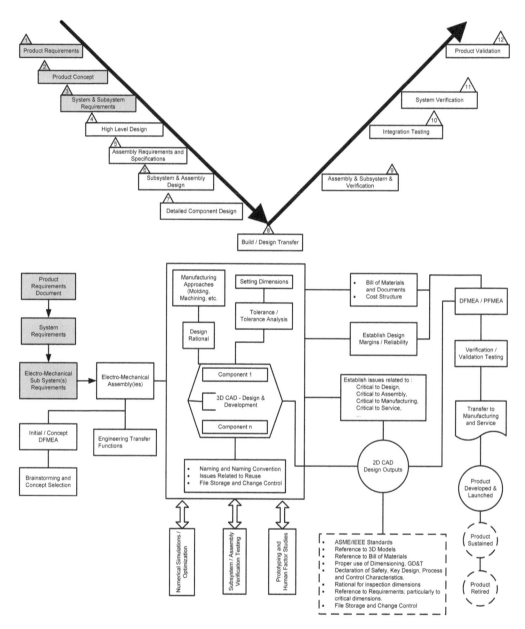

FIGURE 3.11
A typical V-model for product development along with the roadmap. The highlighted area indicates the segments that have already been discussed.

effects analyses (DFMEA) are in place. These documents may need to be updated based on the HALT. Should failure modes lead to effects with high severity, components or assemblies may need to be redesigned.

Once these iterative steps are completed, accelerated life testing may be conducted to establish product life expectancy and evaluate service maintenance schedules. Finally, reliability demonstration tests are conducted to provide objective evidence that reliability requirements can be met.

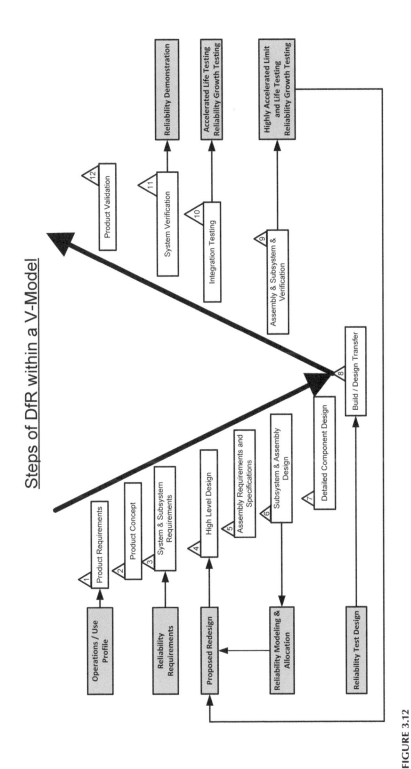

FIGURE 3.12
Design for reliability and the V-model.

Notes

1. Stage-Gate is a registered trademark.
2. The famous "If I had asked people what they wanted, they would have said faster horses" quote, generally attributed to Henry Ford, speaks to this fact.
3. DMADV stands for define, measure, analyze, design, and verify. This tool is used in product design, whereas DMAIC (define, measure, analyze, improve, and control) is used in failure investigations.
4. We use the term "problem statement," but in reality, it may also be called "opportunity statement" if a new market is to be created through this effort. For consistency sake, we will uniformly use "problem statement."
5. Notice that we do not say "fully" defined. We may come back and add to the define phase at just about any point as we learn more about the project.

4

Reliability Requirements, Modeling, and Allocation

Introduction

In Chapter 3, we reviewed the new product development process and various stages of design activities. We learned that in the input and planning phase, user needs and operating conditions play an important role in defining product requirements. In turn, these requirements lead to the system architecture and design concepts. Reliability requirements, being an aspect of design inputs, are defined either based on the business case, competitive edge performance, or safety thresholds,[1] or explicitly by the customer. Reliability requirements should be aligned with the design concept and other system requirements.

There is an interesting aspect of reliability requirements that sets them apart from other types of requirements. Most requirements are true and valid for all stakeholders. For instance, should a car engine have a fuel consumption requirement of, say, 20 miles per gallon, every stakeholder would come to agree with and support this fuel consumption. In the case of product reliability, different stakeholders have different needs, which may cascade to different requirements. For example, the service team would require a certain *mean time to repair* (MTTR), whereas manufacturing may be interested in a metric called *out-of-box failure* (OBF), which is a measure of latent manufacturing defects. Additionally, design engineering may need to allow for a reliability growth of the product because of latent design flaws that may not be discovered in the initial (prelaunch) product testing. There may also be different values of reliability requirements to be satisfied before the product design may move from one phase of development to the next. Therefore, at the product level, various "voices" need to be heard and accommodated in the set of reliability requirements.

Early in the design process, once reliability requirements have been set and design concepts developed, a series of activities called *reliability planning* should take place. These activities, which precede the development of design for reliability (DfR), include reliability analysis, allocation, feasibility, and modeling. They are meant to identify any gaps between the target and the attainable reliability inherent in the system design. Reliability planning is developing the strategy to focus DfR on closing these gaps within the product development cycle.

In the first half of this chapter, we will provide a guideline for understanding various requirements that different stakeholders may have, and how some businesses may select one over another. Additionally, we will provide an overview of how to drive reliability requirements from user needs. The second half of this chapter will focus on reliability allocation and feasibility analysis along with their impact on reliability planning and strategy.

Developing Reliability Requirements

In systems engineering and the world of writing requirements, the acronym *SMART* is quite well known. It stands for specific, measurable, attainable (achievable/actionable/appropriate), realistic, and time-bound (timely/traceable). The objective of having SMART requirements is that no one would ask what they mean. Hence, SMART requirements are easier to understand, to implement, and then to be verified that they have been implemented. Reliability requirements are no exceptions.

If we were to break down SMART requirements for reliability, we would have the following:

- Specific—Requirement should refer to specific and quantitative metrics. Shortly, we will review a few examples of these metrics.

- Measurable—Requirement should refer to a specific and quantitative metric, which may be verified by inspection, analysis, or testing during the product development time frame.

- Attainable—The product functionally should be well defined so that the measurable reliability requirements relate to their intended function. This includes descriptions of customer usage and operating environment profiles.

- Realistic—Allowable product degradation along with a clear definition of failure should be specified. This information is then incorporated into verification tests. It is preferred that components or subsystems are tested to failure as part of the reliability demonstration.

- Time-bound—The allowable degradation should be defined over time. In other words, the design team needs to understand what the design life of the product should be. This will have an impact on components selection and the associated verification activities. Time frame is expressed in a variety of measure, such as hours, years, cycles, distances, and actuations. A clear statement of time and design life should be specified as part of the reliability requirements.

Reliability Metrics

Earlier we mentioned that various stakeholders have reliability requirements that are suited to their needs. Most frequently—though not always—this need is expressed in terms of *mean time between X* or *mean time to X*, where X is an event such as failure or, say, service or repair. For some products, it may be more appropriate to express time in terms of other scales, such as cycles. So the metric of interest for a washing machine may become mean cycles between failures.

In this section, we will review some of the more popular reliability metrics and their implications. These are mean time between failures, mean time to failure, failure rate, mean cumulative function, availability, reliability and life, and probability of success rate.

Mean Time between Failures

A commonly known and used reliability requirement is the mean time between failures (MTBF) metric. MTBF is a practical metric for repairable systems and products.

It is practical because when a device fails, the failed component is either repaired or replaced, then device is placed back in service. Another component of the same device may fail some other time requiring additional repair activities. Note that a common mistake is to believe that MTBF is the time to the first failure of a product or device. This is not the case. Chapter 6 will provide a more in-depth and mathematical view of this metric.

MTBF is defined and calculated as the total time of operation of the entire population of a product in the field divided by all failures of the entire population in that period of time. Realistically, both the population and the failure data are noisy. For this reason, to remove the noise, manufacturers adopt an averaging period of 3, 6, or even 12 months to track and monitor MTBF.

The main business objective of this metric is to plan service resources along with spare parts inventory and predict the cost of service. We need to note that MTBF is not necessarily a constant value over time. In the period following the launch of a new product, there may be undiscovered design flaws along with immature manufacturing processes. These factors may lead to relatively higher failure counts and lower MTBF numbers. As both of these elements mature, the product reliability approaches a steady (albeit, noisy) value. Eventually, as the product population begins to age substantially, the rate at which devices fail increases leading to lower MTBF values once again. This behavior follows a trend that is commonly referred to as the bathtub curve (see Chapter 6).

Mean Time to Failure (MTTF)

Mean time to failure (MTTF) is another reliability metric that is similar to MTBF and is often confused with it. MTTF measures an average time to occurrence of failures. As we will explain in Chapter 6, mathematically, MTBF is the sum of the MTTF and MTTR of the same devices. However, in practice, MTTF is applied to nonrepairable units and systems[2] to differentiate between repairable and nonrepairable devices. For a nonrepairable product with a number of different failure modes, MTTF is sometimes referred to as *mean life*. This is because the product may no longer be used when the first failure occurs.

Similar to MTBF, MTTF does not denote the time to the occurrence of the first product to fail in the population, rather it is an average time of population failure.

Failure Rate

Another reliability metric that is associated with MTBF and MTTF is a product's failure rate. While there is often a discussion on the physical meaning of either MTBF or MTTF, failure rate is an extremely tangible metric. It reflects what the service organization sees in terms of repaired units, or what the finance team experiences in terms of cost of products replaced during the warranty period. For large fleets of devices or products, the average failure rate over a period of time is equal to the reciprocal of either MTBF or MTTF in that period.

What is worth mentioning here is how different stakeholders may view the three metrics introduced so far. For a manufacturer, there is little difference between the implications of MTBF, MTTF, or failure rate. The reason being that these three metrics are interrelated. However, from an end user point of view, none of these metrics are meaningful. An end user may be more interested in how reliable a single unit may be, how long it would last, or if it is dependable enough to be operated whenever the device is put to use.

Reliability and Life

The life metric was initially popular in the bearing industry and called B_x life. B stands for bearing and x is an index pointing to a percent of the units that fail. For instance, $B_{10} = 2$ years would mean that 10% of the bearings fail within a 2-year period. Eventually, this metric was more widely adopted outside the bearing industry as L_x life of nonrepairable products that are typically (though not always) owned and operated by a customer or end user.

Unlike the aforementioned measures, this metric does not assume a constant or average failure rate over time. Rather, it assumes that there is a distinct probability of failure of the design at each time segment. In specifying this metric as a requirement, care must be exercised to understand how the applied stresses in the new design may have an impact on the previously known failure data.

In Chapter 6, we will examine reliability and L_x again and discover their interconnection to MTBF or MTTF.

Availability

Within the context of reliability, *availability* is defined as the proportion of time a system is in functioning mode as expected, or in a specified operable state, when it is called to perform a mission at any point of time. It stands to reason that as a system's reliability increases, its availability also increases. Barlow and Proschan (1975) defined availability as the probability of a system operating at a specified time.

Availability is a function of the relationship between the operable time and the downtime of the design. A common formula for calculating *average availability* is given by

$$A = \frac{\text{Average Uptime}}{\text{Average Uptime} + \text{Average Down Time}}$$

This formula can be expressed in terms of other reliability metrics as follows:

$$A = \frac{MTBF}{MTBF + MDT} \text{ or } \frac{MTBF}{MTBF + MTTR}$$

where MDT is defined as mean down time.[3]

The availability metric is often used for high-throughput and rather complex products that may require high-frequency maintenance and repair activities. For example, fighter aircraft require some type of service activity after each mission. Another example is large equipment in power plants that need to be maintained on a regular basis. Availability of this equipment has a direct impact on whether electric power is available for customers.

Scheduled preventive maintenance becomes an essential aspect of reliability modeling for the type of products which require maximum reliability during the operation time. The model generally provides a schedule to conduct proactive service and repair during a predetermined down time when systems are not in their operating shifts.

Mean Cumulative Function

A lesser-known reliability metric for repairable systems is called *mean cumulative function* (MCF). Simply put, MCF is a measure of the number of allowable failures over either the design or service life[4] of a product. Nelson (2002), however, suggested that MCF is measured as the expected number of failures at a point of time.

The MCF is a nonparametric statistical method that enables simplified methodologies and provides the analyst the ability to show event counts, costs, and maintenance down times (indicator of availability), among other factors. Compared to MTBF and the inherent noises in that metric, MCF provides the population metric of interest for each function. This metric often follows a staircase curve with unequal step rises that denote the behavior of the product in the field. At any given age or time (t), the corresponding distribution of the metric curves has a mean $M(t)$, which is denoted by MCF.

Mathematically, in testing of N systems, the average MCF at any time is the average of the MCF of individual systems calculated as follows:

$$M(t) = \frac{\sum_{i=1}^{N} M_i(t)}{N}$$

where $M_i(t)$ is the number of failures observed on sample i at time t.

Tracking MCF in the field requires knowing the time to failure of each product. This may sound simple in theory but in practice becomes a monumental task. However, MCF as a requirement may be used to develop requirement thresholds for subsystems and components, and enable test designs and studies. Manufacturers may use MCF as a reliability requirement to control cost of service, on the one hand, and, on the other, ensure end user perception of the product reliability and satisfaction.

In the chapters to come, we will explain the methods of calculating and demonstrating the main reliability metrics including MTBF, reliability, L_x life, and probability of success. We need to keep in mind that these metrics are the basic metrics for reliability that can be measured and calculated directly from either test or field data. Other reliability metrics such as availability, failure rate, and mean cumulative failure are all derivative of the aforementioned basic metrics and can be calculated and revised, and monitored in the field over time based on real-life experience.

Probability of Success Rate and Out-of-Box Failures

Up to this point and implicit in our discussion, the reliability metrics have been applicable to products that may be used over and over again. These metrics are used for products where age or repetitive use impacts reliability. On this basis, they are not applicable to disposable or consumable products. For single-use items, a more appropriate metric is out-of-box failure or defect rate. *Defect rate* is probability of a single-use product failing as an end user attempts to deploy the product. In this metric, the basic assumption is that the reliability of each item is totally independent of the success or failure of the previous or next item.

Required Inputs for Developing Requirements

Up to this point in this chapter, we captured a variety of reliability metrics that may be used as requirements. In this section, we will talk about information needed to make the appropriate metric selection based on the needs of various stakeholders. Recall that the needs of stakeholders are often called a *voice*. For instance, the voice of the customers implies what customers want. In developing reliability requirements, one needs to be particularly sensitive to a balance between design, operational, and economical needs expressed either implicitly or explicitly.

Reliability Requirements Based on Voice of Business

The business plan, which reflects the business' needs (otherwise known as the voice of the business), is developed at the start of new product development activities. One of its purposes is to define the required budget for manufacturing, installation, repair, spare parts inventory, and product throughputs among other factors. We cannot overemphasize the significance of these budgets and fund allocations. The manufacturing cost target of a product will influence the quality of selected components in the design, which in turn impacts its reliability. The development budget influences the level of resources that are dedicated to the design and the time needed to flush out design flaws during the development cycle. In Chapter 1, we mentioned that higher product reliability may be achieved through selecting more reliable (and therefore more expensive) parts or implementing redundancy in the design. Alternatively, a higher reliability may be realized with additional development time to identify and eliminate various failure modes. The target cost of a product plays a dominant role in its inherent reliability, and hence, in defining its reliability requirements. In practice, cost of manufacture as a reliability requirement is reflective of the voice of the business.

Another element in defining reliability requirements, from a business point of view, is the target *cost of service* (CoS) for a new design. CoS is particularly important to a business during a warranty period during which it is the manufacturer who pays and not the customer. To account for this factor, we recommend developing a reliability requirement on the basis of CoS during the warranty period as follows:

$$R = 1 - P_f$$

$$R = 1 - \frac{\text{Maximum CoS Target during warranty}}{\text{Number of Parts (Average Part Cost} + \text{Average labor Cost)}}$$

where R is reliability of the product at the end of a warranty period and P_f is the probability of failure. For example, a reliability requirement on the basis of cost of service may read as follow:

> More than 85% of the product population shall survive the first year of operation and more than 75% of the product populations shall survive to the end of the second year of operation.

Our role as reliability engineers is to decipher what this requirement would mean for the design. Later in this chapter, we will discuss reliability allocation and cascade to subsystems and assemblies.

Let's consider another example. A desktop printer manufacturer conducts a cost and revenue analysis to develop a new low-cost, low-margin printer. This organization's business model was to base its revenue on the sale of its ink cartridge and to use the printers as a sales lead. This business would realize substantial profits on the sale of its ink cartridges if it sold at least three cartridges to 75% of its customers. The expectation was that printers would have high failure rates due to the low cost of manufacture. However, they need to survive long enough to use the prescribed three cartridges.

How should we as engineers translate this business need into a reliability requirement? Considering that the function of the printer is to move ink out of the cartridges and on to the paper, the reliability requirement may be based on a "page yield" metric, which is a measure of the number of pages that might be printed with a certain density of printed area. For this organization, the average page yield of each cartridge is 1200 pages with 65% of each page covered by text. The reliability requirement in this case was written as follows:

> 95% of the printers shall be capable of printing at least 3600 pages of a 65% text content to be demonstrated in testing at a 90% confidence, without any failure.

Reliability Requirements Based on Voice of Customer

In Chapter 3, we reviewed how the voice of the customer (VoC) influences product requirements. One example of a reliability VoC was that product design shall be reliable. This is a rather vague expression of a need and should be cast in a more explicit expression using voice of the business or competitive products. In the same chapter, we suggested a product level reliability requirement to be product design shall have a life expectancy of 5 years during which the average failure rate of less than 1.0% per month shall be maintained for the population.

Some customers may express their reliability requirements more explicitly. These may be expressed in terms of productivity, minimum downtime, scheduled service plan, or cost of services, to name a few. For example, a medical analysis lab defines its main performance metrics in terms of daily throughputs to be no less than 36,000 samples per day. It would also plan downtime for repair and maintenance. These metrics are then used to manage resources for operation and sample processing.

In this scenario, the lab management does not really care if the processing equipment has a low MTBF or need scheduled services. Rather, it would want to know how quickly a unit in need of repair may be brought back to service. So long as the total throughput of the equipment is at or above 36,000 samples, units could fail and be maintained at any acceptable rate. In scenarios like this, a customer-based requirement may be expressed as follows.

> The deployed product fleet at each lab shall support a daily throughput of 36,000 samples averaged on a monthly basis.

This requirement speaks to the availability of the product fleet. Later in this chapter, we will discuss how this requirement may be cast in the form of a reliability metric.

Reliability Requirements Based on Use Profile

Failure rate, MTBF, or MTTF are reliability metrics that are aligned with and influenced by the use profile of a fielded product; however, they do not address it directly. Clear and quantifiable use profiles express one or more of the following three elements: product mission, use distribution, and operating conditions. These three elements are discussed next.

Product Mission

In brief, *product mission* defines what functions are performed with a given time duration. An example of such a mission is as follows:

> The pump shall have a reliability of 95% after delivering 1000 fills at an average filling rate of 150 liters per hour for an average of 10 hours operation.

Product Use Distribution

Obtaining the product *use distribution* is a nontrivial and extremely important task. Analysis of this data reveals how a fielded product is used by customers. In Chapter 5, we will discuss this distribution and its uses in some detail. In short, this distribution indicates how average customers along with low- and high-usage customers have operated predicate products in the past. This information may be used to define what the user expectations will be. For instance, for the printer manufacturer mentioned earlier, a use-distribution-based requirement may be as follows.

> The printer shall have a 3-month rolling average MTBF of 10 months when it is used to print up to 500 pages per day with an average text fullness of 60% per page.

Use and Operating Conditions

The *operating conditions* requirement simply states what the reliability of a product should be under a *given set of conditions for a time period*. The caveat is that the reliability should be maintained at the end of the required time under varying stress. This requirement implies that our product should be functional for a known period of time (typically design life) under the stress conditions that are applied with use. For instance, if we operate a device in a desert environment, we anticipate large temperature fluctuations between day and night. For an electromechanical product, this temperature fluctuation would induce thermal fatigue or creep of solder joints leading to lower reliability. Another example is the variety of medical devices deployed on an ambulance. The environment of use will not only be a wide range of temperature variation, but it also includes shock and vibrations associated with vehicle movement. In a scenario like this, accurate distributions of the temperature range over the year, vibration spectrum, and shock profile, as well as any other stressors, should be described in the statement of reliability requirement. We will discuss verification of this type of requirement in Chapter 8, using a test method called cumulative-damage testing.

The reliability requirements may clearly state the expected use and operating conditions and stresses over the design life of the product. An example would be as follows:

> The product shall have a reliability of 25% after 7 years of performing a medical infusion at rate of 150 milliliters per hour for 10 hours on a day at a temperature represented by a normal distribution with a mean of 18°C and a standard deviation of 5°C and a maximum vibration level of 20 G_{rms}.

Use and operating conditions that affect reliability requirements include, but are not limited to, temperature, humidity, vibration, pressure, shocks, altitude variations, operator skill, and mounting position and orientation.

Reliability Modeling and Allocation

Recall Figure 3.12 (see Chapter 3) in which the DfR process within the V-model was depicted. From this model, reliability requirements are developed based on customer needs, business case, and benchmarking with competition. However, the established reliability requirements may not be fulfilled unless the components and selected parts within the design are inherently capable of meeting those requirements. This raises the question of how reliable components should be in order to meet the overall system reliability. To answer this question, we need to allocate and budget reliability values to each subsystem and module. Once this is done, we can then compare allocated values to the inherent reliability of each component. To do this, we need to create a reliability model, which begins with developing the reliability block diagram. Then, the reliability of each block maybe calculated or estimated depending on available data.

Reliability Modeling

Reliability modeling is simply using the system architecture to break down the system into functional subsystems and modules. Often, this breakdown is called the *reliability block diagram* or RBD. In this model, the interrelation of each module to others is represented graphically, and the contribution of each module's reliability to the entire system is calculated mathematically. Each module's contribution to the overall system reliability is the reliability allocation to that module. For those of us who are familiar with network diagrams in either electrical or mechanical engineering fields, the RBD is the network diagram in the reliability discipline.

Similar to any other network diagram, there are two basic elements as shown in Figure 4.1. They are series and parallel networks.

The equivalent reliability of a network of components in series is the product of the reliability of each block. In other words,

$$R_{\text{Series eq.}}(t) = \prod_{k=1}^{N} R_k(t) \tag{4.1}$$

In this equation, N is the number of blocks in series and t is the duration of time of interest. The equivalent failure rate of a network of components in series is the sum of the reliability of each block. In other words,

$$FR_{\text{Series eq.}} = \sum_{k=1}^{N} FR_k \tag{4.2}$$

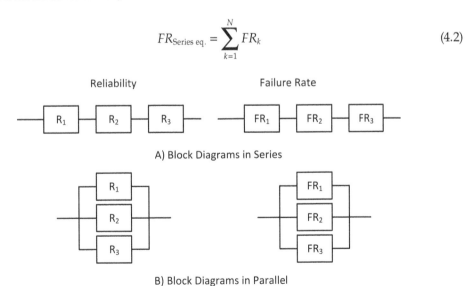

FIGURE 4.1
Elements of reliability block diagrams.

The equivalent reliability of a network of components in parallel is expressed as

$$R_{\text{Parallel eq.}}(t) = 1 - \prod_{k=1}^{N}[1 - R_k(t)] \tag{4.3}$$

Same as before, in this equation, N is the number of blocks in series and t is the duration of time of interest. The equivalent failure rate of a network of components in parallel is as follows:

$$\frac{1}{FR_{\text{Series eq.}}} = \sum_{k=1}^{N}\frac{1}{FR_k} \tag{4.4}$$

We draw your attention momentarily to a topic commonly mentioned when discussing reliability networks. In a series RBD, failure of a single block suggests failure of the entire system. Whereas in a parallel RBD, failure of a single block does not inhibit functioning of the other blocks. The implication is that should two (or more) identical blocks be set up in parallel, in the event that one fails, the entire system can function on the second unit while the first block is being replaced or repaired. A system with this configuration is called a *redundant* system. The failure rate of a redundant system may be calculated as follows:

$$FR_{\text{Parallel eq.}} = \frac{FR}{\sum_{i=(N-q)}^{N}\frac{1}{i}} \tag{4.5}$$

where N is the total count of components, and $(N - q)$ is the required number of components for operation.

For example, suppose that the components shown in Figure 4.1B have equal reliability and failures. Furthermore, only one out of the three components is required for the system to operate. The failure rate of the three-element system is given as follows:

$$FR_{\text{Parallel eq.}} = \frac{FR}{\sum_{i=1}^{3}\frac{1}{i}}$$

$$FR_{\text{Parallel eq.}} = \frac{FR}{\frac{1}{1}+\frac{1}{2}+\frac{1}{3}} = \frac{6FR}{11}$$

Notice that the failure rate of the parallel system is less than the failure rate of a single component, i.e., the reliability of the parallel system improved by having a 1-out-of-3 redundant system. Equation 4.5 is valid under the assumption that all components in parallel are active during operation, however, only $(N - q)$ is required for the success of the mission. More information about failure rate calculation for different in-parallel configurations and assumptions is available in the Reliability Analysis Center's Reliability Toolkit (1993) and MIL-HDBK-338.

In the strictest sense, there are other forms of networks that may be formed. However, a description of them is beyond the scope of this book.

Example

An example of an in-series system is shown in Figure 4.2 where an AC power source operates a pump to fill a tank. Assuming the reliability of the main parts in this system are R_{AC}, R_{PUMP}, and R_{TANK}, respectively, the system reliability is calculated as shown in Figure 4.2, based on the following equation:

$$R_{\text{Series eq.}}(t) = \prod_{k=1}^{3} R_k(t)$$

Suppose that the reliability of the selected components for the AC power source, the pump, and the tank are 0.92, 0.87, and 0.98, respectively, within a 1-month period of operation. Based on this data, the overall system reliability is

$$R_{\text{system}} = 0.92 \times 0.87 \times 0.98 = 0.78 \text{ or } 78\%$$

Note that the overall reliability of the system is lower than the reliability of each individual component. The implication of a 78% overall system reliability of this system is that in a 1-month period of interest, the probability of failure is (100% − 78%) or 22% of the time.

Now suppose that a 22% probability of failure is not acceptable to the business due to high risk of not having enough liquid in the tank. To mitigate this type of risk, a system with redundancy can be used. An example of a redundant system for this example is shown in Figure 4.3. To mitigate power loss, a battery with a reliability of 0.90 is added, and to mitigate the loss of pumping a second identical pump is added.

To calculate the reliability of this system, we first need to find the equivalent reliability of the redundant power subsystem composed of AC power and battery and then calculate the reliability of the two pumps in parallel. Then the system reduces to three blocks in series as shown in Figure 4.4.

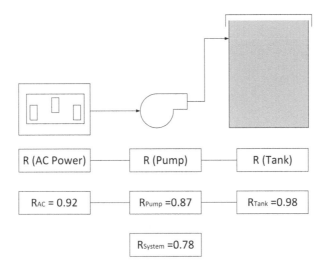

FIGURE 4.2
An example of developing a system reliability model: a tank filling system with no redundancy.

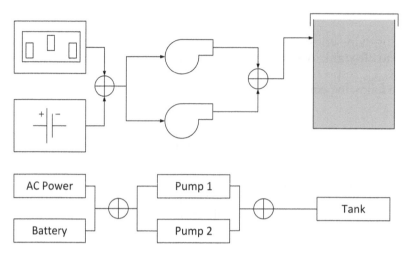

FIGURE 4.3
An example of developing a system reliability model: a tank filling system with redundancy.

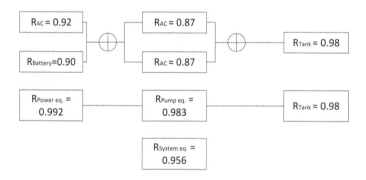

FIGURE 4.4
An example of developing a system reliability model reducing component reliability into a single system value.

To calculate the equivalent reliability of the power subsystem, we have

$$R_{\text{Power eq.}} = 1 - \prod_{k=1}^{2}[1 - R_k]$$

$$R_{\text{Power eq.}} = 1 - (1 - R_{AC})(1 - R_{\text{Battery}})$$

$$R_{\text{Power eq.}} = 1 - (1 - 0.92)(1 - 0.90) = 0.992$$

Similarly, we can calculate the equivalent reliability of pumping

$$R_{\text{Pumping eq.}} = 1 - \prod_{k=1}^{2}[1 - R_k]$$

$$R_{\text{Pumping eq.}} = 1 - (1 - 0.87)(1 - 0.87) = 0.983$$

Now, we can calculate the reliability of the system in series:[5]

$$R_{\text{System eq.}} = 0.992 \times 0.983 \times 0.98 = 0.956 \text{ or } 95.6\%$$

As is demonstrated in this example, redundancy can in fact increase the reliability of the entire system but at the cost of extra components. We should also note that components will continue to operate at the level of their inherent reliability, and redundancy does not reduce component failure rates. It can only be used to contribute to increased system availability. As we will explain later, through redundancy, we improve system (or mission) reliability, but there will be an increase of size, weight, and cost of ownership in terms of manufacturing as well as maintenance and repair.

Reliability Requirement Allocation

Previously, we demonstrated a simple application of RBDs and how to calculate system-level reliability using each block's reliability values. We like to point out that there is not a simple and straightforward relationship between component and system-level reliabilities. Depending on the configuration of the RBD and as demonstrated in the tank-filling example, system reliability may be lower or higher than its components—or even possibly the same.

This poses a problem in cascading reliability requirements from the system level down to subsystems and components. Fortunately, almost all RBDs may be formed in such a way that redundancies are within subsystems or modules. In other words, at a level just below the system, all subsystems are in series. As one might have noticed, in the aforementioned tank-filling example, just below the system, there are three subsystems that are in series, namely, power, pumps, and the tank. It is at the subsystem that redundancies are seen.

Why is this important? It is important because we can cascade system-level reliability requirements to subsystems very easily. Note that for a system in series

$$R_{\text{system}} = R_{\text{subsystem 1}} R_{\text{subsystem 2}} \cdots R_{\text{subsystem } n}$$

If we have a system of identical subsystem reliabilities, we would have

$$R_{\text{system}} = R_{\text{subsystem}}^{n}$$

Historically, the assumption of a subsystem with equal reliabilities led to the method of *equal allocation*. This approach assumes that the target system reliability is achieved by allocating the same (or equal) reliability to all subsystems. We now know that this method is unrealistic. It is seldom to have a system that is composed of subsystems of equal importance, equal functionally, and equal inherent failure rates.

A more rational and realistic approach is to allocate reliability based on design criteria and ranking of relevant subsystems to determine the contribution of each subsystem to the reliability of the system. Shortly, we will discuss what these criteria may be, but first we need to review the concept of allocation and the apportionment table.

In a general sense, reliability may be expressed as an exponential relationship of failure rate (λ) and time (t), i.e., $R = e^{-\lambda t}$. Therefore, we can say that the failure rate of a system is the sum of the failure rate of its subsystems and components—provided that they are in series. In other words

$$\lambda_{system} = \sum_{i=1}^{N} \lambda_i$$

This is known as the *parts count* formulation of reliability and is widely used.

Based on this formula, it would be a straightforward calculation to cascade the system-level reliability requirements into subsystems using an apportionment table that would guide us on what portion of the failure rate would be assigned to each subsystem. Assuming that we have this apportionment table,[6] the steps for this flow down is as follows:

1. First establish the system failure rate (λ_{system}) based on available data or target reliability.

2. Next multiply the percent apportionment for each subsystem with the system failure rate to get the subsystem failure rate:

$$\lambda_{subsystem} = \% \, \text{Apportionment} \times \lambda_{System}$$

3. Cascade the subsystem failure rate further down to lower-level modules using the same approach.

Example

Let us assume that Table 4.1 is the apportionment table for a hypothetical system with seven subsystems.

The requirement of this system is to have an MTBF of 50 months. Using this value, we can calculate the required failure rate:

$$\lambda_{System} = \frac{1}{\text{MTBF}_{System}} = \frac{1}{50}$$

$$\lambda_{System} = 0.02 \text{ or } 2\% \text{ per month}$$

TABLE 4.1

Example of Failure Apportionment for Subsystems

Subsystem	Apportionment
Controls	31.5%
Wireless	18.5%
User interface	28.5%
Hydraulics	10%
Receiving module	7.5%
Wiring	2.5%
Caseworks	1.5%
Total	100%

Note: The apportionment column should add up to 100%.

TABLE 4.2

Example of System to Subsystems Reliability Requirement Cascading

Subsystem	Apportionment	Subsystem Failure Rate (λ)	Subsystem MTBF
Controls	31.50%	$31.5\% \times 0.02 = 0.0063$	$\dfrac{1}{0.0063} = 159$
Wireless	18.50%	$18.5\% \times 0.02 = 0.0037$	270
User interface	28.50%	0.0057	175
Hydraulics	10.00%	0.002	500
Receiving module	7.50%	0.0015	667
Wiring	2.50%	0.0005	2000
Caseworks	1.50%	0.0003	3333
Total	100%	0.02	$\dfrac{1}{0.02} = 50$ at system

Based on this value and the apportionment in Table 4.1, the failure rate and the MTBF (in months) for subsystems may be calculated as shown in Table 4.2.

Developing the Apportionment Table

Now that we see how application of an apportionment table works, we need to understand how to develop such a table using the appropriate logic to budget failure rates that may not be solely on part count. Admittedly, there is not an exact science behind this task. The first and foremost approach is to conduct an analysis of the complexity of each subsystem, their importance, and their design evolution. These elements can be characterized and quantified as follows:

1. Reused components—By using components and modules that have already been used on previous products, we can draw on our knowledge of their failure. This will reduce the risk of unknown reliability or higher failure rate of modules with high state-of-the-art scores.

2. State-of-the-art—Novelty of the technology used in the design, along with engineering team knowledge of associated failure modes, will be a major contributor to failure apportionment.

3. Design complexity—Complexity of each module as a function of part counts and complexity of design involved contribute to the apportionment table.

4. Duty cycle—This is an indication of the mission time at which an item may be vulnerable to wear or degradation. Additionally, it is indicative of the importance of the same component to the overall function or performance of the system.

5. Loading or stress—The higher the stresses or the harsher the environment under which a component in the device is operating, the higher the expected failure rate.

6. Software change required—At times, we may be able to control the stresses on a component through the use of software. For instance, for the thermal management of a device, a fan may be needed. By powering the fan constantly, it may

fail more frequently than by turning it on when internal temperatures exceed certain thresholds. This control is done through software. Should such controls be needed or used, they are usually indicative of higher defects and failure rates.

7. Cost of service—This criterion is used to assign lower desired failure rates to subsystems with the highest cost of service in terms of parts, labor, etc.

8. Mean downtime or repair time—Similarly, this criterion is used to assign lower desired failure rates to subsystems with the longest downtime or repair time to increase availability by increasing uptime.

Reliability Requirements Feasibility and Gap Analysis

Once requirements have been cascaded to subsystems, we need to ask whether the inherent reliability of the subsystem exceeds its requirement. Inherent reliability or failure rate of each subsystem is collected from predicate or similar product actual data. Care must be exercised in using predicate data, as differences in usage, duty cycle, and application may be present.

In cases where gaps exist, we need to examine whether design improvements would close the gaps. In cases that design improvements are not sufficient, maintenance programs must be developed. Regardless of the decision made, impacts on other business matters such as project time lines or required resources must be evaluated.

Case Study

A new product to be developed is composed of four subsystems. It has three phases of operation to complete its mission:

1. Power up and self-test
2. Sample processing and results analysis
3. Cleaning up and shutting down

The reliability requirement has been determined based on customer satisfaction, user needs, and the limitations of technology:

The system shall have a maximum of 1.5 failures per year.

This requirement of the system has to be allocated to the four subsystems. The design of this product includes novel technology, software changes, increased productivity and throughputs identified as major user needs, and cost of service as key parameters in the business case submitted by the marketing department for the new design.

Table 4.3 shows the breakdown of functions, time spent at each function, and the noninvolvement of each subsystem at each function in order to define the duty cycle apportionment of each subsystem. Based on this factor alone, one would expect that subsystem 4 would have the most failures since it has the highest duty cycle of operation in the system. Note that in general, any and all criteria may be included in this table. Table 4.4 accounts for design elements. In this case study, we have subsystem intricacy, novelty of the design and technology, complexity of movements, required changes to software, and firmware involved in each subsystem. These criteria can be used to identify expected faults and failures from each subsystem operation.

TABLE 4.3

Duty Cycle Breakdown per Subsystem and Functions

	Function 1	Function 2	Function 3	
	Power Up and Self-Test	Processing & Results Analysis	Cleaning Up and Shutting Down	**Relative Duty Cycle of Each Subsystem Spent at System Level**
% of System Time Spent at Each Function	**5.00%**	**90.00%**	**5.00%**	
Module/ Subsystem	**Involvement of Each Subsystem at Each Function, e.g., Number of Active Parts to Perform This Function, or Number of Subfunctions Performed by Each Subsystem**			**Sum of %Time × % of Each Function Involvement**
Ssy1	0	1	0%	$0.90 = 0\% \times 5\% + 100\% \times 90\% + 0\% \times 5\%$
Ssy2	0	0.9	0.5	$0.84 = 0\% \times 5\% + 90\% \times 90\% + 50\% \times 5\%$
Ssy3	0.5	0.8	0.75	$0.78 = 50\% \times 5\% + 80\% \times 90\% + 75\% \times 5\%$
Ssy4	1	1	1	$1.00 = 100\% \times 5\% + 100\% \times 90\% + 100\% \times 5\%$
				3.5175

The calculation of the target failure rate allocation in Table 4.4 is the result of the ratio of the multiplication of all the allocated weighted factors for all the criteria for each subsystem. For example, for subsystem 4 we have

$$\text{Total Failure Rate Contribution} = 6 \times 5 \times 1 \times 5 \times 6 \times 5 \times 9 = 40500$$

$$\text{Target Failure Rate Allocation } \% = \frac{40500}{145148} = 0.17 \text{ or } 17\%$$

$$\text{Allocated Failure per Year} = 17\% \times 1.5 = 0.26 \text{ Failures per Year}$$

The output of the reliability allocation is to identify the flow down of system requirements into reliability apportionment, and compare it to the inherent (basic) reliability of the selected parts. This table may be used to identify whether any reliability gaps exist. The next step is to estimate the inherent reliability of the design. This is done based on system architecture, reliability of the part selected and counts, and field data from existing legacy design.

Legacy and Field Reliability Data Analysis

The initial list of components selected for this design is provided in Table 4.5. We have gathered and analyzed available reliability data on all selected parts along with information from legacy or similar products or designs. Our sources of reliability data were the supplier's data based on testing, published databases, and reliability information such as Telcordia (2006) and NPRD (1995). Table 4.5 shows the estimate of the inherent reliability in the design. This table provides the aggregate failure rate of the selected parts as 2.75 per year, while the reliability requirement for this design is equal to or less than 1.5 failures per year.

TABLE 4.4

Reliability Allocation Based on Customer Input and Design Specifications and Limitations

Contributing Factor	A	B	C	D	E	F	G	Total Failure Rate Contribution	Target Failure Rate Allocation % =	Target Failure Rate Allocation (Failure per Year)
	Intricacy (part count)	Novelty	Relative duty cycle (from Table 4.3)	Loading and stresses	Software changes required	Cost of service in failure	Down/repair time	A×B×C×D×E×F	A×B×C×D×E×F×G/System Total	
Rank or Influence on System (from 1 to 10)	10 is higher part count	10 is most novel technology	10 is more time and duty cycle	10 is highest stresses	10 is a complete new software	10 is lowest cost of service	10 is lowest downtime			
Ssy1	5	6	0.9	5	10	5	4	27000	19%	0.28
Ssy2	8	8	0.835	5	9	5	5	60120	41%	0.62
Ssy3	4	2	0.7825	10	8	5	7	17528	12%	0.18
Ssy4	6	5	1	5	6	5	9	40500	28%	0.42
System							25778	145148	100%	1.5

TABLE 4.5

An Example Estimate of Inherent Reliability of the Design

Module	Parts	Data Source	A Identified Failure Count	B Normalized Failure Count	C Stress Change Factor	=B×C	Failure per Year (Prorated Failure per Year)
				=A/Population		=B×C	Sum of B×C for each subsystem
Ssy1	C1	Field	23	0.18	1	0.18	Ssy1 =
	C2	Field	30	0.24	1	0.24	0.88
	C3	Field	29	0.23	1.2	0.28	
	C4	Test	19	0.15	1.2	0.18	
	C5	Field	21	0.17	1.2	0.20	
Ssy2	C6	Field	18	0.14	1	0.14	Ssy2 =
	C7	New	14	0.11	1	0.11	0.92
	C8	Field	17	0.14	1	0.14	
	C9	Field	12	0.10	1	0.10	
	C10	Test	27	0.22	0.9	0.19	
	C11	Test	20	0.16	0.5	0.08	
	C12	Field	22	0.18	0.5	0.09	
	C13	New	17	0.14	0.5	0.07	
Ssy3	C14	Field	18	0.14	1	0.14	Ssy3 =
	C15	New	14	0.11	1	0.11	0.70
	C16	Field	29	0.23	·1	0.23	
	C17	Field	22	0.18	1.2	0.21	
Ssy4	C18	New	17	0.14	0.5	0.07	Ssy4 =
	C19	Field	22	0.18	0.5	0.09	0.44
	C20	Test	18	0.14	0.5	0.07	
	C21	Field	18	0.14	0.5	0.07	
	C22	Test	13	0.10	0.5	0.05	
	C23	Field	21	0.17	0.5	0.08	
Total	Total		461	3.69			2.94

* Number of units in the field = 125 in 1 year.

The output of the inherent reliability of the design is to define the attainable reliability of each subsystem or comments and to be compared to the allocated reliability so that the reliability engineer and the design team are aware of the reliability gaps in the design.

As shown in Table 4.5, C4 had 19 failures in the legacy data over 125 units during 1 year of service; this means that C4 has on average: 19/125 = 0.15 failures per year. Also, we know that C4 will be used at a 20% higher stress in the new product, hence, the C4 predicted failure rate will be $1.2 \times 0.15 = 0.18$ failures per year. This means that the subsystem 1 failure rate will be

$$\text{Failure per Year} = 0.18 \times 1.0 + 0.24 \times 1.0 + 0.23 \times 1.2 + 0.15 \times 1.2 + 0.17 \times 1.2 = 0.88$$

Reliability Gap Analysis

The next step is to conduct reliability requirements feasibility to define reliability gaps and plan the strategy to address them for each subsystem. We do this by comparing the allocated reliability to the inherent or known reliability for each subsystem as shown in Table 4.6.

TABLE 4.6

Reliability Requirements Feasibility, Gaps, and the Strategy to Close Gaps

Module	Inherent (Basic Failure per Year)	Allocated Failure per Year Requirements	Gap	Reduction Task	Rationale for Risk in Closing Gaps
Ssy1	0.88	0.28	0.61	Implement current recommendations from design team	Supported by current design improvement
Ssy2	0.92	0.62	0.30	New development effort	Newly licensed technology with no current improvement plan
Ssy3	0.70	0.18	0.52	Upgrade assembly XXX to handle increased duty cycle	Unable to do this so far, but additional resources and focus should help
Ssy4	0.44	0.42	0.02	Assess newer design	Similar improvements seen in other design refreshes
Aggregate Initial Estimate	**2.94**				
Target		**1.50**			
Aggregate Initial Gap			**1.44**		

The output of the reliability feasibility and gap analysis is used as the input for design for reliability planning and strategy.

Reliability Planning Strategy

In the next chapter we will explain the required inputs to the design for reliability. These inputs are required to create a plan that will define all the necessary tasks and activities to ensure and demonstrate reliability of the design. Reliability modeling and allocation conducted in this chapter is one of those inputs. It helps us to identify what areas of the design require focus and resources to ensure reliability requirements are achieved.

The following guidelines are useful for creating the reliability strategy and the reliability plan for the design:

- For components with unknown failure modes or new technology, exploratory testing such as HALT and design margin testing needs to be planned.
- For components with no previous history on similar applications or supplier test data, a reliability study on a small sample size needs to be performed.
- For reused parts with known failure modes and high reliability, make sure proper stress derating is conducted along with recalculation of new failure rates.
- For components with a "service effectiveness" focus in the reliability plan, ensure the inclusion of service training on actual returns from the field via teardown analysis.
- For components known to have limited life or severe wear out failure modes, make sure to have a life test (test to failure) to establish the actual reliability and cut-off limit of proactive replacement or service/maintenance.

- Based on the identified risk areas, reliability growth and demonstration tests should be planned to ensure any risks are mitigated and minimized.

Other Considerations in Reliability Requirements Development

Reliability requirements are demonstrated through testing and/or reliability analysis prior to market launch. We need to be mindful that the applied stresses during the reliability demonstration period do not reflect the actual stresses that the product experiences in the field. In-house reliability testing demonstrates the capability of the design in a controlled environment. Sample sizes are limited—typically from a single build—and there are minimum use errors or misuse scenarios. Furthermore, these tests do not reflect shipping damages, manufacturing defects, or other adverse influences such as use errors. In short, demonstration testing examines the long-term reliability of a newly developed product.

Typically with any newly launched product, there is a learning curve. Use errors are at their highest level at the launch when the end users are not familiar or well trained. These errors and their associated service events decrease over time with customer's familiarity with the product and improvements in user manuals and troubleshooting guides. Shipping and manufacturing process issues are random and can be triggered anytime when a third party or a supplier has an out-of-control process that impacts specifications. Service and repair-induced failures may be experienced with some parts at the beginning of the product life cycle and with others with wear-out at the end of the service life.

We need to be conscious of these actors and non-design-related failures, and to "translate" their contributions into the reliability requirements and design failure rates. Figure 4.5 illustrates how these additional failure rates add to the design failure rate over the product life cycle. The *bathtub curve* shown in Figure 4.5 is not expected to have a fixed shape or proportion over years of manufacturing lots. Rather, as the product matures in the field, use errors and service-induced failures are expected to drop. With this drop, early life failures should drop as well and the overall failure rates approach a more steady value (constant failure rate). Eventually, components begin to wear. This tends to increase the failure rates once again. At this stage, a latecomer contributor is repair-induced failures.

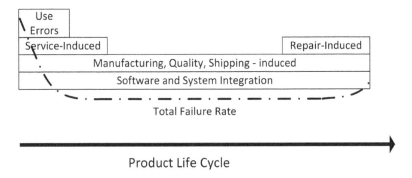

FIGURE 4.5
Composition of failure rate over product service life for a new product in the field.

TABLE 4.7

Recommended Criteria for Writing Reliability Requirements

Criteria	Considerations in Writing Reliability Requirements
Quantitative	The stated requirement is not vague or general but is quantified in a manner that can be verified by one of these four alternative methods: inspection, analysis, demonstration or test.
Measurable	Reliability requirements are stated as probability statements measurable by test or analysis during the product development time frame. They are not goals or nonmeasurable allocated numbers.
Benchmarked	Reliability requirements are benchmarked against target competition, when possible, so that requirements execution exceeds best in class.
Functionality	Functional product requirements are well defined so that the measurable reliability requirements relate to their intended function.
Defined failures	The requirements include a clear definition of product failure, and this definition is incorporated into tests. It is preferred that components or subsystems are tested to failure as part of the reliability demonstration.
Defined mission	A clear statement of time (hours, cycles, therapies, infusion, etc.) and expected service life must be specified as part of the reliability requirements.
Customer usage and operating conditions	The combined customer usage and operating environment profiles are adequately defined in product requirements, and the corresponding tests duplicate actual field failure modes.
Allowable degradation over time	The allowable degradation over time is defined in the requirements and represented by tests.

A Summary of Criteria for Reliability Requirements

Table 4.7 provides a summary of what we need to keep in mind as we write reliability requirements.

Notes

1. We will discuss the relationship between reliability and risk (which is associated with safety) in Chapter 13.
2. Technically, for nonrepairable devices, MTTF and MTBF are numerically the same value. However, the connotations are different.
3. MDT is the average total time when a product is not in an operable state. This time period includes trouble shooting, waiting for spare parts, and the time spent being repaired.
4. Service life is often defined as the period of time when a manufacturer is willing or able to repair a product.
5. We have to admit that strictly speaking, we are assuming that the reliability of any switching mechanism in this system is 100%.
6. We will explain how to develop this table in the section "Developing the Apportionment Table."

5

Reliability Planning

Introduction

Chapter 3 sets the foundation for design for reliability (DfR) with the design process and the V-model at a relatively high level. In Chapter 4, we learned how to develop reliability requirements in preparation for developing the DfR plan. Before we dive into its details, it may be prudent to follow the design process a bit longer. The engineering design stage begins by examining the subsystem and/or assembly along with the interface requirements. Once requirements have been decomposed into design specifications and the manufacturing approach is selected, components and assemblies may be designed using a variety of computer-aided design (CAD) software. This broader process is demonstrated in Figure 5.1.

The process as represented in Figure 5.1 is somewhat deceptive, or should we say oversimplified. It correctly suggests that the design team begins by decomposing assembly (or subsystem) requirements and developing design specifications. The reality is that to develop a comprehensive set of specifications, requirements, and inputs from other functions within the research and development (R&D) department as well as from a variety of departments outside of R&D should be taken into account and incorporated. In some organizations, the outcome of these activities and the decisions made as applicable to manufacture and service of the end product, are documented in a *device master record* document. The outcome and decisions concerning interfacing various functional inputs from within R&D are documented in a set of *interface documents*.

As shown in Figure 5.2 (and in no particular order), organizations that communicate design intent with R&D and their category of inputs are listed next:

1. Marketing as well as compliance (including regulatory) teams are labeling stakeholders. *Labeling* refers to any markings on the product (or shipped along with it) that communicate a message or a series of messages. An example of a label is a decal that provides information required by a regulatory body, such as UL, CE, or TÜV. Another form of labeling is any marketing decal, such the product brand, that may be printed or otherwise placed on the product. Some organizations may consider any manuals (e.g. service manuals) as labeling as well. Product and shipping containers and any manuals also fall within this category. Often, the marketing team designs these documents; however, these designs are treated, maintained, and controlled as any other component design document.

2. The business team is focused on the financial aspect of the product and its impact on the bottom line. Hence, one of the major and early requirements for any product is its projected cost of manufacture. This cost requirement should be properly

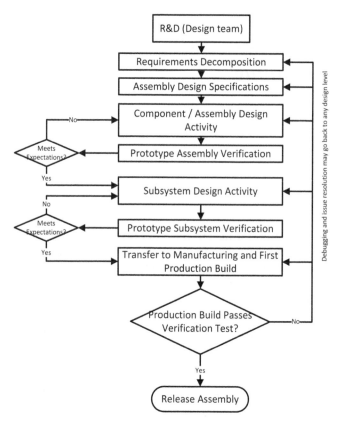

FIGURE 5.1
Steps in developing a model including its production build and verification.

cascaded to each subsystem and ultimately to each component. This factor influences the choice of manufacturing and production to a large extent. Associated with this cost sensitivity are influencers such as reliability. Clearly, the higher the product reliability, the lower the service and warranty costs. In addition to reliability, risk also plays an important role. No product is without risk; it so happens that some products have higher risks than others. It is the responsibility of the design team to identify risk factors, develop a risk benefit analysis of the product, and identify safety-related components within the design.

3. The manufacturing team provides input and feedback on design for assembly (DFA) and design for manufacturability (DFM) concerns. Should the rules of DFA and DFM be properly followed, the overall costs of the product—both direct and indirect—will be lowered and optimized. However, while R&D receives information and input on DFA and DFM, it provides leadership and guidelines on what appropriate manufacturing approaches should be selected, and works closely with manufacturing to establish needed processes and sets needed factors so that manufactured components meet design intent. An example of this R&D input is in injection-molding components. Often, by reducing mold hold time and pressure, an injection-molding operation can save the cost of fabricating components; however, the price to be paid for this saving is higher part-to-part variability. R&D

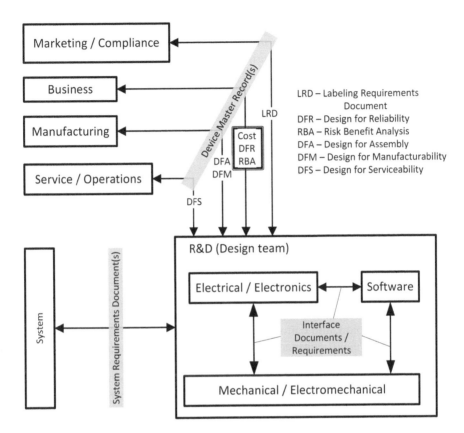

FIGURE 5.2
The interaction of the cross-functional team and its influence on design. The impact and influence of design for reliability (DfR) is highlighted.

and manufacturing should partner to identify optimum process conditions that meet both design intent and manufacturing goals. Once identified, these process parameters should be captured in a related device master record.

4. Finally, service team members have a stake in the design process as well. They bring to table their expertise and know-how in service to arrange assemblies so that the service process is straightforward and time is not wasted to open the product and replace failed components. Similarly, R&D works with the service team to provide effective diagnosis tools and codes such that service personnel can easily and effectively identify a faulty component. As an example, new computers may be considered. Once a malfunction is experienced, these newer units are equipped with diagnosis software that may be executed at the startup of the computer. The system conducts a test of various critical components (such as memory, hard drive, or display) and reports if any issues are observed. Another example is an engine electronics module that is installed on many newer-model automobiles. To identify any malfunctions, the mechanic attaches a diagnosis device to this module and within a short period of time, the issue is identified and located.

As indicated by Figure 5.2, cost and reliability requirements are typically set by the business team as driven by customer needs or industry regulations.[1] Even though the design team might show an appreciation for such requirements, as suggested in Chapter 4, reliability requirements are among the more difficult requirements to be cascaded from system levels to component levels and design specifications.

The intention of this chapter is to explore the means by which a system-level reliability requirement may be achieved and verified.

Design for Reliability Plan Process

Figure 5.3 provides a visual overview of the design for reliability process. Elements of this process that will be reviewed briefly in this chapter and expanded upon later in this book are the business plan, design and development plan, needed inputs, design for reliability plan, and deliverables.

Business Plan

Considering the influence of business and market needs, the DfR process begins with a review of the business plan, and commercial expectations of the product's reliability needs or market demands. Design engineers should be aware of these expectations, interpret these needs in technical terms, and capture that "voice" in the design and development plan. These activities were explained in Chapter 4 and examples of developing requirements based on various voices (i.e., customer, business, etc.) were provided. In the example presented in Figure 5.3, the business plan has suggested that the market expects a 10-year product life. This is a rather vague expression. Does it mean that on the day after the 10-year anniversary, the product will somehow not be functional anymore? Clearly, this is a silly suggestion. Does it mean that on the day after the 10-year anniversary, the product will no longer be supported and repaired should it fail? Well, this may be a plausible explanation. Another explanation may be that the market expects a new product every 10 years. Therefore, the organization plans the design and development cycle so that in 10 years a new product platform is ready to be launched into the market.

Design and Development Plan

Whatever the intentions of the business, the design and development plan (DDP) should provide further explanations of the business needs and the means of achieving them using engineering terms. This DDP document provides an overview of the direction that the product development should take and the overall strategy for design features. For example, in Figure 5.3, the design of the casework should be such that the units are modular and that the display and keyboard are integrated within a removable door. Additionally, the sensors are field replaceable in case they fail. From a reliability point of view, as we will discuss later, this information will be used to ensure that through the life cycle of this product, the display, the keyboard, and even the door closure mechanism do not experience premature failures or wear out.

One may notice that in this example, the business need "Market expects a 10-year life" has been accepted to mean that the product is to be developed with a 10-year design life. Additionally, engineering believes that the product should have a 1-million-hour mean time between failures (MTBF).

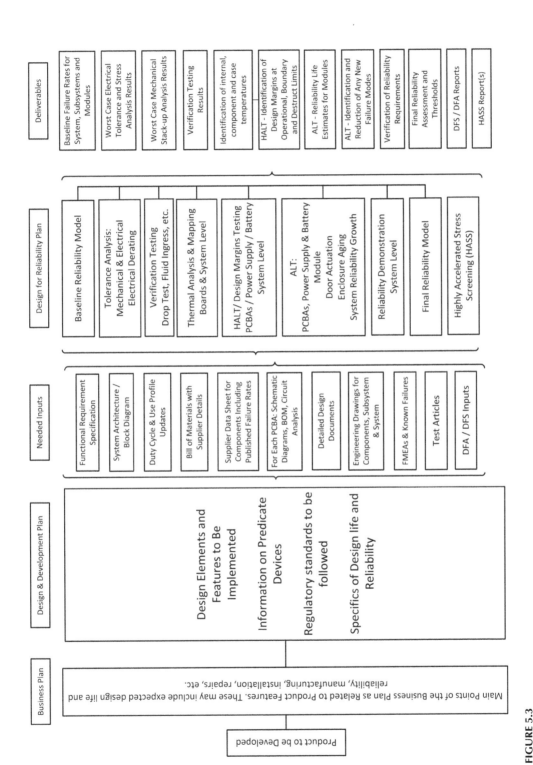

FIGURE 5.3
A visual overview of the design for reliability process.

Often, when our colleagues hear a 10-year design life in the same sentence as a 1-million-hour MTBF, they quickly calculate that the 1 million hours is approximately 110 years! What does this mean they ask. Design life and mean time to failure are two of the most commonly misunderstood reliability terms.[2] For the purposes of our discussions here, the design life of a product is defined as the period of time in which a marketed product maintains a constant failure rate or MTBF.[3] In other words, its components do not begin to wear out during this period and any observed failures are due to random causes.

By and large, the design for reliability plan pivots around market and design development needs and expectations of product life and the product's behavior as expressed in reliability requirements. However, at the onset of the product development cycle, not all the requirements (or needed inputs for a DfR plan) are either fully developed or cascaded. In today's environment and market pressures, the design and development team does not have the luxury of waiting. For this reason, the design for reliability plan should be considered as a living document and updated as additional information becomes available. Once a basic structure is put in place, its details and elements may be filled or updated at a later time during the product life cycle. This statement is true even after the product launch. It is well understood that the requirement documents may not be ready when the design and development plan is written. However, the very fact that its elements are needed will enable the design and reliability teams to work closely with each other, develop better working relationships, and remove any potential "silo" mentalities.

In Chapter 4, we discussed the means of deriving reliability requirements from the voices of customers or the business. This raises one of two questions: Are these voices always provided? or Are the derived reliability requirements reasonable? The answers to these questions may sadly be negative on both accounts. In the first instance, the role of the design (or reliability) engineer is to research competitive products to draw parallels and set reasonable expectations for the business. In the second instance, the role of the design (or reliability) engineer is to provide data and rationale of why the expected life may need to be adjusted to reflect the realities of the product price point, cost, and/or release-date expectations.

Needed Inputs

Once the business as well as design and development plans are understood and developed, the design for reliability process will focus on the details needed to construct an approach to ensure reliability of the product at launch. In summary, these needed inputs are functional requirements, system architecture, bill of materials along with data sheets, schematics (for printed circuit board assemblies [PCBAs]) and detailed engineering drawings, failure modes and effects documents, test articles, and design for assembly and/or service assessments. We will review the need for each of these artifacts next.

Functional Requirements Document and System Architecture

Once the reliability requirements of the product are determined from the design and development plan, Figure 5.3 indicates that the next documents to be consulted are the product requirements document and the system requirements document. These two documents should provide a system view of how the product and the system architecture as well as the functions that a product is to deliver. The system view will provide a pictorial summary of how various stakeholders interact with the system. The system architecture provides an understanding of the interaction between various subsystems and the environment.

To better illustrate this need, Figure 5.4 depicts an example of a system view of an electromechanical device. Figure 5.5 provides an example of a system architecture.

In the example shown in Figure 5.4, the architecture has six subsystems. These are user interface, power, enclosure, mechanical, electrical, and software/applications subsystems. This figure indicates that the device communicates wirelessly with the cloud. This communication enables both software updates as well as online training along with other features. Additionally, the product surveillance team may monitor the product's performance and receive customer complains as needed. In turn, product surveillance may alert the service team to potential failures, which then would take the necessary actions to ensure proper servicing of the product. Other stakeholders depicted in Figure 5.4 are manufacturing and maintenance teams. The last two stakeholders are the end users, and the sales and marketing team. This relationship is important because the sales team becomes the face of both the product as well as the company that produces the product. As depicted in this figure, the sales and marketing teams provide the needed training for end users to enable them to efficiently use the product.

This figure also identifies the environment of use and provides such information as operating temperature, presence of vibration, and daily use as well as the geographical areas where the product will be deployed. This information may be compiled in a *use profile* document. We will discuss the use profile in some detail shortly.

For now, let's consider the system architecture shown in Figure 5.5. This figure is a simplification of a product called Learning Station[4] developed in 1990s. It was designed as a rugged tool in the education of children with cognitive and physical impairments. This architecture shows not only the interaction of the different subsystems but also the

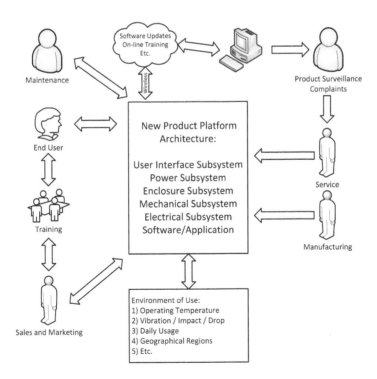

FIGURE 5.4
An example of a system view of an electromechanical product.

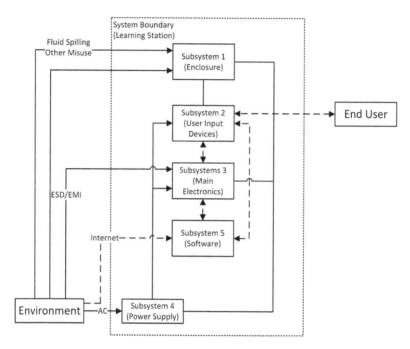

FIGURE 5.5
An example of a system architecture of an electromechanical product.

impact of the environment on the product. Here, the power supply receives 110 VAC from the environment. Additionally, the software may also interact with the Internet and be influenced by it. This figure also identifies other environmental factors and where they may impact.

This information along with the product use profile will be used to design reliability tests and flush out any interactions that may be detrimental to proper functioning of the product.

Use Profile Document

In Chapter 7, we will examine how reliability is impacted by—and in a way driven by—the applied stresses and available inherent strengths. Design engineers have a full grasp of this concept: the higher the stress levels and the lower the strength levels, the higher the probability of failure. The phenomenon that may be overlooked is that in many physical systems, the strength degrades over time. Hence, under a given set of stresses but over time, a product that functions as expected may fail. Fatigue and creep are two well-known examples of these types of progressive failures. In a way, these are addressed in the design and development plan by specifying reliability requirements such as design life to ensure that progressive failures are accounted in the design, and a failure rate to account for random and unexpected failures.

It is relatively easy to account for progressive failures when stress levels are constant. However, stress levels are related to how a product is used; the harder the use, the higher the stresses. It is erroneous to assume that a product is used uniformly across all users and all geographical areas. Some compensate for this with an erroneous mindset that the use of the product (and hence the applied stresses) is governed by a normal (Gaussian)

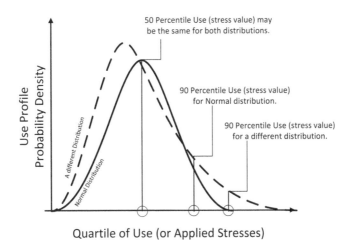

FIGURE 5.6
An example of a use profile distribution.

distribution. Therefore, there is an average "use" and its associate stresses that may be used as the basis for the design.

While evaluation of a use distribution is an appropriate approach, the assumption of normality is often wrong. Figure 5.6 depicts a nonnormal use distribution along with a normal distribution. It is evident that even if the median (50 percentile) of the two distributions may be the same, other use percentiles (e.g., 90 percentile) are different due to the difference in distribution characteristics. From a design point of view, a team needs to decide at which level of use should the product be designed. This decision will impact reliability activities for the product as well. We will discuss this further in reliability modeling and testing sections.

In discussing the percentile use, some may have the misunderstanding that the product fails by that percentage, i.e., for a 50 percentile use, 50% of the parts fail. This is not the case at all. By percentile use, we mean what percent of the population uses a product to that level of stress. For instance, in Chapter 2, we reviewed the example of a minivan under development and the design activities for its brake subsystem. There, we mentioned that a 50th percentile minivan driver applies his brakes just over 320,000 times in a 10-year period, whereas a 90th percentile driver applies his brakes about 530,000 times in the same period (Dodson and Schwab 2006). In Chapter 7, we will provide an example of how use scenarios may be cascaded into design specifications and guidelines.

Detailed Design Documents

As Figure 5.3 indicates, other major inputs into a design for reliability plan are items such as bill of materials, supplier data, schematics for PCBAs, engineering drawings, and theory of operations. These elements are needed not only to act as the basis for theoretical reliability calculations but also to understand what interactions may exist between various components and whether any adverse interactions may exist. An example of the adverse effect is a material mismatch at electrical connections that may lead to galvanic corrosion and failures. It should be noted that these design documents may not be complete at the time the DfR plan is being developed. However, as these documents are developed, we (reliability engineers) should use them to understand the design and develop reliability tests.

Failure Modes and Effects Analysis Documents

Once the use profile and the environment of use of the product under development are understood, the next question to ask is how predicate and similar products have failed. One source of information is the quality records, which provide a detailed investigation of field failures such as corrective action and preventive action (CAPA) records. It may also be possible to study competitive products and evaluate their performance and failures. This may be done by purchasing a few of the competitive product units and subjecting them to a variety of stress tests such as highly accelerated limit testing (HALT).[5]

In some regulated industries, databases may exist that capture failures that may have an end user impact. For instance, the United States Food and Drug Administration (FDA) maintains a database called MAUDE (Manufacturer and User Device Experience), which houses medical device reports submitted to the FDA by mandatory reporters (manufacturers, importers and device user facilities) and voluntary reporters such as health care professionals, patients, and consumers. Its data represents reports of adverse events involving medical devices. By studying MAUDE reports for each product line, prevalent failure modes of various competitive products may be identified. An understanding of how a class or a family of products fail forms the basis of developing the failure modes and effects analyses of a product. A design for reliability plan along with reliability test designs are centered on identified failure modes, their causes, and actions to prevent them. Needless to say, the proper tool for this purpose is a design FMEA matrix. Since this is a design tool, we will review it in Chapter 7, as it pertains to reliability activities during the design process. However, as information becomes available, it will be used to design (or refine) reliability tests.

Test Articles

Depending on the product being developed, the design process may involve a number of engineering builds, each including several iterations and possibly subiterations. Each build or its subiteration is relatively expensive and there is, at times, a tendency to overlook the need for reliability testing. It is paramount to develop a proper estimate of the number of components and/or builds needed for reliability testing and communicate them to the design team as well as the program for proper budgeting.

It is imperative to understand what test articles may be available for reliability testing at which level of product development. Ideally, enough samples should be available to make statistical inference calculations, and, to make proper judgment, units under test should be driven to failure. This may not always be feasible due to cost or size constraints. For instance, for single-use items, for a 95% confidence interval and 90% population, a sample size of 29 units are needed. Now, if the unit under test is a newly developed missile, requesting this number of samples is not realistic. Having said this, the engineer responsible for reliability should understand realistic constraints and work within them.

Design for Assembly (DfA) and Design for Serviceability (DfS)

Other sources of information for reliability considerations are manufacturing and service organizations. Poor manufacturing or service procedures or handling impact infant mortality rates. It is, therefore, important to engage these organizations to ensure that their procedures are verified and validated prior to any system-level reliability testing.

Design for Reliability Plan

Once an understanding of the needed inputs is developed,[6] the draft of the DfR plan may begin. There are typically four segments in a DfR plan. These are the baseline reliability model, design team input, reliability growth testing, reliability demonstration testing, and the final reliability model.

Baseline Reliability Model

In Chapter 6, we will review how to develop reliability models based on theoretical data and information available in various databases such as MIL-HDBK-217F or Telcordia. We have named this theoretical calculation the *baseline reliability model*. The reason is that in lieu of any field or experimental data, it is the best source of information on how the design may behave. Additionally, the design team may understand which subassemblies may underperform from a reliability point of view. It is generally well understood that the baseline model is an expectation of a system's reliability and not a reflection. The baseline model may be used to assess the reliability budget for each subsystem.

Design Documents

There are DfR activities that are truly outside a pure "reliability" realm. Nevertheless, any shortcomings in them will indeed lead to a lack of robustness and increased failures. Tolerance analysis, or derating of components, is among such activities. Our role as reliability engineers is to flush out areas of concern based on design documents and bring these areas to the attention of the design team for further analysis.

Similarly, verification of requirements is a responsibility of the design team in general and of the systems engineers in particular. This information forms an aspect of the design for reliability and through good communication any observed failures during verification should be shared with the reliability team.

Reliability Growth, Evaluation, and Demonstration Testing

Two elements of the DfR—and in fact the lion's share of reliability activities—focus on activities to grow the reliability of a product and eventually demonstrate via objective evidence that the product meets its reliability requirements. Elements of reliability growth include HALT. Often HALT is associated with an environmental chamber by the same name. Here, by HALT, we mean any accelerated testing that would flush out failure modes of the product. The advantage of HALT is that in a relatively short period of time, it identifies how a product may fail. The disadvantage of it is that no timing may be associated with the failure. In other words, after a HALT event, no one may claim that the product may last a month or a year in the field. To obtain "life" information, accelerated life testing (ALT) is required. Though the advantage of ALT is to identify when a failure may take place in the life of a product, the time frame to develop this knowledge may take months to years.

Once the product design and development is concluded and the design is frozen, the systems engineering team typically begins verification activities. From a reliability point of view, system-level reliability requirements should be verified.

Final Reliability Model

As test data is collected and a better picture of failures are formed, the baseline reliability model may be augmented and updated using statistical techniques such as Bayesian methods. We call this combination of theoretical calculations with test data (such as reliability growth or demonstration) a *blended reliability model*. It may be possible to combine the blended model with field data of predicate device data and generate a *final reliability model*. This model may be used to predict fleet reliability or be used as a means of trending field failures.

Highly Accelerated Stress Screening

Ensuring the reliability of the product does not end with the end of the design cycle and completion of the final reliability model. At the onset of production, there are a number of defects that could adversely impact product reliability. We need to continue to safeguard product reliability by monitoring and detecting these defects. One of the commonly used tools is called the highly accelerated stress screening (HASS) test. It takes place at elevated stresses based on the output of HALT. Typically HASS runs for a few hours using a combination of thermal cycling and vibration and it may be applied on 100% of the production lots. In alignment with the product development process and as an output of the DfR activities, we will discuss it in Chapter 10 in more detail.

Design for Reliability Deliverables

As with any plan, a list of deliverables should be identified. The deliverables for a DfR plan include the following elements:

1. Baseline failure rates of the components, subassemblies, and the final design.
2. A summary of worst-case analyses for both electrical and mechanical designs with an understanding of their impact on reliability.
3. Identification of internal component and case temperatures.
4. Identification of design margins through the use of HALT.
5. Identification of issues and design flaws through the use of HALT or ALT.
6. Reliability life estimates for subassemblies and modules as well as any required preventive maintenance schedules.
7. Verification of reliability requirements.
8. Final reliability model with its assessments as well as thresholds for field monitoring.
9. Highly accelerated stress screening reports, which include the test conditions, and, for early production units, screening and criteria for continued testing.
10. Optionally, the DfR report may include a summary of design for assembly and service assessments.

Now that we have reviewed the elements of a DfR plan, we begin to expand and explain the elements of DfR in the following chapters. Starting with Chapter 6, we will first discuss common statistical tools used in reliability as well as development of baseline reliability models. Additionally, Chapter 6 will include tools needed to analyze data that is generated through reliability testing. We will then follow in Chapter 7 with a discussion of predictive reliability tools in engineering design. Chapters 8 and 9 will provide guidelines on conducting component-, subsystem-, and system-level testing. Chapter 10 will provide a fuller view of what proper reliability outputs need to be. The work of the reliability engineer continues even after the launch of a product. Chapter 11 provides guidelines on how to analyze field data, particularly when the data is noisy. It also includes a description of the DMAIC approach to failure investigation.

Since risk and reliability are intertwined, it is important that we develop an understanding of risk and its relationship to reliability. This topic is covered in Chapters 12 and 13. Chapter 14 provides a list of standards and guidelines on reliability.

Notes

1. In regulated industries such as medical devices, the FDA and other regulatory agencies insert their influence on the reliability of products, particularly when reliability of a product may tie directly to patient safety.
2. This confusion is partly due to the fact that there are three terms with similar meanings that are mistakenly interchanged. These terms are *design life*, *useful life*, and *service life*.
3. Under the assumption of a constant failure rate, MTBF is the mathematical inverse of the failure rate.
4. Ali Jamnia developed this product as the director of engineering at Airtronic Services.
5. We will discuss HALT and other reliability tests in more detail, particularly in Chapter 8.
6. We need to recognize that the needed inputs may not be readily available when the design for reliability plan is being drafted.

6

Reliability Statistics

Introduction

In Chapter 1, we introduced the questions that product development teams need to consider and answer. These questions included *How frequently does the product need to be repaired or maintained?* and *How reliable is my product?* In the same chapter, we defined reliability as the probability that an item will perform its intended function for a specified time under specified conditions. Hence, we need data to ascertain the probability of failure of a product over its intended use period and to determine the frequency with which it needs to be repaired or maintained.

Good data along with proper statistical analysis are needed in order to accurately calculate the reliability of any product. Basic reliability calculations are simple, but they can become quite sophisticated once advanced statistical theories are employed. Herein, the most basic approaches will be presented in order to enable product development teams to evaluate their design from a reliability point of view as well. We will summarize the most commonly used distributions to model reliability data. The focus will be on the use and application of the statistical analysis techniques. These methods may either be employed as a part of the design process to model and predict reliability, or they may be used to understand component and system reliability from either testing or field failures.

For a deeper understanding of theories of statistical and probabilistic analysis of reliability data, a number of good resources has been published. The interested reader is encouraged to consult these references. *Practical Reliability Engineering*, 5th edition (O'Connor and Kleyner 2012), *AT&T Reliability Manual* (Klinger et al. 1990), *Practical Reliability of Electronic Equipment and Products* (Hnatek 2003), *Reliability Improvement with Design of Experiments* (Condra 2001), and MIL-HDBK-338B (1988) are recommended.

While we are on the subject of recommending books and references, we would like to point to one difficulty that we (as authors of such books) face. Oftentimes, *real-world data* is noisy, not well behaved, and at times ambiguous. As a result, most authors tend to use academic data so that a lesson can be taught! The outcome is once the individual tries to apply these models to the so-called real world, it would be difficult to interpret the results.

In Chapter 11, we try to tackle this problem by providing examples of these types of data and the approaches of dealing with them. In this chapter, our main concern is to introduce the statistical models that are commonly used in reliability.

Basic Definitions

Before we begin, we need to become familiar with some common probabilistic and statistical terms used with the context of reliability analyses.

In the context of reliability modeling, *probability density function* (pdf) is the distribution of the time to failure of a product (component or system), expressed as $f(t)$. The value of $f(t)$ is the probability of the system failing precisely at time t. As illustrated in Figure 6.1, $f(t)$ can have different values for the same failure mode on the same component at different times, hence, $f(t_1) \neq f(t_2)$.

A *cumulative distribution function* (cdf) is the probability of a system's first failure at or before time t. It is generally denoted as $F(t)$ and is measured as the area under the curve of the probability density function from 0 to time t, as shown in Figure 6.2.

$$0 < F(t) < 1$$

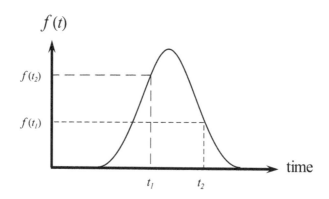

FIGURE 6.1
Probability density function distribution of time to failure.

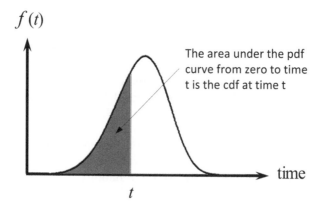

FIGURE 6.2
Cumulative distribution function of time to failure.

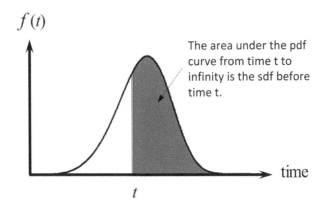

$f(t)$

The area under the pdf curve from time t to infinity is the sdf before time t.

time

t

FIGURE 6.3
Survival distribution function of time, t.

Also

$$F(t) = \int_{-\infty}^{t} f(t)\,dt$$

Survival distribution function (sdf), also known as the reliability at time t, is the probability that a product does not fail at time t—only sometimes after. It is therefore the complement of the cdf. $R(t)$ is measured as the area under the curve of the probability density function from time t to ∞, as shown in Figure 6.3.

$$S(t) = R(t) = 1 - F(t)$$

Hazard rate, $\lambda(t)$, is the instantaneous rate of failure of a population that have survived to time t. It is also defined as the *probability of failure* at any time interval, Δt, given that it survived up to this time interval. To an end user this means "what is the reliability when I use the product *now*, given its history so far."

Hazard rate is calculated as follows:

$$\lambda(t) = \frac{f(t)}{R(t)}$$

If we were to graph the behavior of the hazard rate over the product life cycle, the outcome would often have a bathtub shape and is often referred to by the same name. We will discuss the bathtub curve in some detail shortly.

Hazard rate is commonly expressed in units of failure in time (FIT). One FIT is equal to 10^{-9} per hour. In other words, there is a 10^{-9} probability of failure in the next hour.[1] For most engineers, the hazard rate has no tangible meaning. Rather, the term *failure rate* is more familiar. More specifically, a *failure rate per month* or *failure rate per mission* are terms that are more tangible.

Mean time to failure (MTTF) is a term commonly used in reliability calculations and its meaning may be self-evident. Mathematically, it is the area under the survivor (reliability) distribution function as time goes to infinity:

$$\text{MTTF} = \int_0^\infty R(t)\,dt \ \text{or}\ \text{MTTF} = \int_0^\infty S(t)\,dt$$

It may be shown that for a constant hazard rate (λ), we have

$$\text{MTTF} = \frac{1}{\lambda}$$

MTTF is a term that is typically used in association with components of a system. This is because components are not typically repaired and are replaced in repairable systems. For repairable products, it is more customary to use the term *mean time between failures* (MTBF). It is the average[2] time that a system functions between two consecutive failures. It should be noted that once a system fails, it has to be repaired before it is placed back in service. Thus, MTBF is the sum of MTTF and the mean time to repair (MTTR):

$$\text{MTBF} = \text{MTTF} + \text{MTTR}$$

We like to make two points here. First, most individuals use the terms MTBF and MTTF interchangeably. One could argue that the reason is MTTR is substantially smaller than MTTF and as such may be ignored. The second is more of a mathematical nature. It is to note that the four variables $F(t)$, $S(t)$, $R(t)$, and $\lambda(t)$ are interrelated, and if one is known, then the other three can be calculated. Furthermore, time (t) is generally, though not always, expressed in terms of hours.

Continuous Distributions and Reliability Analysis

Many of us are familiar with the *normal* distribution or curve. This curve, also known as the *bell curve*, is a statistical model. We hear of this model so frequently that we may develop the impression that just about any data distribution follows this model. Although the normal distribution has its uses in reliability calculations, we need to bear in mind that much of the data obtained in reliability does not fit this distribution. Accurate calculations of reliability metrics of any product are incumbent on a correct characterization of its failure-in-time distribution model. This characterization is done by fitting an appropriate equation to a set of test or field data. An example of this curve fitting is shown in Figure 6.4.

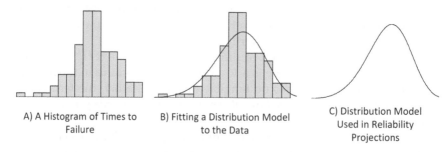

A) A Histogram of Times to Failure

B) Fitting a Distribution Model to the Data

C) Distribution Model Used in Reliability Projections

FIGURE 6.4
The progression from a histogram of times-to-failure data to a mathematical representation model.

Once the data is represented by a mathematical equation, we can use the equation to run various analyses and eventually make the needed engineering decisions.

Analysis of the failure data on a variety of parts, components, and systems in the last 60 or so years along with attempts to fit mathematical equations to these distributions have shown that the failure-in-time behavior of many engineering components and systems actually follows only a handful of equations. In this chapter, these common distribution models and their applications, namely, normal, exponential, gamma, chi-square, lognormal, and Weibull, are discussed.

Normal Distribution Model

Normal distribution is commonly used to represent distribution of an engineering measurand (Johnson and Kotz 1972). Typically, physical variables identified in a manufacturing environment are candidates for this class of modeling. These include variables such as process temperature, pressure, flow rate, and part dimensions.

The probability density function of a variable, t, is given by

$$f(t) = \frac{1}{\sigma_t \sqrt{2\pi}} \exp\left[-\left(\frac{1}{2}\right)\left(\frac{\ln t - \mu_t}{\sigma_t}\right)^2\right]$$

where μ_t is the mean value of variable t, and σ_t is its standard deviation. In the context of reliability, μ_t may be thought of as the MTTF.

Most of us have come to know about the normal distribution through its association with the *standardized* normal distribution and the Z score. In this sense, the standardized normal distribution is a method used to calculate the probability of a variable being greater than, less than, or equal to a target value. Z score is calculated using the following formula:

$$Z = \frac{T - \mu}{\sigma}$$

where T is the value that is being standardized, μ is the mean value of the distribution that represents T, and σ is the standard deviation of the same distribution. Table 6.1 provides a small section of the Z score data. This information is easily retrievable from the Internet or a reference book on statistics.

The following example will help clarify the use of the Z score.

TABLE 6.1

A Portion of a Table of the Standard Normal Cumulative Distribution

Z	0	0.01	0.02	0.03
2.0	0.9773	0.9778	0.9783	0.9788
2.3	0.9893	0.9896	0.9898	**0.9901**
2.7	0.9965	0.9966	0.9967	0.9968
2.8	0.9974	0.9975	0.9976	0.9977
3	0.9987	0.9987	0.9987	0.9988

Source: NIST 2012 (https://www.itl.nis t.gov/div898/handbook/eda/sect ion3/eda3671.htm).

Example

Suppose that a light bulb manufacturer has developed a new line of light bulbs. For marketing reasons, this manufacturer needs to provide a warranty period of 20,000 hours. The design team has tested a large number of the new bulbs and has concluded that this design has a mean life of 79,650 hours and a standard deviation of 21,146 hours. How can this team calculate the probability of this new product line surviving the warranty period?

Since a large sample set has been tested, and the mean life and its standard deviation been determined, one may assume that the distribution of the times to failure may be modeled using a normal distribution and the Z score:

$$Z = \frac{T - \mu}{\sigma} = \frac{20,000 - 79,650}{21,146} = -2.82$$

The probability that Z is greater than –2.82 is the same as probability that Z is less than 2.82.[3] From Table 6.1, we obtain the probability of Z less than 2.82 to be 0.9976. This means that the light bulbs will have a 99.76% reliability that they will survive their warranty period. The probability of failure for the same period is the probability of Z being greater than 2.82. Hence,

$$P(Z > 2.82) = 1 - 0.9976 = 0.0024 \text{ or } 0.24\%$$

This calculation implies that 0.24% of light bulbs will fail during the warranty period. Armed with this information, the business decides that a failure of 1% may be tolerated if a marketing advantage is gained by increasing the warranty period. Hence, the task of the design team is to calculate the new warranty period.

A 1% population failure is the same as a reliability of 99%, which is a different way of saying that the probability of survival is equal to 0.99. Based on Table 6.1, this probability corresponds to a Z score of –2.33 (note the negative sign). With the same mean value and standard deviation, we can calculate the new warranty period:

$$Z = \frac{T - \mu}{\sigma} = \frac{T - 79,650}{21,146} = -2.33$$

$$X = -2.33 \times 21,146 + 79,650$$

$$T = 30379.82 \text{ hours}$$

Based on this calculation, the warranty period may be increased to 30,000 hours for marketing purposes. Finally, it is worth mentioning that normal distribution is used to model time to failure of wear-out mechanisms. Another common use of normal distribution is to model variation in the manufacturing process.

Exponential Distribution Model

In the *exponential distribution* model, it is assumed that the hazard rate (λ) is constant and the probability density function varies exponentially with time (Hnatek 2003). This distribution is usually used to model reliability of electronic components, since they do not typically experience wear-out-type failures.[4] This model is also called memoryless, as the failure (or hazard) rate at any time interval along the life of the design is constant and not impacted by the previous run time. The probability distribution of failure at any time, t, is expressed as

$$f(t) = \lambda e^{-\lambda t}$$

Also, the probability of failure up to and before any time, t, becomes

$$F(t) = \int_{-\infty}^{t} f(t)\,dt = \int_{-\infty}^{t} \lambda e^{-\lambda t}\,dt = 1 - e^{-\lambda t}$$

Thus the reliability (or survivor function) becomes

$$R(t) = 1 - F(t) = e^{-\lambda t}$$

Mean time to failure is defined as

$$\text{MTTF} = \frac{1}{\lambda}$$

Example

A device has a 250 FIT hazard rate. Determine the MTTF, reliability, and the probability of failures for the first year. As mentioned earlier, 1 FIT is equal to 1×10^{-9}, hence, we have

$$\lambda = 250 \times 10^{-9} = 2.5 \times 10^{-7}$$

$$\text{MTTF} = \frac{1}{\lambda} = \frac{1}{2.5 \times 10^{-7}}$$

$$\text{MTTF} = 4,000,000 \text{ hours}$$

To calculate the reliability of this device, one may note that there are 8760 hours in a year assuming the device is operating around the clock, therefore,

$$R(t) = e^{-\lambda t}$$

$$R(8760) = e^{-8760(2.5 \times 10^{-7})}$$

$$R(8760) = 0.998$$

$$F(8760) = 1 - R(8760) = 1 - 0.998$$

$$F(8760) = 0.002$$

Thus, for this device the reliability (or survival probability) is 99.8% for the first year and only 0.2% failure.

Example

A device has a MTTF of 250,000 hours. What is the first-year probability of failure assuming that a constant hazard (or failure) rate is the failure-in-time behavior?

$$\lambda = \frac{1}{\text{MTTF}} = \frac{1}{250000}$$

$$\lambda = 4 \times 10^{-6} \text{ or 4000 FITs}$$

$$R(8760) = e^{-8760(4 \times 10^{-6})}$$

$$R = 0.965$$

$$F = 1 - R = 1 - 0.965$$

$$F = 0.0344 \text{ or 3.44\%}$$

In other words, a MTTF of 250,000 hours represents a 3.44% first-year failure! The significance of this example is to illustrate that a seemingly high MTTF may still lead to significant first-year failures.

Exponential distribution is suitable for calculating reliability and MTBF for systems that consist of many components, each of a different failure pattern over time. For example, a complex device may experience some electronic assembly failure at the early life stage due to manufacturing or shipping defects. It may have some random hardware and software failures due to marginal design or lack of robustness against variation of loading and strength. Finally, wear-out failure modes on mechanical components and moving parts due to aging and wear out may be exhibited. Each of these different failure modes and patterns can be represented by distinct distributions that would reflect increasing failure rates over time. However, for the entire system, the time between multiple failure modes of different components tends to be random, i.e., arriving at a constant rate for the system. We do like to emphasize that the exponential distribution at the component level is appropriate if and only if the component does not exhibit a wear-out failure mode.[5] We will revisit this topic again in this chapter under the subject of the bathtub curve.

Gamma Distribution

Normal distribution belongs to a family of statistical models that are defined by two parameters. In the case of the normal distributions, these two parameters are the mean (μ) and the standard deviation (σ). Another commonly used two-parameter model is called *gamma distribution*. The two parameters associated with this distribution are the shape parameter β and the scale parameter α. The probability distribution of failure at any time, t, is expressed as (Leemis 1995)

$$f(t) = \frac{1}{\beta^a \Gamma(a)} \exp[-t/\beta]$$

where $\Gamma(a)$ is the gamma function calculated at a:

$$\Gamma(a) = \int_0^\infty x^{a-1} e^{-x} dx$$

In the context of reliability, β may be thought of as the MTTF and a is akin to an event rate.

Application to Cumulative Damage

An important use of this distribution in reliability is to model cumulative damage caused by the random application of a number of excessive loads (such as, say, mechanical shocks) on a system. Any one of the excessive loads may not cause a catastrophic failure by itself; however, after an accumulation of k instances, the system fails and should be repaired. The time to failure, T_f, is the sum of the times associated with the application of each excessive loads up to the kth (and final) application. In other words:

$$T_f = T_1 + T_2 + T_3 \ldots + T_k = \sum_{i=1}^k T_i$$

We use the gamma distribution to find the time to the first failure after a series of excessive loads.

Reliability can be calculated as

$$R(t) = \sum_{k=0}^{a-1} \frac{(t/\beta)^k \exp(-t/\beta)}{k!}$$

β is the average time between partial failures. a is an integer and represents the number of applications of excessive loads, i.e., the kth event leading to failure. For an integer a, $\Gamma(a)$ the gamma function reduces to

$$\Gamma(a) = (1-a)!$$

A specific engineering application of this approach is in calculating the period between recalibration of equipment. We often experience degradation in the performance of a certain class of hardware that can actually be repaired simply by adjustment or recalibration. These hardware degrade, drift, and accrue certain levels of cumulative damage. Eventually there comes a time when it is impossible to retain proper functionality through adjustments or recalibration. Gamma distribution may be used to define the probability of failure after k adjustments and the proper time to replace the hardware.

Case Study: Pressure Sensor First Failure after Partial Physical Damages

In a portable medical device, a pressure sensor monitors the pneumatic pressure value in a pumping mechanism to control the volume of the medicament infused in the patient. A basic configuration of this sensor is depicted in Figure 6.5. In this sensor, the external pressure causes a deflection of the diaphragm, which in turn reduces the gap (d) between the diaphragm and the sensing element. This change causes changes in the electrical signals

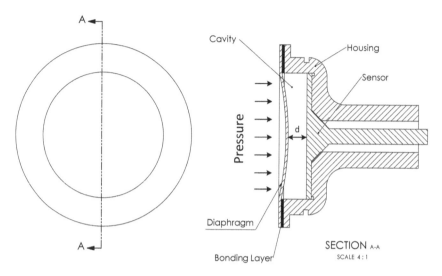

FIGURE 6.5
Pressure sensor structure diagram.

that are then conditioned and correlated with the pressure values. Clearly, any unintentional variation in the gap will cause errors in the calculation of the pressure leading to variations in the amount of the drug that has been administered to a patient. From a risk point of view, these variations need to be understood and managed.

The gap and its variations depend on the interactions between the diaphragm, the bonding layer, the sensing element, the housing, and their variations.[6] For the purposes of the current discussion, we need to keep in mind that the proper functioning of these elements is adversely impacted by shocks (such as drops) during the product handling and use. A major and somewhat imperceptible impact of drops is sensor drift, which eventually leads to a drift of the sensor. For this reason, to mitigate the risk of a false reading, the patient is instructed to follow a simple procedure to recalibrate the device monthly. Eventually, the damage caused by mishandling becomes severe enough that a recalibration does not bring it back to proper readings and it needs to be replaced. It is the manufacturer's responsibility to create a timetable for replacing these sensors to mitigate the risk of misreading pressure and harming patients. Clearly, replacing these sensors sooner would lower the risk to the patient but it would increase cost. These devices are typically returned to a service center every 14 months for a general cleaning and maintenance. The question that the design team asks is whether these sensors should be replaced every 12 months or every 24 months.

Data from the service department indicates that most products that are returned to service require some type of recalibration. Additionally, collected and test data indicate that the sensor needs to be recalibrated after every three consecutive drops.

Based on this information, we can develop our model using the gamma distribution. The scale parameter (α) is equal to 3 because the sensor requires recalibration after the third drop, and the shape parameter (β) is 14, because every 12 months the sensors require calibration in the service center. Thus, we have

$$\alpha = 3$$

$$\beta = 14$$

The probability that the sensor will not fail (need to be fully replaced), with a 14-month recalibration period, is calculated by

$$P(t) = \sum_{k=0}^{2} \frac{(t/14)^k \exp(-t/14)}{k!}$$

For this scenario, we run the model using $t = 12$ and 24:

$$P(12) = \frac{\left(\frac{12}{14}\right)^0 \exp\left(-\frac{12}{14}\right)}{0!} + \frac{\left(\frac{12}{14}\right)^1 \exp\left(-\frac{12}{14}\right)}{1!} + \frac{\left(\frac{12}{14}\right)^2 \exp\left(-\frac{12}{14}\right)}{2!}$$

$$P(12) = 0.4244 + 0.3637 + 0.1559 = 0.944 \text{ or } 94.4\%$$

This means that the probability that a sensor will survive a continuous 12-month functioning is 94.4%, i.e., 5.6% of the fielded devices will need to be repaired. The probability that the sensor will survive a continuous 24-month functioning is calculated by

$$P(12) = \frac{\left(\frac{24}{14}\right)^0 \exp\left(-\frac{24}{14}\right)}{0!} + \frac{\left(\frac{24}{14}\right)^1 \exp\left(-\frac{24}{14}\right)}{1!} + \frac{\left(\frac{24}{14}\right)^2 \exp\left(-\frac{24}{14}\right)}{2!}$$

$$P(24) = 0.1801 + 0.3087 + 0.2646 = 0.7534 \text{ or } 75.34\%$$

This means that the 24-months replacement interval will reduce the reliability to 75.34%. In other words, over 24.65% of fielded devices fail unexpectedly. This may be considered a high-risk failure.

Chi-Square Distribution

The *chi-square distribution* (often notated as χ^2) is not used to model physical phenomena, rather, it is employed to calculate a population metric based on observations made on a limited sample quantity and by specifying a statistical confidence interval. This model is the most widely used distribution for referral statistics.[7] In the field of reliability, we use the chi-square model to ascertain whether the mean time to failure of a set of samples under test is greater than or equal to a particular value.

Mathematically, the chi-square distribution is a special case of the gamma distribution, where $\beta = 2$ and the scale parameter $\alpha = r/2$, where r, being a positive integer, is called the *degree of freedom*. The degree of freedom is the number of *observations* (or pieces of information) in the data that are free to vary. We will discuss the degree of freedom again shortly. For now, let us keep in mind that the shape of the chi-square distribution varies with the value of the degree of freedom. The probability density function of the chi-square distribution is given by

$$f(t) = \frac{1}{\Gamma(r/2)2^{r/2}} t^{\frac{r}{2}} e^{-\frac{t}{2}}$$

The most commonly known application of the chi-square distribution is in calculating the MTBF (or failure rate) in a reliability test using three variables of test duration, number of observed failures and the desired confidence interval (Nelson 1982; Cox 1953; Lawless 1982).

This model enables one to calculate either the lower bounds or the two-sided confidence bounds under either time-bound or failure-bound test conditions. In reliability testing, it is customary to test either to achieve a predetermined number of failures regardless of the length of time or for a predetermined length of time regardless of the number of observed failures. The first category is called *failure truncated* and the second is called *time truncated*.

Failure-Truncated Test

The two formulae for a failure-truncated test are as follows:

$$\text{MTBF} \geq \frac{2\,T}{\chi^2_{(c,\,2r)}}$$

and

$$\frac{2\,T}{\chi^2_{\left(\frac{c}{2},\,2r\right)}} \leq \text{MTBF} \leq \frac{2\,T}{\chi^2_{\left(1-\frac{c}{2},\,2r\right)}}$$

where T is the time duration, c is the acceptable risk of error (i.e., confidence interval), r is the number of failures that have been prescribed, and MTBF is the mean time between failures. The first formula is used to calculate the lower bound of the MTBF, and the second formula enables one to understand both the upper and lower bounds of the same metric.

Standard tables of values of the χ^2 variables can be found in multiple sources, including the MIL-HDBK-338. It can also be calculated directly using standard formulae listed in many commercial applications such as Microsoft Excel.

Time-Truncated Test

The two formulae for a time-truncated test are as follows:

$$\text{MTBF} \geq \frac{2\,T}{\chi^2_{(c,\,2r+2)}}$$

and

$$\frac{2\,T}{\chi^2_{\left(\frac{c}{2},\,2r+2\right)}} \leq \text{MTBF} \leq \frac{2\,T}{\chi^2_{\left(1-\frac{c}{2},\,2r+2\right)}}$$

Note that since both the χ^2 and exponential distributions are subsets of the gamma distribution, the failure rate is the inverse of the MTBF. In other words:

$$\lambda = \frac{1}{\text{MTBF}}$$

Example

A system reliability demonstration test was conducted on seven different samples (devices). As each unit under test failed, it was repaired and placed back in the test. This test produced 17 different failure events. The accumulated test times on the seven devices were recorded as 174, 345, 234, 432, 189, 456, and 267 hours. To clarify, a unit with 267 test hours, might have operated 106 hours prior to the first failure, an additional 50 hours prior to the second failure, and finally 111 hours prior to testing being stopped. We like to calculate the lower bound of the mean time between failures with a 90% confidence level. In other words, the acceptable risk of error is 10% (= 100% − 90%). Since this test was time-truncated, we can use the following formula:

$$\text{MTBF} \geq \frac{2\,T}{\chi^2_{c,\,2r+2}}$$

In this situation, T is the total sum of the test time on all seven systems. Thus, we have

$$\text{MTBF} \geq \frac{2 \times (174 + 345 + 234 + 432 + 189 + 456 + 267)}{\chi^2_{0.10,\,2 \times 17 + 2}}$$

$$\text{MTBF} \geq \frac{2 \times (2097)}{47.21} = 88.83 \text{ Hours}$$

This tells that the mean time between failures of a larger population (possibly the fielded units) is minimally 88.83 hours. In comparison, had we calculated the MTBF solely based on a point estimate (i.e., total time divided by the number of failures, or $2097/17 = 123.3$ hours), we would have reported a much less conservative value.

Just as relevant is the question of what is the upper bound of the MTBF that we can expect. This is calculated as follows:

$$\frac{2(2097)}{\chi^2_{\left(\frac{0.1}{2},\,2 \times 17 + 2\right)}} \leq \text{MTBF} \leq \frac{2(2097)}{\chi^2_{\left(1 - \frac{0.1}{2},\,2 \times 17 + 2\right)}}$$

$$\frac{2(2097)}{50.998} \leq \text{MTBF} \leq \frac{2(2097)}{23.269}$$

$$82.24 \leq \text{MTBF} \leq 180.24$$

The results indicate that for this product, 90% of the time, the MTBF of the fielded units will not be lower than 82.24 hours, nor will it be greater than 180.24 hours.

Lognormal Distribution

Hnatek (2003) provides this definition of *lognormal distribution*: "If the natural logarithm of a function is found to be distributed normally, then the function is said to be lognormal." Although Bethea et al. (1991) suggest that there may not be a universally accepted form of the lognormal distribution, a generally accepted form is as follows (Klinger et al. 1990; Bethea et al. 1991; Condra 2001):

$$f(t) = \frac{1}{\sigma t \sqrt{2\pi}} \exp\left[-\left(\frac{1}{2}\right)\left(\frac{\ln t - \mu}{\sigma}\right)^2 \right]$$

where σ, called the shape parameter, is the standard deviation of the natural logarithm of the failure time; and μ, known as the location parameter, is the mean of the natural logarithm of the failure time.

The cumulative distribution function (cdf) becomes

$$F(t) = \frac{1}{\sigma\sqrt{2\pi}} \int_0^t \frac{1}{x} \exp\left[-\left(\frac{1}{2}\right)\left(\frac{\ln x - \mu}{\sigma}\right)^2\right] dx$$

The probability of reliability is

$$R(t) = 1 - F(t)$$

and the hazard rate is

$$\lambda(t) = \frac{f(t)}{R(t)}$$

It should be noted that unlike the exponential distribution, there is no longer a simple inverse relationship between the MTTF (or MTBF) and the hazard rate λ. Furthermore, conducting back-of-the-envelope calculations using the lognormal distribution becomes exceedingly difficult. Having said this, we should note that there is a relationship between the lognormal and normal distributions. Should we calculate the natural logarithm of each data point in the set, the resulting data would fit a normal distribution; hence, normal distribution characteristics such as average and standard deviation would apply to this new data set.

Lognormal distribution well represents a population of mechanical components with wear-out failure modes. It can be used to calculate the time needed to repair a product prior to a catastrophic failure due to wear. Additionally, it is applicable to cycles to failure due to fatigue, and material strength and loading variations.

Weibull Distribution

Another commonly used tool in the analysis of reliability data is the *Weibull distribution* model. This distribution can be used to represent a variety of failure data distributions and trends over time. We will explain later that normal, lognormal, and exponential distributions are all special cases of the Weibull distribution model.

A general form of the Weibull probability density function is as follows:

$$f(t) = \frac{\beta}{\eta}\left(\frac{t-\gamma}{\eta}\right)^{\beta-1} \exp\left(-\left(\frac{t-\gamma}{\eta}\right)^\beta\right)$$

where β is called the shape parameter; η is called the scale parameter or characteristic life (the life at which 63.2% of the population has failed); and γ is called the location or time delay, also known as failure free time (O'Connor and Kleyner 2012).

Based on this formula, the cumulative distribution function is

$$F(t) = 1 - \exp\left(-\left(\frac{t-\gamma}{\eta}\right)^\beta\right)$$

and the reliability becomes

$$R(t) = 1 - F(t) = \exp\left(-\left(\frac{t-\gamma}{\eta}\right)^{\beta}\right)$$

The hazard rate λ is

$$\lambda(t) = \frac{f(t)}{R(t)}$$

or

$$\lambda(t) = \frac{\beta}{\eta}\left(\frac{t-\gamma}{\eta}\right)^{\beta-1}$$

It is noteworthy to recognize the versatility of this formula. For $\gamma=0$, $\eta=1/\lambda$, and $\beta=1$, Weibull's distribution reduces to the exponential distribution. Additionally, by adjusting the three parameters of γ, η, and β, almost any distribution may be represented.

For $\beta<1$　　　Weibull reduces to the gamma distribution.

For $\beta=2$　　　Weibull reduces to the lognormal distribution.

For $\beta=3.5$　　Weibull approximates the normal distribution.

For this reason, Weibull distribution is widely used to analyze test and/or field data.

A Modified Weibull Distribution

Klinger et al. (1990) have created a special version of Weibull distribution to describe early life behavior (up to 10,000 hours). In this formulation, the following assumptions are made:

$$\gamma = 0$$

$$\beta = 1 - a$$

$$\eta^{\beta} = \frac{1-a}{\lambda_o}$$

where $0 < a < 1$ is the new shape parameter, and λ_o is an initial (or scale factor) hazard rate. With these assumptions, Weibull's probability density function takes the following form:

$$f(t) = \lambda_o t^{-a} \exp\left(-\frac{\lambda_o t^{1-a}}{1-a}\right)$$

Based on this model, the reliability and cumulative distribution functions and the hazard rate become

$$R(t) = \exp\left(-\frac{\lambda_o t^{1-a}}{1-a}\right)$$

$$F(t) = 1 - R(t) = 1 - \exp\left(-\frac{\lambda_o t^{1-a}}{1-a}\right)$$

$$\lambda(t) = \lambda_o t^{-a} \qquad \lambda_o > 0, \quad 0 < a\langle 1, t\rangle 0$$

It should be noted that $\lambda_{\circ}$ is a scale parameter and is the failure (hazard) rate at 1-hour device usage. In addition, the long-term hazard rate is assumed to be achieved by 10,000 hours of operation. This model reduces to the exponential distribution for $\alpha = 0$.

Example

A device has a 250 FIT long-term hazard rate. Using the modified Weibull model, determine the reliability of the device in the first 6 months. Assume that $\alpha = 0.75$.

$$\lambda = 250 \times 10^{-9} = 2.5 \times 10^{-7}$$

This is the long-term hazard rate. To recover the initial rate, use the following equation:

$$\lambda(t) = \lambda_{\circ} t^{-\alpha}$$

$$2.5 \times 10^{-7} = \lambda_{\circ} 10000^{-.75}$$

$$\lambda_{\circ} = 0.25 \times 10^{-4}$$

$$R(t) = \exp\left(-\frac{\lambda_{\circ} t^{1-\alpha}}{1-\alpha}\right)$$

$$R(4380) = \exp\left(-\frac{(0.25 \times 10^{-4}) \times 4380^{(1-0.75)}}{1-.75}\right)$$

$$R = 0.9992$$

$$F(t) = 1 - R(t)$$

$$F = 0.0008$$

Therefore, for this device the failure rate in the first 6 months is only 0.08%.

The Weibull distribution is well suited for calculating and projecting reliability metrics. In particular, it can be used for anticipating the number of failures for a fielded product that has a population with different age ranges. The Weibull distribution can also be used to project conditional reliability, i.e., if a product has survived a certain period of operation, what would be the expected reliability of the product beyond that period.

Selecting the Right Distribution for Continuous Variables

There is no simple answer to the question of what is the right distribution to choose. The point to consider is this: the distribution models of which most commons were just introduced should represent the failure behaviors of the product of interest. For instance, if the product of interest is a purely mechanical system that experiences a steady-state level of vibration, there is a good chance that a lognormal model represents its reliability better than any other distribution. For electronic equipment with no mechanical components, the exponential model may describe its reliability best.

The dilemma is that most real-life products are a mixture of both electrical and mechanical components that exhibit failures caused by chance and wear. Bazovsky (2004) has demonstrated that the exponential distribution model is suitable for electromechanical systems exhibiting a degree of component wear. The shortcoming of the exponential model is that it cannot predict whether the infant mortality period for a new product has come to an end or whether the wear-out portion of the bathtub curve has begun. The Weibull distribution enables the data analyst to determine whether the product reliability belongs to the infant mortality, steady-state, or wear-out phases.

From a practical point of view, we need to ask how accurate do we need to be. If either a back-of-the-envelope estimate of reliability or even a more in-depth probabilistic reliability prediction is needed, then the only choice is exponential distribution. However, as we move into using deterministic reliability predictions involving life data (i.e., both test and field data), then either lognormal or Weibull distributions may be more appropriate. The choice of model should be backed up by goodness-of-fit tests to ensure that the correct model selection has been made.

The Exponential Distribution

The exponential distribution is probably the most widely used distribution. It is the basis of what is called the "parts-count" calculations in MIL-HDBK-217 and Bellcore TR-332 (now Telcordia). There are a number of advantages to the using of this distribution:

1. Ease of use—a single reliability metric; namely, λ.
2. It has a wide applicability and is mathematically traceable.
3. It is additive in the sense that the sum of a number of independent exponentially distributed variables is exponentially distributed.

To use this distribution, the following assumptions should hold:

1. The failure rate is constant, not age dependent.
2. The "infant mortality" region has been eliminated through a postproduction environmental stress screening test.
3. There is no or very little redundancy in the device.

This distribution is often used to develop an understanding of the reliability of a product design at its early stages when test data is not readily available. The exponential distribution is very suitable when tracking reliability testing for an electrotechnical system that consists of multiple modules and has multiple failure modes with different rates and trends over time. The exponential distribution can be used to model the mean time between failures for a population of fielded products with mixed age ranges and brackets. It is also useful for the service department in any organization to plan for required service capacity.

The Lognormal Distribution

The lognormal distribution is suitable for a reliability analysis of semiconductors (Klinger et al. 1990) and life degradation due to progressive damage such as fatigue or wear of mechanical components (Condra 2001). This distribution is also commonly used in maintainability analysis (a topic beyond the scope of this work).

This distribution is better suited for reliability predictions when either field or test data is available.

The Weibull Distribution

The Weibull distribution is widely used in reliability work since it is a general distribution, which, by adjustment of the distribution parameters, can be made to model a wide range of life distribution characteristics of different classes of engineered items. It is a powerful tool particularly to identify either infant mortality or wear-out sections of the bathtub curve.

Similar to the lognormal distribution, this distribution is better suited for reliability predictions when either field or test data is available.

The data fed into the Weibull analysis should contain the actual time to failure and time to survival. However, it is to be noted that the most accurate application of Weibull distribution is when the data analyzed is for a specific failure mechanism that leads to a certain failure mode. Mixing failure mechanisms usually produces misleading results and leads to wrong conclusions about the failure trend and any following corrective actions.

In Chapter 11, we will review—in some detail—fitting of reliability data into statistical distribution models and the use of these models to project and calculate reliability characteristics.

Up to this point, we have only covered continuous distributions. However, a class of reliability problems is addressed and modeled using discrete distributions. In the next section we will discuss some of the most common statistical distributions used to model discrete variables in engineering applications.

Discrete Distributions and Their Applications to Reliability

Up to this point in this chapter, we have considered the failure of repairable systems. There are, however, a large number of products that are either single use (i.e., disposable) or have a limited mission time. Examples of these devices are car airbags, fire extinguishers, and many disposable medical devices, such as scalpels, syringes, and urine test strips. The reliability of these single-use items is not represented by the probability of repeat use without failure nor by time to failure, but rather by the probability of success of mission of the first and only use of the device.

There are several models that can be used to predict the probability of success. Of these, we will review and discuss three models. They are the binomial, the geometric and hypergeometric, and the Poisson distributions.

Binomial Distribution

The first model that we might use to represent the probability of success or failure of a single-use item is the *binomial distribution*. It expresses the probability of exactly x failures occurring in n trials. This probability is given by

$$P(x) = \binom{n}{x} p^x q^{(n-x)}$$

where $P(x)$ is the probability of x failures in n trials. p is the defect rate, i.e., the percentage of defective products in a population. q is the probability that one sample is defect-free, i.e., equal to $1 - p$. n is the number of trials or samples.

We can formulate the binomial distribution to calculate the probability of occurrence of r failures in n usages (samples). Conversely, the same may be used to calculate the probability of failure-free usages. This formulation is given by

$$P(r) = \sum_{x=0}^{r} \binom{n}{x} p^x q^{(n-x)}$$

Example

A pneumatic manifold contains 32 valves. From bench testing, it is known that the mechanical failure probability of a single valve after 30,000 cycles of operation is 0.05 (5%). Refurbishing manifolds is considered to be a high cost saver; however, the service department can only refurbish a manifold with a maximum of three bad valves. What is the probability that these pneumatic manifolds can be refurbished for use on another device after 30,000 cycles of use?

Although we are interested in the reliability of the manifold after 30,000 cycles, we consider the probability of failure of each valve to be independent of the failure of other valves in the same manifold. Therefore, we can calculate the probability of x valves failing out of the total of n valves in the manifold by applying the binomial distribution.

Since a manifold can be salvaged if no more than three valves have failed, the probability of repairing and reinstalling the manifold is the probability of no more than three valve failures, i.e., $x = 0$, 1, 2, or 3. In this scenario, $r = 3$, $n = 32$, $p = 0.05$, and $q = 1 - 0.05 = 0.95$.

Thus, by applying the binomial formula, we have

$$P(r) = \sum_{x=0}^{r} \binom{n}{x} p^x q^{(n-x)}$$

$$P(3) = \sum_{x=0}^{3} \binom{n}{x} 0.05^x 0.95^{(n-x)}$$

$$= \binom{32}{3} 0.05^3 0.95^{(32-3)} + \binom{32}{2} 0.05^2 0.95^{(32-2)}$$

$$+ \binom{32}{1} 0.05^1 0.95^{(32-1)} + \binom{32}{0} p^0 q^{(32-0)}$$

$$= \frac{32!}{3!29!} 0.05^3 0.95^{(32-3)} + \frac{32!}{2!30!} 0.05^2 0.95^{(32-2)}$$

$$+ \frac{32!}{1!31!} 0.05^1 0.95^{(32-1)} + \frac{32!}{0!32!} 0.05^0 0.95^{(32-0)}$$

$$P(3) = 0.1401 + 0.2661 + 0.326 + 0.1937 = 0.9262.$$

Thus, the probability of having no more than three valve failures is 92.62%.

Geometric Distribution

Another distribution related to the reliability of single-use items is the *geometric distribution*. This distribution enables the calculation of the probability of the required number of independent trials, k, to the first event as follows:

$$P(k) = \left(1 - p\right)^{k-1} p$$

p is the probability of an event in a single trial. This is a constant for each trial. k is the total number of trials before the first event occurs. This can be, for instance, the size of inventory at a customer's site.

This distribution is particularly useful for calculating the probability of success from the point of view of an end user (customer) of a product. An important application of this analysis may be situations when a customer has not fully adopted a new line of products and is using them in order to develop better familiarity and gain confidence. In these scenarios, early failures may trigger rejections by the customer.

Example

A chemical plant uses sterilized water produced by a water purification device. However, prior to bringing a new batch on line, a conductivity sensor measures its conductivity level in order to ensure that it meets the sterilization requirements. Once the measurements are done, the data is communicated to a control center via a Bluetooth module. The Bluetooth connection has a 3% probability of failure at every trial of communication. If the communication failed with the control center for more than 5 minutes after the batch is completed, the system has to be rebooted, and another measurement has to be performed. This action causes delay, or the produced batch may lose sterility. What is the probability that there will be no communication in the first, second, and third connectivity trials?

$$P(k = 3) = (1 - 0.03)^{3-1}(0.03) = 0.028 \text{ or } 2.8\%$$

Based on this calculation, there is a 2.8% chance that a communication failure may be observed during the first three trials.

Hypergeometric Distribution

A *hypergeometric distribution* is another model that can be used in calculating the probability of failure of single-use items. For example, small shipments of disposable medical kits to patients in rural and remote areas can have no more than the allowable number of defected sets in each shipment. Suppose that a manufacturer has started a new line of products that exhibit 5% defect in each batch of product being shipped to customers. On the one hand, if the customer uses five units every week, it is possible that five defective units are encountered. On the other hand, it is quite likely that the five units are all functioning as expected. Hypergeometric distribution enables one to calculate the probability of encountering a defective unit in the selected set.

This model is expressed as follows:

$$P(k) = \frac{\binom{D}{k}\binom{N-D}{n-k}}{\binom{N}{n}}$$

where $P(k)$ is the probability of encountering k number of events. N is the population size or the total number of outcomes. K is the number of observed events or outcomes. D is the number of defected events in the population. n is the number of executed trials before k events or outcomes are observed.

Notice that this distribution is different from the binomial distribution in that the probability of an undesired event is changing with every trial. What we mean is that once a sample is selected (whether it includes any defects or not), it is not returned back to its population. As a result, the unit pool gets smaller as more and more samples are drawn.

Example

A service technician conducts maintenance of vending machines at different customer sites. Sometimes, he has to install a new rechargeable lead acid battery on these machines if the existing battery is not at its proper remaining capacity. This activity is time-consuming, as it requires a partial disassembly of the housing.

The technician's truck has a compartment to store ten batteries. However, on average four out of ten batteries lose their charging capacity due to age or weather conditions. The only time that this defect is discovered is when the battery is installed and the unit is tested. On average, the technician changes seven batteries a week. On a weekly basis, he replenishes his truck's inventory of parts to be replaced including the batteries.

What is the probability that he will pick only one defective battery during the first 4 days of the week?

To answer this question, we first need to identify the input parameters to the hypergeometric distribution model. We note that the total number of units in the truck's inventory is ten. Thus, $N=10$. Furthermore, the expected total number of batteries to be used in the first 4 days is also four (one per day). So, $n=4$. We only need to use one battery a day, therefore $k=1$. Now, we know that four out of ten batteries are expected to be defective. This information leads to $D=4$. We have summarized this information in Table 6.2.

The probability of installing only one bad battery in the first four machines of services is

$$P(k) = \frac{\binom{D}{k}\binom{N-D}{n-k}}{\binom{N}{n}}$$

TABLE 6.2

Counts of Batteries Installed and Remaining Batteries

	Total	Selected	Not Selected
Bad batteries	$D=4$	$k=1$	$D-k=3$
Good batteries	$N-D=6$	$n-k=3$	$N+k-n-D=5$
Total batteries	$N=10$	$n=4$	$N-n=6$

$$P(1) = \frac{\binom{4}{1}\binom{6}{3}}{\binom{10}{4}}$$

$$P(1) = \frac{\dfrac{4!}{1! \times 3!} \times \dfrac{6!}{3! \times 3!}}{\dfrac{10!}{4! \times 6!}}$$

$$P(1) = 0.3809 \text{ or } 38\%$$

This calculation implies that there is a high chance (38%) that one bad battery will be selected by the technician during his maintenance process in the first 4 days of the week.

Poisson Distribution

Poisson distribution describes the probability of occurrence of certain numbers of an event in a continuous domain such as fixed time distance or length. It can be used in a manufacturing facility to model the expected number of defects in a given process. The main assumption is that the occurrences of these events are totally independent of each other, i.e., the occurrence of this event is random and has a constant rate. Also, events occur at certain intervals, and no two events happen at the same instant. The probability of a certain number of events in a fixed time interval is given by the following equation:

$$P\left(k \text{ events in an interval } t\right) = e^{-\lambda} \frac{\lambda^k}{k!}$$

where λ is the average number of events per interval, i.e., an occurrence rate. k is the number of events.

This formulation of the Poisson distribution function can be rewritten to model a number of events in time in terms of the time rate of an event, r, as follows:

$$P\left(k \text{ events in interval } t\right) = e^{-rt} \frac{(rt)^k}{k!}$$

where r is the rate of event in units of 1/time, t is the time interval of interest, and k is the number of events.

Example

Remember the water purification device of a previous example. In-house testing revealed that Bluetooth disconnection between the chemical plant and the water device occurs on average every 50 hours. What is the probability that three disconnections will occur in 1 day?

Based on a 50-hour mean time between failures, we can calculate the failure rate as follows:

$$r = \frac{1}{50} = 0.02$$

Now we can calculate the probability of three disconnects a day:

$$P(3 \text{ events in 24 hours}) = e^{-0.02 \times 24} \frac{(0.02 \times 24)^3}{3!}$$

$$P(3 \text{ events in 24 hours}) = 0.0114 \text{ or } 1.14\%$$

Case Study: New Product Launch

A medical device manufacturer was about to launch a new product called "PainFree" on a limited scale. The new product was to be distributed to selected customers for a 3-month clinical trial period. This device performs medical therapy using a single-use tubing set to transfer and control medicament to patients. One of the major innovations in this new product was the new tubing set used; however, although this new design made the therapy easier, the manufacturing defects were substantially higher (about 10%) due to the complexity of the molding and sterilization processes. Clearly, the design and manufacturing teams needed to solve this problem even after the limited launch. However, a simple solution outcome did not seem likely because teams believed that the defects represented inherent issues with the molding and sterilization processes.

Due to market pressures, management needed to authorize the limited launch, but at the same time, patient risk had to be assessed as well. The major areas of concern stemmed from the fact that the product would be shipped to patients who live in rural areas. To keep the cost of therapy reasonable, the tubing sets would be shipped to them every 3 months. The set consumption is one tubing set per day per patient; hence, a 10% defect would mean that a patient in a rural area could miss—on average—9 days of therapy every 3 months. One way of mitigating this risk was to ship additional sets to replace any potentially defective sets; hence, a set of 100 tubing sets could be shipped to cover for the expected 3-month consumption of 90 sets with the known defect rate. While this decision may alleviate the patient risk superficially, there are two other concerns:

1. From a marketing point of view, should patients encounter three defective sets in the first 2 weeks of the 3-month trial, the organization providing therapy (i.e., the customer[8]) may withdraw from the contract due to high failure rates. The reason is that this product line is new and novel, and unfamiliar to the customer or the patient. In lieu of perceived excessive failures, customers may prefer safety and revert to the old product with which they are more familiar.

2. From a manufacturing point of view, the defect rate is an average of 10% per lot. However, individually shipped batches may have higher defect counts. Should this be the case, the risk model allows for the patient losing no more than five therapies. In other words, the maximum tolerable number of defects per shipment should not be greater than 15 sets.

To address the marketing concern, we can calculate the probability of the three defective sets appearing in the first 2 weeks of receiving therapy, i.e., in the first 14 sets used. To do this, we apply the hypergeometric distribution for the following reasons:

- Each patient will have 100 tubing sets, not to be renewed before 3 months.
- Every time a defective set is used, the number of defective sets is decreased by one.

Recall the hypergeometric distribution formula:

$$P(k) = \frac{\binom{D}{k}\binom{N-D}{n-k}}{\binom{N}{n}}$$

Here, $P(k)$ is the probability of encountering k defective sets. N is the nonreplaceable population size. In this scenario, it is equal to 100. k is the number of observed (adverse) events. For us, it is equal to 3. n is the number of draws/trials before k events is observed. It is equal to 14. D is the expected number of defects in the population. In this situation, it is equal to $100 \times 10\% = 10$.

$$P(3) = \frac{\binom{10}{3}\binom{100-10}{14-3}}{\binom{100}{14}}$$

$$P(3) = \frac{\binom{10}{3}\binom{90}{11}}{\binom{100}{14}}$$

$$P(3) = \frac{\dfrac{10!\,90!}{3!\,11!}}{\dfrac{100!}{14!}}$$

$$P(3) = 0.113$$

The hypergeometric distribution calculation may be done using most statistical software including Minitab or can be executed using Microsoft Excel. The result implies that 11.3% of patients are at an increased risk in the first 2 weeks of therapy. This may be a high rate of failure and the organizations providing therapy may refuse to sign a contract even though the new product may have many added benefits.

To address manufacturing's concern of the probability of one set shipment of 100 sets to have more than 10 defected sets, we apply the binomial distribution, where the probability of "good" parts shipped is

$$P(r) = \sum_{x=0}^{r}\binom{N}{x}p^x q^{(N-x)}$$

and the probability of shipping "bad" parts is

$$F(r) = 1 - \sum_{x=0}^{r}\binom{N}{x}p^x q^{(N-x)}$$

Here, $F(x)$ is the probability of encountering x defective sets. N is the number of usages. In this scenario, it is equal to 100. r is the number of failures. For us, it is equal to 15. p is the average defect rate per batch. In here, it is 10%. q is the expected number of good parts in the population, i.e., $1 - p = 90\%$.

Alternatively, the binomial distribution may be formulated to calculate the probability of occurrence of r failures in n usages (samples):

$$F\left(r = 15 \text{ in } 100 \text{ sets}\right) = 1 - \sum_{x=0}^{15} \binom{100}{15} p^x \left(1-p\right)^{\left(100-x\right)}$$

$$= 1 - 0.968 = 0.032 \text{ or } 3.2\%$$

The results indicate that 3.2% of the patients may receive shipments of 100 sets that have more than 15 defected sets. This defect rate may cause five delays of therapy in 3 months. The leadership team has to discuss the option of shipping an additional 4% or 5% tubing sets to each customer in every shipment. This decision has to be weighed against profit margin, manufacturing capacity and the logistics, human factors for the truck drivers, redesign of the tubing sets packaging, etc.

Case Study: Expected Service Calls

After the full launch of the aforementioned PainFree medical device, the manufacturer received a number of complaints that were attributed to a gasket. Studies had indicated that the gasket would fail the self-diagnosis of the device at a rate of 1 per 10,000 therapies. The system software algorithm runs its self-diagnosis test multiple times during a therapy. It triggers an alarm after three consecutive self-test failures. When this happens, the customer support team deploys a service technician to replace the gasket. The cost of this service call is $400. Assuming one therapy per day per patient and for a population of 10,000 devices, we need to determine the expected number of service calls per year.

The probability of encountering three consecutive self-test failures may be calculated using the geometric distribution formula as follows:

$$P(k) = \left(1 - p\right)^{k-1} p$$

where p is the probability of encountering a defective gasket. Here, it is 0.0001 (= 1/10,000). k is the number of defects before the first success. It is equal to three in this scenario.

$$P(3) = \left(1 - 0.0001\right)^{3-1} \times 0.0001$$

$$P(3) = 9.99 \times 10^{-5} \text{ per therapy}$$

The annual probability of service calls and part replacements due to this event is

$$\text{Probability of Service call per year} = 9.99 \times 10^{-5} \times 365 = 0.036 \text{ or } 3.65\%$$

Thus, for a population of 10,000 devices, it is expected to have 365 devices repaired annually due to repeat gasket self-test failures.

A Discussion on Reliability Metrics and the Bathtub Curve

Now that we have reviewed reliability statistics, let us focus on a widely used and commonly discussed metric, namely, mean time between failures.[9]

Is MTBF the best metric for reliability? The answer is that it depends. MTBF is a good metric for measuring, predicting, and calculating cost of service for large product populations. It is also a good measure for a manufacturer (or even for a customer with large inventories) to plan operation, downtime, preventative maintenance, etc.

From the point of view of the end user, MTBF may not be an appropriate metric. No end user keeps track of the statistical analysis of the failure trend or rate of the entire population of the product line; rather, the user tracks when and where his/her own specific item has failed. An everyday example is the car that we drive. On a cold and rainy day, as we sit to drive our car, our expectation is that it will start and transport us to our destination. Or, when we buy a car (new or used) how long we will be able to use it trouble-free.

A common mistake in the business and engineering communities is that the MTBF is time to the first failure, or the time at which all the fielded products will fail. Let's examine what this term means. As mentioned earlier, a very common model for failure rate distribution for many engineering systems is the exponential distribution. This distribution assumes that the failure rate is constant over time.[10] The reliability $R(t)$ and the probability of failure $F(t)$ at any time over the product life t are calculated based on the exponential distribution formula as follows:

$$R(t) = e^{-t/\text{MTBF}}$$

$$F(t) = 1 - R(t) = 1 - e^{-t/\text{MTBF}}$$

Now, let us calculate the failure of product when $t = \text{MTBF}$. We have

$$F(t) = 1 - R(t) = 1 - e^{-\text{MTBF}/\text{MTBF}} = 1 - e^{-1} = 1 - 0.37 = 0.63$$

The results indicate that 63% of the product would have failed when an operating time equal to the MTBF is reached. So, when we discuss or review a MTBF (or MTTF) metric, we are really talking about a time frame when over 60% of products have failed. Do we really need to talk about a time frame when so much of our population fails? Or, are we trying to communicate something else?

Often, in the same breath that MTBF is talked about, there is a mention of the *bathtub curve*. This curve, which is graph of a product's failure rate as a function of time, is shown in Figure 6.6. $\lambda(t)$, defined as the hazard (or failure) rate, is the instantaneous rate of failure of a population that have survived to time t.

The main point of this graph is to draw attention to the fact that reliability failures may be observed in three different time intervals in a product's life cycle. These three time intervals are identified as three regions of the bathtub curve:

1. Infant mortality region—The initially high but rapidly decreasing hazard rate corresponding to inherently defective parts. This portion is also called early mortality or early life.
2. Steady-state region—The constant or slowly changing hazard rate.

3. Wear-out region—This generally occurs in mechanical electronic parts (such as relays) or when component degradation and wear out exist. In this stage reliability degrades rapidly with ever-increasing hazard rates.

A single end user of any repairable product could experience failure incidents in a pattern following the bathtub curve shown in Figure 6.6. High rates of failure in the early life of using the product may be expected because of manufacturing imperfections, shipping damages, or software bugs. Failure rates drop in the useful life or steady-state period in the life of the product. Then in the wear-out period of the product life, the failure rate increases due to the end of life of moving parts and mechanical and structural components. In other words, once an individual buys a product, the probability of encountering manufacturing defects is high. As the product is used without defects, and as time goes by, the probability of failure drops to a steady-state value caused by random failures. Eventually, over time, the probability of failure increases as various components age and wear out. This curve and probability of wear-out failures may be reduced, should a preventive maintenance be instituted. See Figure 6.6.

For a product manufacturer, the product population in the field is a mix of devices at different phases of their life cycles. Manufacturers experience an average of the failure rates associated with each phase of the product life cycle (i.e. early life, steady state, and wear-out period, as shown in Figure 6.6). Figure 6.7 illustrates this average failure rate experienced by the manufacturer, $\lambda_{average}$, by superimposing a number of bathtub curves associated with the lives of each fielded product. Since $\lambda_{average}$ may be considered to be a constant, the assumptions of the exponential distribution apply here and the MTBF is the inverse of the average failure rate.

The end user would be interested in the probability of failure at each phase of the life cycle of a single device he/she owns, whereas the manufacturer would be interested in the number of services per month of the entire product population to plan for factors such as cost of service, spare parts, technical support personnel, etc.

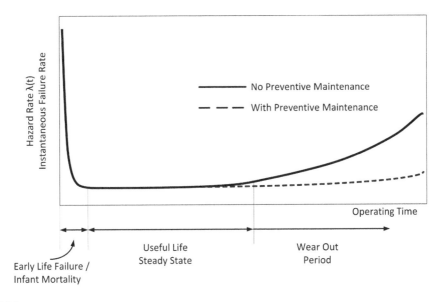

FIGURE 6.6
Hazard rate versus operating time. The influence of preventive maintenance has been shown.

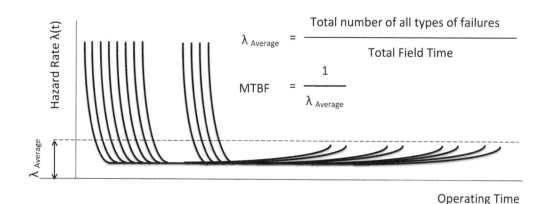

FIGURE 6.7
Average failure rate for population mix and MTBF for a product.

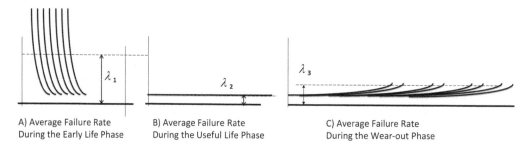

A) Average Failure Rate During the Early Life Phase

B) Average Failure Rate During the Useful Life Phase

C) Average Failure Rate During the Wear-out Phase

FIGURE 6.8
Average failure rate of fielded products experienced by a manufacturer during various stages of the product life.

During the early launch of a new product, manufacturers usually experience higher failure rates due to early life reliability issues or design immaturities. A lower MTBF should be expected when compared to a mature product during the useful life period when manufacturing and quality issues have been resolved. In this period field failures occur due to design robustness and random failures. This is depicted in Figure 6.8.

Notes

1. Military standards are different in that they use *percent failing per 1000 hours of operation*.
2. We use the term *average* here to convey the message. But, there is an inaccuracy associated with the word average as it implies 50% of the population. A better word might be the centroid of distribution, but who would know what we are talking about? Just keep in mind that depending on the distribution of failure times, the average time may not be as we expect.
3. Standard normal distribution tables typically provide probability measures for positive values of Z.
4. Wear-out failures reflect themselves in increasing failure rates.
5. This comment raises a potentially interesting point. Should a component have several failure modes, theoretically, it may be possible that each mode may have a different distribution model with one of them being exponential. Though, in practice, it would be very rare to encounter.

6. In Chapter 7, we will take a more in-depth look at reliability in design.
7. In other words, we draw conclusions for the entire population based on a sample quantity that has been tested.
8. Remember that the customer and end user do not have to be one and the same entity.
9. Or MTTF (mean time to failure) for components or nonrepairable systems.
10. This is a valid assumption for most engineering systems that consist of multiple parts, each with a different failure rate trend over time.

7

Predictive and Analytical Tools in Design

Introduction

In Chapter 6, we discussed and reviewed the major reliability models for both continuous and discrete variables. In this chapter, we will demonstrate how some of these models may be used to predict component or system failure rates, and predict a product's reliability. Recall that in Chapter 5, we outlined the design for reliability (DfR) planning process and pointed out that developing a reliable product begins with an understanding of business needs and the strategy along with how the design team will implement those goals. In this sense, reliability activities begin with the design of the product from the beginning. In our experience, many design engineers who are unfamiliar with reliability and its role in the design process assume that the reliability team tests components that are specified and/or purchased to ensure that they meet reliability requirements and needs. Or, they assume that the reliability engineering team begins to test assemblies, subsystems, and eventually the system once the design work is completed and prototypes are developed.

This assessment might have held true at some point but it is no longer valid. Reliability evaluation and analysis begins the moment that design activities begin. In fact, the US Food and Drug Administration has suggested that the role of reliability engineering is to understand the product and its failures, whereas the role of other engineering disciplines is to apply their own knowledge to particular aspects of the product.[1] This being the case, the focus of this chapter is to develop an understanding of how we can apply the principles of engineering to understanding how the product may fail and through this understanding develop a better—more robust and reliable—design. When we speak of reliability and robustness in design, we mean taking systematic steps to understand what it is that we are designing and how this design may behave without necessarily resorting to laboratory testing.[2]

In this chapter, we will review elements of design strengths and their interactions with induced stresses, and how failures may be induced or introduced. A note of caution: this chapter relies to a large extent on the reader's previous knowledge and background, since detailed treatment of each topic would be impractical.

Stress versus Strength

We define failure as the inability of a product to meet its requirements or function as expected. However, no one designs a product to not meet its requirement. The fact is that when a product fails prematurely, it is often due to oversight by the design team.

A product is designed with a certain degree of strength to withstand an assumed degree of applied stress. Once the applied stress exceeds the designed strength, a component within a product fails and the intended function is disrupted. One common oversight is that designed strength relative to applied stress is not fully understood. Hence, prior to the start of design activities, we need to understand how a product is expected to be used in the field by various users and what this usage means in terms of applied stress.

Cascading the Use Profile into Component Specifications

In Chapters 4 and 5, we discussed the product use profile and use environments. Once a basic design concept has been agreed upon, it is the responsibility of both the design and reliability teams to ask how reliability requirements such as, say, product design life cascades down to component selection.

Once the stress levels associated to use cases are known, they may be cascaded to various subassemblies as design specifications on the one hand, and on the other hand, they may be used to design reliability tests. This flow down is done on the basis of design details. For instance, in the brake system mentioned in Chapter 5, a cam, a cam follower, and a spring are needed to actuate the brake. So, for each application of the brake, we need to know how many times the cam, its follower, and the spring are actuated. On the basis of this relation, their usage may be calculated. To clarify this further, let us consider the following case study.

Case Study

A product delivers a certain level of torque to a fluid delivery system. This is an electromechanical system comprised of a motor, gearbox, cam, cam follower, and their internal springs. Additionally, this device has a door that opens to couple the device with the fluid delivery system. The door houses a display and keypad to enable the user to specify application settings such as duration and volume of the fluid to be delivered. The connection of the display and keypad to the main body is via a flex cable. This product is powered by either a 110 AC or 220 AC outlet. The product is to have a 10-year design life.

Table 7.1 provides the usage of this product for both 50% and 95% of the population. In this situation, the duty cycle for 50 percentile is 0.55 and for 95 percentile is 0.75. Other

TABLE 7.1

Device Use Profile

Device Usage	Usage Percentile	
	50%	95%
Life (years)	10	10
Duty use	0.50	0.75
No. of applications in a day	2	5
No. of use hours in a day	6	14
No. of keypad presses per application	5	7
No. of door actuations per application	2	2
No. of power adapter connections per day	2	5
Volume level setting*	100	250

* One may inquire as to the units of torque. For the sake of this example, no particular units have been used.

relevant information is the number of applications in a day, which are 2 and 5 for 50 percentile and 95 percentile, respectively. Additionally, 50% of the population uses the product no more than 6 hours a day, whereas 95% of the population uses the product for a maximum of 14 hours a day.

We need to pay particular attention to the duty cycle. Table 7.1 suggests that the 50 percentile usage is only 0.55, meaning that the product is used 55% of the time in a year or 6.6 months. At other times, it sits idle. Similarly, the 95 percentile usage is 0.75; in other words, the product is used only 75% of the time or 9 months out of a year.

Table 7.2 provides design details in that it provides the relationship between various components and the delivered output. For instance, the fluid volume moved per actuation is 0.200 and the number of cam rotations per actuation is 0.5.

Table 7.3 provides the flow down of the product use profile down to various components. For argument sake, let us review how the usage for some components were calculated:

$$\text{AC outlet connectors} = \text{No. of power connections/day}$$

$$\times 365 \text{ days/year} \times \text{duty cycle} \times \text{life (years)}$$

TABLE 7.2

Device Design Details

Design Details	
No. of cam follower rotations per actuation	0.5
No. of cam rotations per actuation	1
Fluid volume per actuation*	0.200
No. of spring movements per actuation	1
Gear ratio	4:1

* One may inquire as to the units of torque. For the sake of this example, no particular units have been used.

TABLE 7.3

Flow Down of Product Use Profile to Subassemblies and Components to Design Life

Component	Usage Percentile	
	50%	95%
AC outlet connectors	3,650	13,688
LCD display (life in hours)	10,950	38,325
Door flex cable	7,300	27,375
Keypad (lights)	10,950	38,325
Keypad (actuations)	18,250	95,813
Door springs	7,300	27,375
Door latches	7,300	27,375
Door hinges	7,300	27,375
Internal springs	5,475,000	47,906,250
Cam followers	2,737,500	23,953,125
Motor + bearings	10,950,000	95,812,500
Mechanism gear rotations	10,950,000	95,812,500

$$\text{AC outlet connectors @ 50\%-ile} = 2365 \times 0.50 \times 10 = 3650$$

$$\text{LCD display life (hrs)} = \text{daily use (hr)} \times 365 \times \text{duty cycle} \times \text{life (years)}$$

$$\text{LCD display life @ 50\%-ile} = 6 \times 365 \times 0.50 \times 10 = 10950 \text{ hrs}$$

Calculations of the motor and bearings life may be a bit more involved:

$$\text{Motor and bearing revolutions in life} = (\text{No. of motor revolutions/year}) \times \text{life (years)}$$

$$\text{No. of motor revolutions per year} = (\text{No. of cam rotations/year}) \times \text{gear ratio}$$

$$\text{No. of cam rotations/year} = (\text{no. of actuations/year})$$
$$\times (\text{No. of cam-follower rotations/actuation})$$

$$\text{No. of actuations/year} = \frac{\text{Fluid volume/year}}{\text{Fluid volume/actuation}}$$

$$\text{Fluid volume/year} = \text{duty cycle} \times (\text{use hours/day}) \times \text{volume level setting} \times 365$$

To calculate the motor and bearing life, we begin by calculating the fluid volume per year:

$$\text{Fluid volume/year @ 50\%-ile} = 0.50 \times (6/\text{day}) \times 100 \times 365$$

$$\text{Fluid volume/year @ 50\%-ile} = 109500$$

$$\text{No. of actuations/year} = \frac{109500}{0.200} = 547500$$

$$\text{No. of cam rotations/year} = 547500 \times 0.5 = 273750$$

$$\text{No. of motor revolutions per year} = 273759 \times 4 = 1095000$$

$$\text{Motor and bearing revolutions in life} = 1095000 \times 10 = 10950000$$

Similarly, the required life of other components and assemblies may be calculated and used as a design parameter for part and component selection. Once we understand how a product is used and the duration of its use, we can begin the design process.

Uncertainty in Strength and Stress: Single-Point Solutions versus Distributions

An aspect of the design process is to calculate the stresses endured by its components and the inherent strength of the design. In conducting numerical simulations, often single-point calculations are made, in other words, loading and material properties under a given set of conditions are used as inputs for numerical simulations and a single deterministic output is calculated. However, in reality, these input values are subject to variations. Should these variations be ignored a great deal of information may be lost. Two examples may provide more clarity.

Example

In the design of cabinets and racks for electronic systems, a design criterion that may be overlooked is buckling. A simple approach to account for buckling is to identify components under compressive loads, and then calculate these components' load-carrying capacities. In this case study, a rack system is made of a number of slender truss members. A simple slender member would be able to carry a compressive load so long as it is below P_{max}:

$$P_{max} = \frac{\pi^2 EI}{L^2} \tag{7.1}$$

where E is Young's modulus of elasticity, I is the second moment of inertia, and L is the column's length. Should the compressive load in a given member exceed this value, it would lose its load-carrying capacity and buckle. The member under study has a 0.5 in×0.5 in cross-section (12.5 mm×12.5 mm) and a length of 45 inches (1.143 m). The material is steel with a modulus of elasticity of 30,000,000 psi (206.8 GPa). Based on these values, a single-point calculation gives us the following load-carrying capacity for this member:

$$P_{max} = 761.5\,lbs\,(3.39\,KN)$$

We need to remember that in reality, the material along with geometrical properties have inherent variations, measured from either part to part or even from one lot to another. The variation of steel's Young's modulus used for this member has been shown to have a normal distribution with a tolerance of $\pm 1.5 \times 10^6$ (in other words, $E = 30 \times 10^6 \pm 1.5 \times 10^6$). The component has the following nominal and tolerance dimensional values:

$$(0.5 \pm 0.01\,inch) \times (0.5 \pm 0.01\,inch) \times (45.0 \pm 0.5\,inch)$$

The dimensional distribution of this part supplied is uniform (i.e., any component may have a dimension within the minimum and maximum values).

Figure 7.1 depicts the result of the Monte Carlo simulation conducted using Apogee (SDI Tools 2010) and Equation 7.1. It shows the distribution of the expected column buckling loads for the member of interest. Note that the minimum load-carrying capacity of the components is no longer 761.5 pounds, rather it is about 710 pounds. This is reflective of a 6.8% reduction, which is the expected "average" strength of the member. Additionally, by means of this example, we have demonstrated that strength is not a single point but a distribution.

Example

Now consider a different a problem where an injection-molded box as shown in Figure 7.2 is to designed with snap-fits as locking features into a mating part. Additionally, the part is to be sourced from two different suppliers. The molding process at supplier A is such that a normal distribution of various geometric dimensions may be maintained. The process at supplier B, however, only allows a uniform distribution of the same dimensions.

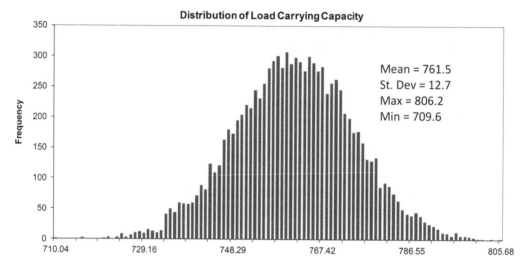

FIGURE 7.1
Impact of input variations and distributions on expected load-carrying capacity of a column.

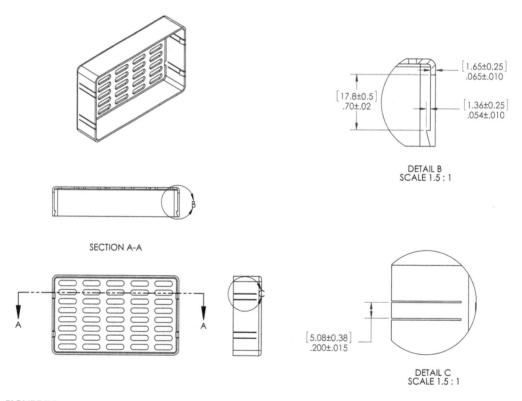

FIGURE 7.2
An injection-molded enclosure with snap-fit features.

The material selected for this enclosure is a PC/ABS blend with a Young's modulus of elasticity of 392 (±50) ksi and a yield strength of 4900 psi.[3] The maximum stress induced at the base of the snap-fit feature is calculated based on the assumption that the snap-fit feature is approximated as a beam:

$$S = \frac{3Ec\delta}{L^2} \tag{7.2}$$

where E is Young's modulus of elasticity, c is half the thickness ($c = h/2$), δ is the maximum deflection, and L is the length of the beam at the point of maximum deflection. Table 7.4 summarizes single-point calculations of maximum stress at nominal dimensions as well as the maximum and minimum values of both geometrical and material properties.

Clearly a nominal value of stress has been calculated along with upper and lower bounds. Furthermore, this material may likely fail at a stress level of 4900 psi. Considering that the upper stress bound is 6882 psi, is a redesign necessary to avoid failures? The fact is that the upper stress bound is only realized if the maximum values of E, h, and δ are aligned with the minimum value of L in the same part. What is the probability of this combination occurring? A more appropriate question may be to ask, What is the probability of the stress values being greater than 4900 psi? Will the parts from supplier B have more failures than supplier A? To answer these questions, two Monte Carlo runs were conducted, as shown in Figure 7.3.

There are some interesting points to be noted here. First, the average stress value of the Monte Carlo population is lower than the calculated nominal value (3759 or 3753 compared to 4212). This is due to the random combination of feature dimensions. Second, the failure rate of parts (i.e., parts that fail due to stress levels above 4900 psi) produced by supplier A is only 0.2% because of the process control and a normal distribution of various feature dimensions. This failure rate increases to 2.86% for supplier B whose components may experience higher stresses.[4] This is due to the fact that supplier B does not have the same process controls, and hence, can only provide parts that meet the "print."

Interaction of Strength and Stress Distributions

In the foregoing two examples, it is easy to recognize that the first calculates the strength of a slender member under buckling conditions. It suggests that so long as the applied

TABLE 7.4

Single-Point Stress Calculations

	Nominal	Tolerance	Min	Max
δ	0.054	0.01	0.044	0.064
h	0.065	0.01	0.055	0.075
L	0.7	0.02	0.68	0.72
E	392000	50000	342000	442000
S	4212		2395*	6882*

* Care must be exercised in calculating maximum and minimum stress values.

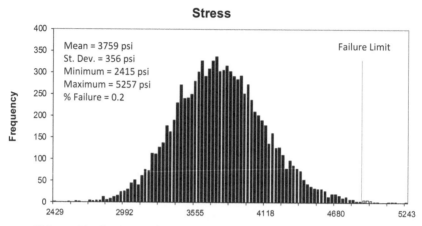

A) Stress Distribution in The Snap-Fit area of Parts Provided by Supplier A

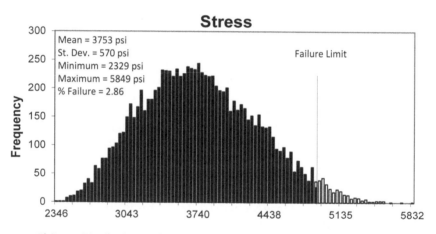

B) Stress Distribution in The Snap-Fit area of Parts Provided by Supplier B

FIGURE 7.3
Impact of dimensional variation and distributions on part stresses.

loads remain below a value, as calculated and shown earlier, the member will function as expected. In contrast, the second case study focuses on the applied stresses and indicates that so long as these stresses remain below a certain threshold, the snap-fit arm functions as projected. If we were to generalize these case studies, we would say that any product has an inherent strength that is represented by a probability distribution function. Similarly, the same product when used or in operation undergoes a degree of stress represented by another probability distribution function. Should these two distributions overlap, the product will fail. The probability of failure is determined by the area under the overlap (Sadlon 1993; Dudley 1999). This concept is shown graphically in Figure 7.4.

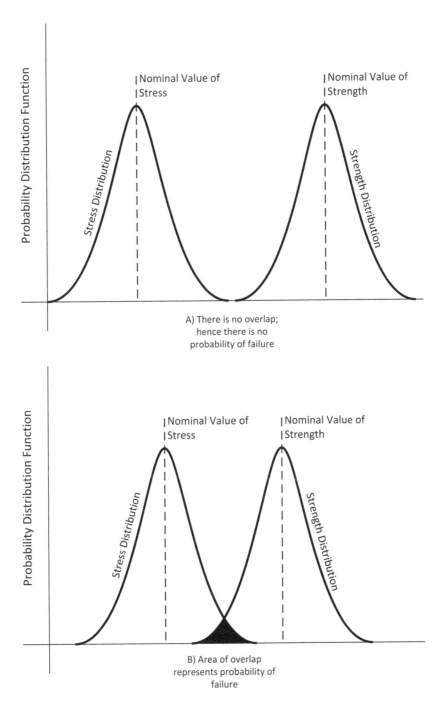

FIGURE 7.4
The relationship between stress and strength and their potential overlap.

Safety Factor or Design Margin

Often, we, as engineers, either design for a factor of safety (also known as design margin) or calculate it for a design that has been developed to ensure that there are no shortcomings. We define this factor as the ratio of the nominal value of strength and the nominal value of stress:

$$\text{Factor of Safety or Design Margin} = \frac{\text{Nominal Value of Strength}}{\text{Nominal Value of Stress}}$$

It should be clear that unless the distribution of both the stresses and strengths of a design is known, simply providing or calculating a design margin purely on the basis of a nominal value may not provide much of a protection against unexpected failures.

Another often forgotten factor is that the strength of a design may be reduced by a number of different physical and/or environmental influences. Fatigue, creep, and even corrosion impact strength reduction. In other words, even though a product might have the appropriate strength to withstand the applied loads at the point of manufacture, as times goes by, this strength maybe lessened due to conditions beyond the control of the design team. Eventually, there comes a time when stresses overlap strengths, and the product begins to fail.

At times, this loss of strength is called *wear out*.[5] Understanding the design margin of any product along with its particular wear-out mechanism and probability of failures is among the primary tasks of both the design and reliability engineers. For this purpose, engineering analysis tools including numerical simulations may be employed.

A Statistical Approach

Should both strength (S) and stress (s) be normally distributed, reliability can be calculated using standard normal variant Z:

$$Z = \frac{\overline{S} - \overline{s}}{\sqrt{\sigma_S^2 + \sigma_s^2}}$$

where $\overline{S}$ and $\overline{s}$ represent the mean values of strength and stress, respectively, and σ represents their standard deviations.

In case all statistical characteristics of the random variables are known, reliability (or probability of failure) of the design may be more accurately calculated using the probability distribution of the safety margin between the design strength and stresses.

Example

Many complicated engineering systems may be reduced to simple configurations for the purposes of calculating the induced stresses and estimating whether the design has allowed for proper levels of strength. In the system shown in Figure 7.5, a cantilever beam model is used to evaluate whether the supporting structure pipe has the proper load-carrying capacity. In this design, the chassis connectors at the wall float. As a result, the entire weight of the equipment is carried by the airflow pipe. The idealized cantilever model is, hence, loaded at the end of its span with a concentrated load, *P*, as shown in Figure 7.5, representing the weight of the equipment.

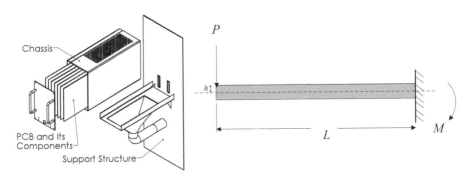

FIGURE 7.5
Depiction of the free body diagram for a cantilever beam.

Our goal is to calculate the probability of failure due to variations in the strength of material as well as applied stresses.

Acting stress at the end of the beam is given by

$$S_y = \frac{PLh}{I} \tag{7.3}$$

where $I = \pi h^3 t$ is the second moment of inertia of the cross section of the cantilever beam (pipe), L is the beam length, and h is the vertical distance from the point at which the stress is calculated on the surface of the beam to the neutral axis of the cross section of the beam. It is equal to the radius of the pipe. t is the pipe-wall thickness.

We are aware that every input value is subject to its own variations. Atua et al. (1996) have shown that the reported yield strength of materials is in reality the mean value of a distribution. Often engineers fail to inquire of the standard deviation of the material property that they use. In this example, the cantilever beam model has the following configuration as given in Table 7.5. Note that the variations of pipe-wall thickness has been ignored.

Based on the work done by Ayyub and McCuen (1997), the mean and the standard deviation of maximum pipe stresses (μ_{S_y}, σ_{S_y}) due to the weight of the equipment (i.e., load) can be calculated from Equation 7.3 using the probability distribution of related variables listed in Table 7.5. In other words,

$$\mu_{S_{Load}} \approx \frac{\mu_P \mu_L \mu_h}{\mu_I}$$

$$\mu_{S_{Load}} \approx \frac{50 \times 30 \times 1}{0.0942} = 15915.49 \text{ psi}$$

TABLE 7.5

Configuration of the Cantilever Beam Model

	Distribution Type	Mean Value, μ	Standard Deviation, σ
P	Normal	50 pounds	0.70 pounds
h	Normal	1.0 inches	0.01 inches
t	A constant pipe-wall thickness of 0.03 inches has been assumed.		
L	Normal	30 inches	0.2 inches
I	Normal	0.094 cubic inches	0.0027 cubic inches
S_{yield}	Normal	20,000 psi	2100 psi

And the standard deviation may be calculated from the following equation. See Ayyub and McCuen (1997) for more detail.

$$\sigma_{S_{Load}}^2 \approx \left(\frac{\mu_L \mu_h}{\mu_I}\right)^2 \sigma_P^2 + \left(\frac{\mu_P \mu_h}{\mu_I}\right)^2 \sigma_L^2 + \left(\frac{\mu_P \mu_L \mu_h}{\mu_I^2}\right)^2 \sigma_I^2 + \left(\frac{\mu_P \mu_L}{\mu_I}\right)^2 \sigma_h^2$$

$$\sigma_{S_{Load}}^2 \approx 49,647 + 17,590 + 214632 + 25330$$

$$\sigma_{S_{Load}} = 554.25 \text{ psi}$$

In order to calculate the reliability of the cantilever beam under the specified loading, the limit state function of the reliability relating the acting stress, S_{Load}, to the yield stress, S_{yield}, of the cantilever is defined as

$$SM = S_{yield} - S_{Load}$$

Since both S_y and S_{Load} are independent variables with normal distribution, the probability of failure can be found using the area under the interaction of the probability distribution function (pdf) of both variables. This can be done using the area under the cumulative distribution function of the *standard normal distribution*, representing the difference between these two variables (S_{yield}, S_{Load}), which is the stress margin in the limit state function given earlier as follows:

$$Z = \frac{S_{yield} - S_{Load}}{\sqrt{\sigma_{yield}^2 + \sigma_{Load}^2}}$$

$$Z = \frac{20,000 - 15915.49}{\sqrt{2,100^2 + 554.25^2}}$$

$$Z = 1.88$$

By looking up the area under the standard normal distribution charts (see Chapter 6) as shown in the sample Table 7.6,[6] we obtain the area above a Z value of 1.88 (1.88 standard deviations):

$$P(Z \geq 1.88) = 3\%$$

TABLE 7.6

A Portion of a Table of the Cumulative Standard Normal Cumulative Distribution

Z	0	0.01	0.02	0.03
1.8	0.9641	0.9649	0.9656	0.9664
1.9	0.9713	0.9719	0.9726	0.9732
2.0	0.9773	0.9778	0.9783	0.9788
2.3	0.9893	0.9896	0.9898	0.9901
2.7	0.9965	0.9966	0.9967	0.9968
2.8	0.9974	0.9975	0.9976	0.9977
3	0.9987	0.9987	0.9987	0.9988

Source: NIST/SEMATECH, Cumulative distribution function of the standard normal distribution, *NIST/SEMATECH e-Handbook of Statistical Methods*, retrieved on May 5, 2012, from https://www.itl.nist.gov/div898/handbook/eda/section3/eda3671.htm.

Therefore, the pipe used to support the electronic equipment has a reliability of 97%, provided the system is not subject to any vibration or shock loads.

Engineering Analysis and Numerical Simulation

Once a component and assembly concept has been chosen and the three-dimensional models developed, the design team should inquire whether the virtual product meets the intended design. Traditionally, this question is answered by fabricating a prototype and testing (or evaluating) it. In the days when the paper design was flat, that is, when designs were two-dimensional projections of someone's mental image, it was understandable that a design–build–test–redesign was justifiable. In today's fast pace product development cycles and the advances in computational fields and 3D solid modeling software, this traditional approach is no longer efficient or effective.

In today's world, much of this old approach is (and should be) replaced by engineering calculations and evaluations. The reason is that it may be next to impossible to remove hidden design flaws through testing alone. However, what-if scenarios maybe conceived and their outcome evaluated via numerical simulations.

There are several applications for the use of computational methods in the design process. The first, as we mentioned, is a numerical what-if game play. For instance, what if we use aluminum as opposed to steel, or what if we made the rib thinner? A second application of numerical simulations is to evaluate if the intended components or assemblies meet particular design intents. Failure analysis belongs to this category of use. Questions such as how much deflection or how hot or what frequency are also answered. Other types of failure analyses focus on the future performance of the product or the performance of the entire population. The questions to be answered are how long will this part last or what percent of the population may fail. Thus, it is prudent to understand possible failure modes and develop means of accurately predicting those failures while the design is still on paper and changes are inexpensive.

Before providing more details on these three classes of simulations, we suggest that numerical simulations may be divided into three types. The first are the closed form solutions that may be solved longhand. These are simple applications of engineering "handbook" formulae. The solution may be developed by using a hand calculator with advanced functions. The second type of numerical simulations involves a set of equations that may not have readily accessed close-form solutions. Up to the late 1990s, some simple software programs were needed to solve these equations and provide solutions in tabular or graphic format. In today's environment, a program such as MATLAB® or Microsoft Excel may easily be used to solve a number of these types of equations and converge to a solution. The third type of numerical simulation is when continuous engineering systems are discretized, and techniques such as finite element analysis (FEA) are used to solve complex engineering problems.

What-If Scenarios

One use of numerical simulations is to answer the what-if questions. As depicted in Figure 7.6, two different design configurations of a clip are subjected to the same loading conditions and the resultant Von Mises stresses are compared. This type of analysis helps the design team make decisions on the direction that the product design and/or its configuration might take.

Similarly, numerical simulations may be used to develop the transfer functions between the input values and the expected output(s). Figure 7.7 depicts the magnetic scalar potential and the flux intensity fields. By employing a Hall-effect sensor, the magnetic field strength may be measured. Considering that magnet and the magnetic paths and leakages must

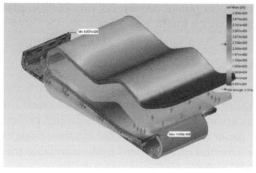

| A) Von Mises Stresses For Configuration A | B) Von Mises Stresses For Configuration B |

FIGURE 7.6
Numerical simulations are used to set design and/or configuration decisions.

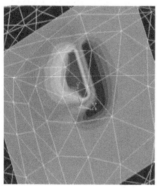 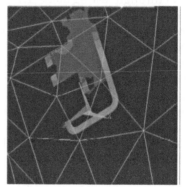

| A) Magnetic Scalar Potential | B) Flux Intensity Field |

FIGURE 7.7
Numerical simulation of a magnetic field to develop the transfer function.

be designed so that the output signal is monotonic and relatively linear, a trial-and-error approach may prove extremely ineffective in bringing such a sensor to market in a timely manner. The transfer function may be developed based on variations of the geometric and material properties of the input values. The numerical simulation would provide an insight into magnetic leakages and their impact on output values may be easily evaluated.

Another aspect of what-if scenarios is to compare the expected results under one set of conditions to another set of conditions. Figure 7.8 shows fan and impedance curves at sea level and at an altitude of 5000 feet.

We iterate that the purpose of the what-if numerical analyses is to develop an optimum balance between stresses and strengths by first understanding the failure modes and mechanisms and then by making changes to various design features if need be.

Stress Derating

Through the right what-if game play, we can develop an understanding of how a product functions. We can ascertain whether the applied stresses are near the strength thresholds. In other words, are there sufficient design margins or factors of safety.

Openning	Area	N	K
inlet	9.31	0.50	1.30E-03
Initial Stage	10.20	0.50	1.09E-03
Fan Tray Area	49.28	1.00	9.31E-05
Flow Thru PCB's	15.75	2.50	9.13E-05
90 deg bend (right)	7.50	1.50	6.03E-03
90 deg bend (top)	37.50	1.50	2.41E-04
90 deg bend (left)	7.50	1.50	6.03E-03
Outlet (right)	2.93	0.50	1.32E-02
Outlet (top)	10.08	0.50	1.11E-03
Outlet (left)	2.93	0.50	1.32E-02

Total K = 0.042
Air Density = 0.0765
Total Flow = 33.000

Flow Rate	Pres. Loss	Fan Curve	Pres. Loss @ 5000 ft	Fan @ 5000
0	0.00	0.360	0.00	0.31
5	0.01	0.348	0.01	0.30
10	0.02	0.335	0.02	0.29
15	0.06	0.323	0.05	0.28
20	0.10	0.310	0.09	0.27
25	0.16	0.298	0.13	0.26
30	0.22	0.285	0.19	0.25
35	0.30	0.273	0.26	0.23
40	0.40	0.260	0.34	0.22
45	0.50	0.248	0.43	0.21

at sea level or 0.0659 at 5000 feet
at sea level or 33.00 at 5000 feet

number of fans = 3

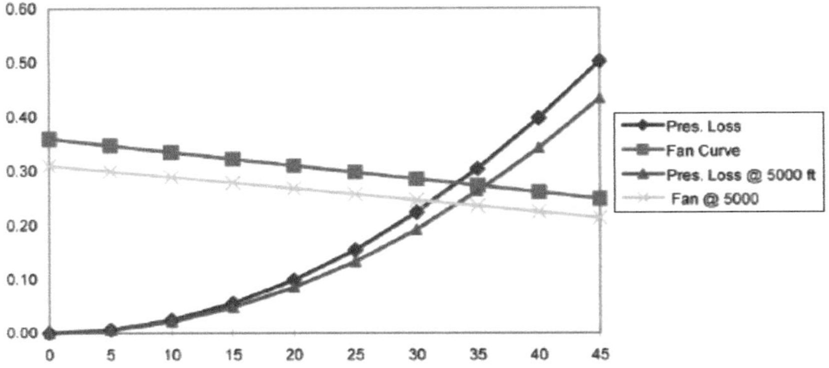

At Sea Level

Delta T = 17.33	deg. F
Input CFM = 33.00	CFM
VA = 245.96	Watts
Power Factor = 0.75	
Barrometic Pressure = 29.92	inches Hg
Ambient Temperature = 60.00	deg. F

At 5000 Feet Altitude

Delta T = 20.03	deg. F
Input CFM = 33.00	CFM
VA = 245.96	Watts
Power Factor = 0.75	
Barrometic Pressure = 24.90	inches Hg
Ambient Temperature = 40.00	deg. F

Legend: Pres. Loss — Fan Curve — Pres. Loss @ 5000 ft — Fan @ 5000

FIGURE 7.8

Comparison of outcome under two different environmental conditions.

Typically, though not always, the term *factor of safety* applies to mechanical systems. For electrical/electronics the term *stress derating* is more commonly used. Dudley (1999, p. 1) defines derating as "the practice of limiting electrical, thermal and mechanical stresses on electronic parts to levels below their specified rating." What this means is that should a component, say, a capacitor, is rated for 2 amps and 85°C, then the applied current and the environmental temperature should be much less than these values for the part to operate properly and for a fairly long time. Derating may be achieved by one or a combination of the following means (Dudley 1999):

1. A stronger component may be selected. For instance, in place of the 2 amps and 85°C capacitor, one with 3 amps and 105°C rating may be chosen. The drawback of this approach is that the overall cost will increase, not only in financial terms but potentially in terms of size and weight budgets as well.

2. The design may be changed to reduce stress levels. Once a stress analysis is conducted on a circuit, it may be possible to optimize the design to reduce current loads on components. The foundation of the stress analysis is to calculate the currents as they pass through various components. Typically, for critical components the ratio of the applied current to the rated current should not be larger than 60%, and for all other components should not be larger than 75%.

3. The third approach is to minimize stress variations. On electronic components, two sources of stress variations are environmental temperature and humidity. Though, in general, controlling these variations may be costly, in extreme cases active thermal management measures, such as localized cooling and heating, may prove effective in preventing failures. This practice was commonplace for computer centers in the 1980s and 1990s where large air conditioning units were built specifically to keep these giant computers cool.

4. The last approach is to minimize the variations in strength. Components have not only a rated value (e.g., 3 amps), but they also have a tolerance level. Some components have a ±10% tolerance, others a ±5% tolerance. If necessary, a component with a tighter tolerance (i.e., smaller variation) may be specified. Needless to say, selection of these components translates into higher costs.

Tolerance Analysis

The next logical area to focus on for reliability and robustness in design is tolerance analysis. Part-to-part variations are unavoidable, but they are controllable. One way that design engineers control part-to-part variations is to specify tolerances on the engineering drawing. Once part variations are defined via specified tolerances, the design should be analyzed to ensure that the specified components and their specified variations would, indeed, form the intended assemblies or function as expected. This activity is called *tolerance analysis*. Typically, there are three common techniques for this purpose. These are worst-case analysis, the root sum of squares method, and Monte Carlo analysis. We will be discussing these methods briefly.

Worst-Case Analysis

Once nominal values of various design elements (or features) have been determined and their expected variations (or tolerances) specified, a tolerance analysis should be conducted. The simplest form of this analysis is called *worst-case analysis* (WCA). WCA may be applied to mechanical assemblies to ensure that all components may be correctly assembled. It may also be applied to electrical circuits to ensure that the circuit can function as expected.

This approach is based purely on the influence of the maximum and/or minimum value of various design elements and their impact on the ultimate feature to be calculated. For instance, in a mechanical assembly, there are typically two questions that concern the design team:

- Will the parts assemble as expected, i.e., is there excessive interference between some or all components?
- Once assembled, will there be excessive play (i.e., gaps) between some or all components?

To illustrate this point, consider the assembly as shown in Figure 7.9. Three blocks, namely, blocks A, B, and C, are assembled within the U-channel L. The success or failure of this assembly not only depends on the nominal sizes of L, A, B, and C but also on their tolerances.

Mathematically, the maximum and minimum gaps may be calculated by using the largest opening (i.e., the width of the U-channel) and smallest blocks. This is expressed as

$$\text{Gap}_{\text{maximum}} = (L + \text{tol\_L}) - \left[(A - \text{tol\_A}) + (B - \text{tol\_B}) + (C - \text{tol\_C})\right]$$

Rearrange the terms to obtain

$$\text{Gap}_{\text{maximum}} = \left[L - (A + B + C)\right] + \left[\text{tol\_L} + (\text{tol\_A} + \text{tol\_B} + \text{tol\_C})\right]$$

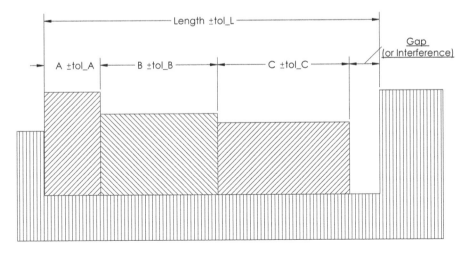

FIGURE 7.9
An assembly block for tolerance calculation demonstration.

Alternatively, the minimum gap is calculated by using the narrowest opening (i.e., the smallest width of the U-channel) and the largest blocks, as shown next:

$$Gap_{minimum} = (L - tol\_L) - [(A + tol\_A) + (B + tol\_B) + (C + tol\_C)]$$

Rearrange the terms to obtain

$$Gap_{minimum} = [L - (A + B + C)] - [tol\_L + (tol\_A + tol\_B + tol\_C)]$$

In this scenario, one could summarize the nominal and tolerances on the gap as follows:

$$Gap_{nominal} = [L - (A + B + C)]$$

$$Gap_{tolerance} = -[tol\_L + (tol\_A + tol\_B + tol\_C)]$$

Note that a gap with a negative value suggests that an interference between components exists.

Electrical Circuits

To illustrate the applicability of this technique to electronics, consider the RCL circuit as shown in Figure 7.10. Equation 7.4 holds for this circuit:

$$z = \sqrt{R^2 + \left(\omega L - \frac{1}{\omega C}\right)^2} \tag{7.4}$$

where z is the total impedance of the circuit, and ωL and $(1/\omega C)$ are the inductive and capacitive reactance, respectively. The maximum current is calculated using Equation 7.5:

$$I = \frac{V}{Z} \tag{7.5}$$

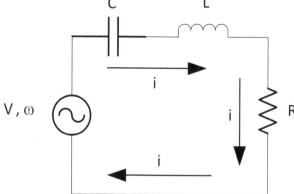

FIGURE 7.10
An RCL circuit for tolerance calculation demonstration.

The same approach may be applied to calculating the worst-case values of z and I_{max} using various tolerances:

$$z_{max} = \sqrt{(R + \text{tol}\_R)^2 + \left[(\omega + \text{tol}\_\omega)(L + \text{tol}\_L) - \frac{1}{(\omega + \text{tol}\_\omega)(C + \text{tol}\_C)}\right]^2}$$

$$z_{min} = \sqrt{(R - \text{tol}\_R)^2 + \left[(\omega - \text{tol}\_\omega)(L - \text{tol}\_L) - \frac{1}{(\omega - \text{tol}\_\omega)(C - \text{tol}\_C)}\right]^2}$$

$$I_{max} = \frac{V + \text{tol}\_V}{Z_{min}}$$

$$I_{min} = \frac{V - \text{tol}\_V}{Z_{max}}$$

It is clear that the nominal values of Z and I as well as their tolerance bands are calculated as follows:

$$z_{nominal} = \frac{z_{max} + z_{min}}{2}$$

$$z_{tolerance} = \pm\left(\frac{z_{max} - z_{min}}{2}\right)$$

Similarly,

$$I_{nominal} = \frac{I_{max} + I_{min}}{2}$$

$$I_{tolerance} = \pm\left(\frac{I_{max} - I_{min}}{2}\right)$$

Root Sum of Squares Method

There are two drawbacks of WCA. The first is that although the worst conditions are analyzed, the distribution of the assembly variable is not determined. Should the worst-case conditions become unacceptable, there are no metrics providing evidence of what percentage may be rejected. The second drawback is that all dimensions occur at their maximum (or minimum) condition at the same time. This is clearly a rare or improbable case. More realistically, assembled components are randomly picked and their sizes could be at any location on the tolerance band.

In the *root sum of squares* (RSS) method, the nominal value of the assembly variable is calculated based on nominal values of each contributing component, however, the tolerance band of the assembly variable is calculated using the square of the standard deviation (also known as variance):

$$s = \sqrt{\sum_i (s_i)^2}$$

Since the tolerance band is directly related to the standard deviation (often tolerance band is equal to three times the standard variation), the following relationship also holds:

$$y_{\text{tolerance}} = \sqrt{\sum_i (x_{\text{tol},i})^2}$$

where $y_{\text{tolerance}}$ is the expected tolerance of the assembly variable due to the specified tolerances of the individual components. $x_{\text{tol},i}$ is the specified tolerance of the ith component. It should be noted that the nominal value of y is calculated based on nominal values of x_i.

Sensitivity Analysis

The process of evaluating the impact of component dimensions on the assembly variables is called sensitivity analysis and is often presented as a percentage contribution to $y_{\text{tolerance}}$. In general it is expressed as

$$P_i = \frac{(x_{\text{tol},i})^2}{(y_{\text{tolerance}})^2}$$

where P_i is the contribution of the ith component to the tolerance of the assembly variable.

Sensitivity analysis provides information on which variable contributes the most to the assembly variable. In turn, this knowledge may be used to redesign the component so that the overall variability may be reduced. For more information, see Creveling (1997) and Ullman (2010).

Example

Recall the assembly as shown in Figure 7.9. Now, suppose that each block has a specified dimension and tolerance band as shown in Figure 7.11.

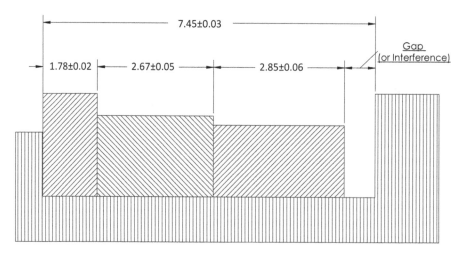

FIGURE 7.11
Assembly block example for tolerance calculations.

TABLE 7.7

Calculation of Nominal Values of the Gap in Figure 7.11 Using WCA

	Nominal	Tolerance	Min	Max
A	1.78	0.02	1.76	1.8
B	2.67	0.05	2.62	2.72
C	2.85	0.06	2.79	2.91
L	7.45	0.03	7.42	7.48
Gap	0.15	0.16	0.31	−0.01

TABLE 7.8

Calculation of Nominal Values of the Gap in Figure 7.11 Using RSS

	Nominal	Tolerance	tol²	P
A	1.78	0.02	0.0004	5.4%
B	2.67	0.05	0.0025	33.8%
C	2.85	0.06	0.0036	48.6%
L	7.45	0.03	0.0009	12.2%
		sum	0.0074	100.0%
Gap	0.15	0.09	0.24	0.06

Sample Calculations
Gap nominal $= 7.45 - (1.78 + 2.67 + 285) = 0.15$
Gap tol. $= (0.0004 + 0.0025 + 0.0036 + 0.0009)^{1/2} = 0.086$
$P_A = 0.0004 / 0.0074 = 5.4\%$
Note that calculations have been done using computer accuracy; however, results have been rounded post computation (e.g., 0.086 is rounded to 0.09).

Tables 7.7 and 7.8 provide the nominal values of the gap as well as its tolerance band using WCA and RSS. Although the nominal value for the gap is identical between the WCA and RSS methods, it is clear that the WCA approach predicts a much wider tolerance band (even an interference) than the RSS method. RSS method is accurate within a three-sigma variation, whereas WCA assumes that all components are at their respective maximum or minimum values.

In addition to the nominal and tolerance values of the gap, Table 7.8 provides a sensitivity analysis. It shows that component C is the largest contributor to the gap variation followed by component B. Should a redesign be warranted, these two components are likely candidates for redesign.

It should be noted that in RSS, it is assumed that component variations and the resulting assembly variable follow normal distributions. Hence, it is possible to calculate the proportion of assembly variables (e.g., gap or interference) that falls within a given range or falls above (or below) a certain threshold.

Monte Carlo Analysis

The *Monte Carlo analysis* technique is a relatively simple technique that was initially developed to study (or model) gambling outcomes (Bethea and Rhinehart 1991). This approach can be used effectively to model the outcome variations based on input changes and variations. In the case of tolerance analysis, input values are the nominal values of each design feature along with its expected tolerance band. Additionally, component variation distributions, if known, may be used as inputs.[7]

Once nominal values for each variable along with specified distribution curves for each variable are specified, then random values for each variable are selected and the outcome is calculated. Clearly, to develop a proper distribution of the assembly variable, a large number of calculations (often in thousands or tens of thousands) should be carried out.

Table 7.9 provides a set of sample calculations for the example shown in Figure 7.11. It contains 15 trials (for demonstration purposes), where random sizes of parts A, B, C, and L were selected and the gaps calculated. Based on these 15 trials, the average (or nominal) gap size is 0.16 and its standard deviation is 0.05.[8] Should the process capability be 3σ (i.e., the tolerance band is three times the standard deviation of the components produced by the process), then this table tells us that maximum gap is 0.32 and the minimum gap is 0.0. As we noted, for a Monte Carlo analysis, the number of trials should be in hundreds if not thousands. As the number of trials increases, the average (nominal) gap and its standard deviation converge. For 50 trials, the nominal value is 0.14 and the standard deviation is 0.05; for 100 as well as 600 trials, the nominal value and the standard deviation remain at 0.15 and 0.05, respectively.[9]

We can use this information to calculate the probability of scrapping assembled product with a gap outside certain acceptable specifications. This can be done using the cumulative standard normal distribution function. If the threshold limit of the gap is to be no less than 0.12 inches, then

TABLE 7.9

Sample Calculations in a Monte Carlo Analysis

Trial	A	B	C	L	Gap
1	1.77	2.64	2.80	7.43	0.22
2	1.80	2.68	2.84	7.43	0.12
3	1.77	2.72	2.87	7.44	0.08
4	1.77	2.65	2.79	7.47	0.26
5	1.80	2.70	2.81	7.44	0.13
6	1.79	2.66	2.89	7.45	0.11
7	1.79	2.72	2.82	7.44	0.11
8	1.78	2.64	2.86	7.47	0.20
9	1.79	2.65	2.84	7.44	0.17
10	1.80	2.63	2.84	7.44	0.18
11	1.79	2.68	2.83	7.42	0.13
12	1.77	2.65	2.90	7.46	0.15
13	1.76	2.66	2.90	7.43	0.11
14	1.78	2.63	2.86	7.44	0.18
15	1.76	2.66	2.80	7.46	0.24
				Average	0.16
				Std. Dev.	0.05
				Max	0.32
				Min	0.00

Notes: Values of A, B, C, and L are generated using a random number generator. Maximum and minimum values are based on 3×Standard Calculation Range.

$$Z = \frac{\text{Gap Threshold} - \mu_{\text{Gap}}}{\sigma_{\text{Gap}}}$$

$$Z = \frac{0.12 - 0.16}{0.05} = -0.8$$

The probability of scrapping if the gap is less than 0.12 inches is expressed by the probability that $Z = -0.8$, i.e., we are now interested in the area under the standard normal distribution of the region below a Z value of -0.8, which is equal to the area under the chart of the region above the Z value of $+0.8$. This is calculated[10] as

$$P(Z \geq 0.8) = 0.2118$$

This means that there is a probability of 21.18% that assembled products will be scrapped because the gap is too tight and below the acceptable specifications.

Unintended Consequences

In engineering design the so-called law of *unintended consequences* is always at play because a design is often required to satisfy conflicting requirements. An example of such conflicting requirements may be found in Figures 7.12 through 7.14. Figure 7.12 depicts a design of several linkage arms that are covered by a housing. A bolt passes through holes in the

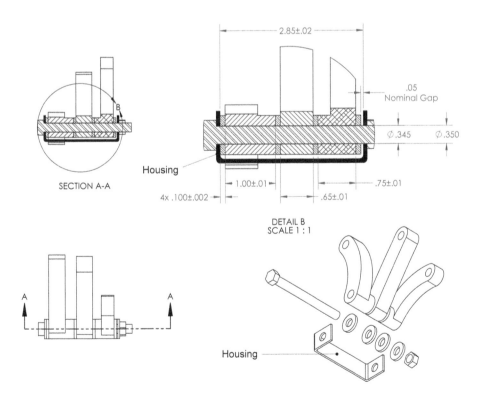

FIGURE 7.12
An example to illustrate various tolerance analysis techniques.

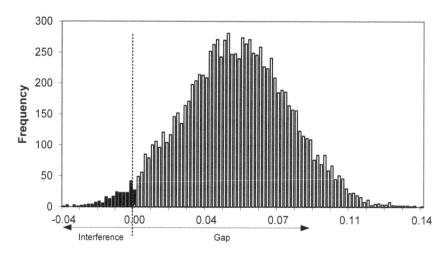

FIGURE 7.13
Monte Carlo analysis results for the gap size in Figure 7.11. This analysis indicates an interference of 0.04 to a maximum gap of 0.14 may exist.

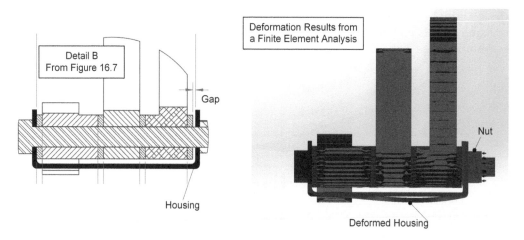

FIGURE 7.14
An often-overlooked issue: deformation of housing due to tightening of the nut on the right-hand side.

linkages as well as the housing. The bolt not only provides an axis of rotation, but also provides a way to assemble the housing.

Due to the possible wide tolerances on the parts used in the assembly, the target dimension may vary from loose to interference (interference occurs when the target dimension is negative, indicating that two parts are trying to occupy the same space). Based on the dimensions shown in Figure 7.12, a Monte Carlo analysis is conducted, as shown in Figure 7.13. It is realized that for a portion of population of assemblies, there will be interference between the housing and the rest of the assembly leading to scraped assemblies.

The team decided to modify the component sizes so that the probability of scraps is reduced. The unintended consequence of this change, however, is shown in Figure 7.14. This figure shows that as the assembly is made and the bolt is tightened, the housing begins to bow out due to excessive gap values.

Another unintended consequence is the impact of environmental conditions such as temperature, pressure, and humidity. It is important that the impact of temperature and material behavior be evaluated when tolerance analysis is conducted. The impact of humidity or other factors, though just as important, may be more difficult to evaluate.

One last thought: whenever an analysis is done, it is helpful to adjust the tolerances and process capability values (C_p) on various parts to see how the stack up is affected.

Functional Tolerance Concerns

In a previous work (Jamnia 2016), it has been pointed out that failures typically occur when design engineers overlook certain (and possibly critical) factors. One of these often-forgotten factors is an investigation of how the delivered function of an assembly or subsystem is affected by various part-to-part variations. Often, engineers conduct worse-case tolerance analyses on both mechanical and electrical components; however, they do not investigate these variations beyond stack-up.

Part-to-part variations of mechanical as well as electronic components along with property variations of the electronic components may have a profound impact on the ability of a design to perform its intended function. At times, this function may be expressed in a closed-form mathematical solution. Equation 7.1 is an example of such a closed-form solution. More often, a closed-form equation is not available. In these cases, careful experiments should be set up to develop an empirical relationship between input variables and the intended outcome, called *response surface*. A robust approach to developing the response surface is to employ *design of experiments* (DOE). While a detailed treatment of the response surface and DOE are beyond the scope of this work, for completeness sake, we will present a brief description as well as a case study in the next section.

Design of Experiments

Design of experiments (DoE), or experimental design, is a technique used during the design phase or in investigations to develop a so-called transfer function. The transfer function is a mathematical relationship between input variables and an expected outcome. For instance, brightness (lumens) of a light bulb is a function of the resistance present in a circuit and the voltage and current that is supplied. If we were to develop a mathematical relationship between these three variables, we have determined its transfer function.

The outcome of this activity is to determine which design parameter has the most impact on a specific performance of the design. The transfer function is used to optimize the design and make it more robust against variations of the input variables. It may also be used to evaluate and understand the interaction between key design parameters. On the one hand, design optimization and robustness contribute to higher product reliability. On the other hand, understanding the impact of the design parameters to the performance, or in some cases, the underperformance of the design, helps to discover root causes of failures, which may lead to design improvements.

Design of experiments is a structured technique to study the impact of a set of factors (inputs or X's) on the response of a system (outputs or Y's). In a nutshell, the basis of DOE is to run a set of experiments (or simulations) on the basis of nominal and/or maximum/minimum values of its inputs. Once this is done and data is collected, regression (i.e., curve fitting) tools are used to develop a mathematical model that relates input variables to the response(s). This technique is well studied and documented. See, for instance, Del Vecchio (1997), Hicks (1973), Morris (2004), and Mathews (2005).

A full description of DoE is beyond the scope of this work. The interested reader is encouraged to follow one of the references mentioned here. As such, in the following section we will use the DOE techniques in a case study involving the analysis of an air-in-line sensor. It will serve as an example to demonstrate how to develop a *response surface* from a set of input variables and a series of structured experiments.

Case Study

A number of medical devices use an air-in-line sensor to mitigate the risk of harming a patient. The device is comprised of an ultrasonic transducer, an ultrasonic receiver, electronic circuitry, and the housing, which keeps the transmitter and receiver apart and places the medium to be sensed in the middle. A simplified configuration is shown in Figure 7.15.

The theory of operation of this sensor is rather simple. One of the transducers sends ultrasonic waves through the medium between the transducer and receiver. The signal strength at the receiver depends on the medium at which the waves travel. Therefore, when air bubbles pass between the transducer and receiver, the signal strength varies indicating their presence in the fluid.

We have determined that variations in the housing may have the most significant impacts on the performance of the sensor, however, we do not know which variables or their interactions may be more significant. Figure 7.16 illustrates the air-in-line sensor and observed variations in the housing. Parameters of interest are the nominal distance between the transducer and the receiver, vertical shift in the alignment of the transducer and the receiver, and the angle in the horizontal level between the transducer and the receiver. These are designated as *d*, *x*, and *α*, respectively.

The approach in DOE is relatively simple to explain: First, the inputs to DOE should be identified. For the air-in-line sensor, these input variables are *d*, *x*, and *α*. One may suggest that electrical power and circuitry may also be considered as input variables. Although this statement holds true, it may be suggested that the electronics and mechanical design in this particular example do not interact and may be decoupled.

Second, the degree of linearity (or nonlinearity) of the model should be decided. Typically, in DOE, one of the linear, quadratic, or cubic models are used. The simplest model is the

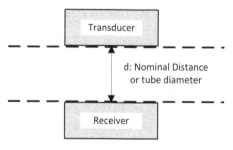

FIGURE 7.15
A simplified configuration of an air-in-line sensor.

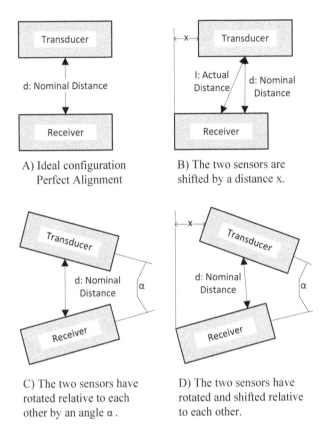

A) Ideal configuration
Perfect Alignment

B) The two sensors are
shifted by a distance x.

C) The two sensors have
rotated relative to each
other by an angle α.

D) The two sensors have
rotated and shifted relative
to each other.

FIGURE 7.16
Air-in-line sensor design critical parameters and DOE factors.

linear model. For a three-variable input such as the ones of the air-in-line example, a linear model is expressed as

$$Y = A_0 + A_1X_1 + A_2X_2 + A_3X_3 + A_4X_1X_2 + A_5X_1X_3 + A_6X_2X_3 + \epsilon \qquad (7.6)$$

The X_i's are independent variables (i.e., d, x, and α) and Y is the transfer function. ϵ is the error term that results from casting a general equation into a specific form, as in Equation 7.6. A_i coefficients will be determined from the experiments that will be conducted.

We understand that Equation 7.6 is the transfer function. In other words, it represents the relationship between the ultrasonic response, Y, and the critical parameters or factors X_1, X_2, and X_3. Notice that the response also includes the first-order interaction between the factors, i.e., X_1X_2, X_1X_3, and X_2X_3. The higher-order interaction between all design parameters has been ignored.

DoE Test Design

The third step is to conduct m number of experiments (or numerical simulations, if running a finite element analysis) using different values of X_i and obtain corresponding values of Y. To ensure that there are an appropriate number of equations to solve for the unknowns (i.e., A_j's), m should be equal to 2^n, where n is the number of independent variables. For the three-variable example ($i = 3$, $m = 2^3 = 8$), this results in the following set of equations:

$$Y_1 = A_0 + A_1 X_{11} + A_2 X_{21} + A_3 X_{31} + A_4 X_{11} X_{21} + A_5 X_{11} X_{31} + A_6 X_{21} X_{31} + \epsilon_1$$

$$Y_2 = A_0 + A_1 X_{12} + A_2 X_{22} + A_3 X_{32} + A_4 X_{12} X_{22} + A_5 X_{12} X_{32} + A_6 X_{22} X_{32} + \epsilon_2$$

$$\cdots$$

$$\cdots$$

$$Y_8 = A_0 + A_1 X_{18} + A_2 X_{28} + A_3 X_{38} + A_4 X_{18} X_{28} + A_5 X_{18} X_{38} + A_6 X_{28} X_{38} + \epsilon_8$$

In matrix form, this set of equations may be written as

$$
\begin{Bmatrix} Y_1 \\ Y_2 \\ \cdots \\ Y_m \end{Bmatrix}
=
\begin{bmatrix}
1 & X_{11} & X_{21} & X_{31} & X_{11}X_{21} & X_{11}X_{31} & X_{21}X_{31} \\
1 & X_{12} & X_{22} & X_{32} & X_{12}X_{22} & X_{12}X_{32} & X_{22}X_{32} \\
\cdot & \cdot & \cdot & \cdot & \cdot & \cdot & \cdot \\
1 & X_{18} & X_{28} & X_{38} & X_{18}X_{28} & X_{18}X_{28} & X_{18}X_{28}
\end{bmatrix}
\begin{Bmatrix} A_0 \\ A_1 \\ A_2 \\ A_3 \\ A_4 \\ A_5 \\ A_6 \end{Bmatrix}
+
\begin{Bmatrix} \epsilon_1 \\ \epsilon_2 \\ \cdots \\ \epsilon_8 \end{Bmatrix}
$$

The general form of this equation may be expressed in the matrix notation

$$\{Y\} = [x]\{A\} + \{\epsilon\} \tag{7.7}$$

Finally, this matrix equation should be solved by minimizing the error term $\{\epsilon\}$. The least squares method is typically used to minimize this term and calculate the regression coefficient vector $\{A\}$ (Myers and Montgomery 2001):

$$\{A\} = \left([\mathbb{X}]^t [\mathbb{X}] \right)^{-1} [\mathbb{X}]^t \{Y\} \tag{7.8}$$

The transfer function (or the response) may be calculated as follows:

$$\{\hat{Y}\} = [\mathbb{X}]\{A\} \tag{7.9}$$

Morris (2004) has developed a template that conducts these calculations in Microsoft Excel. In addition, he provided means of calculating the goodness of fit as well as other statistical measures such as the confidence level associated with each value of the regression coefficient. Needless to say, many analysis software such as Minitab solve these types of equations.

Test Runs and Output Signal Strength

In selecting the values of the input factors (or independent variables) to run the experiments, it is customary to choose the maximum or minimum values, if a linear relationship is assumed. For nonlinear models, additional points are required. Considering the air-in-line example, suppose that the design team anticipates the following parameter ranges:

$$1.5 \le d \le 4.0,$$

$$0 \le x \le 1.0,$$

$$0 \le a \le 10.0$$

Since there are three variables, 8 ($=2^3$) experiments need to be set up. Table 7.10 depicts these values along with the corresponding signal strength measured from the hypothetical experiments.

Mathews (2005) warns against forming the $[\mathbb{X}]$ matrix on the basis of the actual input values. It is possible that the resulting matrix is ill-conditioned, leading to erroneous results. Instead, he recommended transferring the variables into a *coded* space. In this space, variables range from –1 to +1. The transformation equation is as follows:

$$C_i = \frac{2}{X_{i\,max} - X_{i\,min}} X_i + \left(1 - \frac{2}{X_{i\,max} - X_{i\,min}} X_{i\,max}\right)$$

Hence for the air-in-line example:

$$C_1 = \frac{2}{4-1.5} d + \left(1 - \left(\frac{2}{4-1.5}\right)4\right)$$

$$\Rightarrow C_1 = 0.8d - 2.2$$

$$C_2 = \frac{2}{1-0} x + \left(1 - \left(\frac{2}{1-0}\right)1\right)$$

$$\Rightarrow C_2 = 2x - 1$$

$$C_3 = \frac{2}{10-0} a + \left(1 - \left(\frac{2}{10-0}\right)10\right)$$

$$\Rightarrow C_3 = 0.2\,a - 1$$

As a result of this transformation, Table 7.10 turns into Table 7.11; note that Y's have not changed.

TABLE 7.10

Experiment Set with Input Values and Corresponding Measurements

Experiment	Input Factors (X's)			Transfer Function (Y)
	$d\,(X_1)$	$x\,(X_2)$	$a\,(X_3)$	Signal Strength
1	1.5	0.0	0.0	0.47237
2	4.0	0.0	0.0	0.13534
3	1.5	1.0	0.0	0.28650
4	4.0	1.0	0.0	0.10688
5	1.5	0.0	10.0	0.41992
6	4.0	0.0	10.0	0.12912
7	1.5	1.0	10.0	0.26608
8	4.0	1.0	10.0	0.10246

TABLE 7.11

Experiment Set with Coded Input Values and Corresponding Measurements

	Input Factors (C's)			Transfer Function (Y)
Experiment	$d\ (C_1)$	$x\ (C_2)$	$\alpha\ (C_3)$	Signal Strength
1	−1	−1	−1	0.47237
2	+1	−1	−1	0.13534
3	−1	+1	−1	0.28650
4	+1	+1	−1	0.10688
5	−1	−1	+1	0.41992
6	+1	−1	+1	0.12912
7	−1	+1	+1	0.26608
8	+1	+1	+1	0.10246

TABLE 7.12

The [𝕏] for the Air-in-Line Sensor Example in the Coded Space along with the Calculated Coefficients

Intercept	d	x	α	$d\,x$	$d\,\alpha$	$x\,\alpha$		
C_0	C_1	C_2	C_3	C_1C_2	C_1C_3	C_2C_3		
1	−1	−1	−1	1	1	1	A_0	0.2398
1	1	−1	−1	−1	−1	1	A_1	−0.1214
1	−1	1	−1	−1	1	−1	A_2	−0.0494
1	1	1	−1	1	−1	−1	A_3	−0.0104
1	−1	−1	1	1	−1	−1	A_4	0.0356
1	1	−1	1	−1	1	−1	A_5	0.0078
1	−1	1	1	−1	−1	1	A_6	0.0042
1	1	1	1	1	1	1		

Now the $[\mathbb{X}]$ may be set up in the coded space and the coefficient vector may be calculated using Equation 8.3. The results are presented in Table 7.12.

There are two additional metrics that need to be calculated and evaluated prior to accepting the coefficient values (or A_j). The first is a metric indicative of goodness of fit or R^2. This metric ranges between 0 and 1; 0 means that the data fit was very poor and 1 indicates perfect fit. The second metric is the confidence level $(1-p)$. This value is typically chosen at 0.9 or higher (90% confidence or better). Most available statistical software packages provide this information. For developing an algorithm in Excel, see Morris (2004). Table 7.13 provides this information for the previous example.

The $R^2 = 0.9992$ shows that the calculated curve fits the data very well. However, the confidence levels of A_3, A_5, and A_6 are low—much lower than the typical 90% or 95%. These coefficients are associated with α, the angle between transducer and receiver. A low confidence level is indicative of a low degree of influence of the variable on the response. For this reason, the contribution of α may be ignored. Finally, the response equation or the transfer function in the coded space may be written:

$$Y = 0.2396 - 0.1214C_1 - 0.0494C_2 + 0.0356C_1C_2$$

TABLE 7.13

The Confidence Level and R^2 Values
Associated with the Calculated Coefficients

	p-Value	Confidence Level	R^2
A_0	0.010	99.0%	0.9992
A_1	0.020	98.0%	
A_2	0.049	95.1%	
A_3	0.221	77.9%	
A_4	0.067	93.3%	
A_5	0.288	71.2%	
A_6	0.464	53.6%	

In the physical space the response surface for the signal in air may be written as

$$Y_{\text{Air}} = 0.635 - 0.126\, d - 0.255\, x + 0.0569\, d\, x \tag{7.10}$$

This equation indicates that the signal strength is a function of not only the distance and the pitch between the transducer and receiver but also the product of the two. Hence, both d and x and their interactions should be controlled in a robust design. A designed reliability test would then examine their impact on the overall system behavior at a worse (or corner) case scenario.

Physics of Failure

When we speak of physics of failure, the first impression that may form is that of understanding the fundamentals of the mechanism of failure. This section focuses on exploring what failure means. It tends to explore various mechanisms by which failures may happen at system or below system (i.e., subsystem and assembly) levels. Having said this, we acknowledge that a thriving disciple within reliability is called *physics of failure* (PoF), which is the study of the same phenomenon however at an electronic-component level. It has been accepted that methods such as *part count* and probabilistic data as explained in MIL-HDBK-217 F, Bellcore, or Telcordia do not as a general rule of thumb provide accurate prediction of a system's reliability field behavior. The goal of PoF is to improve component failure predictions by including elements such as environmental and use (particularly load) conditions, and to allow the failure rates to be influenced by the physical conditions that affect them. As mentioned, the goal of this section is not to take a deep dive at the PoF but to introduce the concept at a macro level. For more details on PoF, consult Thaduri (2013), White and Bernstein (2008), or McLeish (2018).

Failure Classifications

Reliability is, in a way, a study of failure; hence, it is important to define failure first. Failure is defined as the inability of a system, subsystem, assembly, or component to meet its design specifications. It may be that it was poorly designed and never met its objectives, or that it initially met its design objectives but after some time it failed. In these two scenarios, clearly, there have been certain overlooked factors.

Many factors may lead to component and/or assembly failures. Some may be due to sloppy design and a lack of verification activities, whereas others may simply be due to the designer's lack of insight of potential factors affecting the system; yet, others may be attributed to misuse of the product. In general, failures of electromechanical systems may be categorized as reversible failures, irreversible failures, sudden failures, or progressive failures. These apply equally to both mechanical and electrical systems. A fifth topic concerns itself with chemical failures that are associated with either restriction of hazardous substances (RoHS), or corrosion or electrochemical material migration.

First, we will examine them in generalities and then provide more detail as they relate to either electrical or mechanical systems.

Reversible Failures

From a mechanical point of view, these failures are caused by elastic deformation of a member in the system. In general, once the failure-causing load is removed, the system would function normally once again. In electronic systems, often some excess heat will interrupt functionality but when the unit cools down sufficiently, operations are resumed. Another example of this type of failure is resonant vibration of relays or chatter of printed circuit board assemblies (PCBAs), as shown in Figure 7.17. Once vibratory excitations are removed, the assembly works as expected.

Irreversible Failures

These failures are caused by incremental application of loads beyond a certain (reversible) limit. This behavior is typically observed in systems with nonlinear (softening) behavior. In mechanical components, the applied loads and the associated stress factors create stress fields that are beyond the proportional (yield) limit and in the neighborhood of the yield point. In electronic systems, there are component damage, say, due to arching.

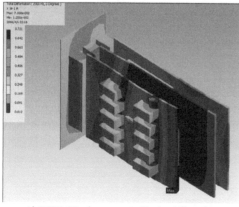
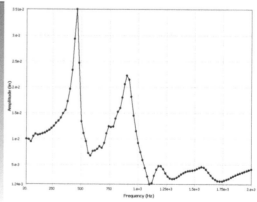

A) PCBA Displacements Under Vibration   B) PCBA Vibration Response

FIGURE 7.17
An example of PCBA chatter and its response.

An example of this phenomenon in an electromechanical device may be the PCBA shown in Figure 7.17. Should the magnitudes of vibration pass a certain limit, the PCBA's chatter may lead to an electric short causing an irreversible failure. Another example is overloading a motor; some overload conditions may be tolerated and others may cause significant degradations.

Sudden Failures

Sudden failures are caused by application of excessive loads where loads have surpassed their ultimate values. A blown fuse is an example of this type of failure. Cracking of ceramic boards is another example. Numerical simulations of sudden failures are rather simple: once the load distribution is calculated, load (or stress) values at desired locations are compared to permissible levels. This type of analysis is also known as the worst-case analysis.

Progressive Failures

These failures are the most serious of failures because initially a device passes all its verification requirements, yet after some time in the field, it begins to fail. Mechanical creep and fatigue belong to this category. Typically, failures of electrolytic capacitors or lead–acid batteries may also be due to this class of failures.

Prediction of progressive failures is more challenging than other types of failures. On the one hand, the expected failure mode should be known and well understood. On the other hand, failure time is not deterministic and a certain level of uncertainty and probability exist; hence, along with an understanding of physics of failure, a knowledge of statistics and probabilities is also needed to analyze progressive failures and create predictive models.

Chemical Failures

The final topic in failure categories that the design team needs to account for in reliability consideration in design is RoHS. A household word within the electromechanical product development community, RoHS was the result of a European Union directive for restricting uses of certain hazardous substances in electronic and electrical equipment in the waste from electrical and electronic equipment (WEEE). The outcome has been a ban on using some materials, including lead, mercury, and cadmium.

In particular, complying with RoHS requires using lead-free solders, and, as a result, a special soldering process. A lead-free soldering process operates at a higher melting point than the traditional tin–lead solder. Higher process temperatures may adversely impact sensitive components on the board, or it may cause warping of the board leading to stressed solder joints.

Use of lead-free solder is also associated with whisker growth. Whiskering is a form of material migration that is a process by which material moves from one area to another area. It happens as a result of internal solder stresses.

Whiskers are needle-like structures with diameters typically up to 3 μm and can grow to over 1 mm in length (Asrar et al. 2007). Their growth has been primarily attributed to the compressive stresses in the tin plating as evidence in a bright tin finish, and as a result bright finishes have been prohibited in microelectronic packages (Xu et al. 2007; Zhang et al. 2007). Whiskers break due to vibration and are a source of debris. They have been known to cause complete failures of several on-orbit commercial satellites (Asrar et al. 2007).

The growth of whiskers is related to the materials and processes used to plate and solder the components on the printed circuit board (PCB) as well as the environment to which the equipment is subjected. It does not depend on the components used, and a variety of components such as diodes, transistors, integrated circuits, microcircuit leads, and even PCBs have been affected. Many metals have been known to whisker, and as such, designers and engineers should be mindful to avoid components plated with materials as well as processes that would facilitate whisker formation.

Failure Modes and Mechanisms

Failure modes and effects analysis (FMEA) helps the design team to clearly identify sources of failure and indicate whether the solutions may be sought in mechanical, electrical, or other causes. In general, there are four causes of mechanical failures. These are as follows:

1. Failures by elastic deflection—These failures are caused by elastic deformation of a member in the system. Once the load causing the deformation is removed, the system functions normally once again.

2. Failures by extensive yielding—These failures are caused by application of excessive loads where the material exhibits a ductile behavior. Generally, the applied loads and the associated stress factors create stress fields that are beyond the proportional limit and in the neighborhood of the yield point. In these scenarios, the structure is permanently deformed and it does not recover its original shape once the loads are removed. This is generally a concern with metallic structures such as chassis and racks.

3. Failures by fracture—These failures are caused by application of excessive loads where the material exhibits a brittle behavior, or in ductile materials where stresses have surpassed the ultimate value.

4. Progressive failures—These failures are the most serious because initially the system passes most, if not all, test regiments and yet after some time in the field, it begins to fail. Creep and fatigue belong to this category.

In electronics and in general, PCBAs and electronic (sub)systems exhibit three modes of failures. These are open circuits, short circuits, and intermittent.

1. Open circuit—When an undesired discontinuity forms in an electrical circuit, it is said that the circuit is open. This is often caused by solder cracks due to a variety of thermomechanical stresses. Another contributor to developing this failure mode is chemical attack (such as corrosion).

2. Short circuit—When an undesired connection forms between two electrical circuits or between a circuit and ground, it is said that the circuit is shorted. Major contributors that short circuits are based on chemical attacks such as dendrites and whiskers, or contaminants left behind from the manufacturing process combined with humidity and high temperatures. Short circuits may also occur under vibration if the PCBAs are not properly secured.

3. Intermittent failure—An intermittent failure is a microscopic open (or a short) circuit that is also under other environmental influences. For instance, a crack in the solder may be exasperated under thermal loads leading to an open circuit failure; however, as the board is removed from its environmental conditions and cools down, the gap closes, and it would work as intended.

The failure mechanism in electronic packages has been well studied over the years. Failure mechanisms may be attributed to

1. Design inefficiencies
2. Fabrication and production issues
3. Stresses such thermal, electrical, mechanical, and environmental factors such as humidity

In addition to the two main mechanical and electrical influences on failure, a third and, to some extent, less considered influence, is chemical and electrochemical factors. This may be divided into two broad categories:

1. Corrosion is a natural two-step process whereby a metal loses one or more electrons in an oxidation step resulting in freed electrons and metallic ions. The freed electrons are conducted away to another site where they combine with another material in contact with the original metal in a reduction step. This second material may be either a nonmetallic element or another metallic ion. The oxidation site where metallic atoms lose electrons is called the anode, and the reduction site where electrons are transferred is called the cathode. In electromechanical systems, corrosion is often caused by galvanic cells, which are formed when one of the following two criteria is satisfied: two dissimilar metals exist in close proximity or the metal is a multiphase alloy in the presence of an electrolyte.

2. Migration is a process by which material moves from one area to another area. If migration happens between two adjacent metals such as copper and solder, it is called diffusion. If it happens as a result of internal stresses and in the absence of an electric field, it is called whiskering. Finally, if it happens in the presence of an electric field and between similar metals, it is called dendritic growth. Two factors have given this topic relevance in PCB and system package design. The first factor is that as the PCBs are more densely populated, it becomes exceedingly difficult to clean and wash away all the processing chemicals. The presence of the pollutants along with an electric field provides an ideal environment for electromigration and dendritic growth. The second factor is more related to legislation and a lead-free environment. Pure tin solder, which is a natural replacement for a leaded solder, has a tendency to grow conductive needles, known as whiskers, in an out-plane direction (z-axis).

Common failure modes and mechanisms in electromechanical devices have been reviewed to some extent by Jamnia (2016).

Failure Modes and Effects Analysis

After the system and subsystems architecture, and a physical concept of the new product have been developed, an early stage design/concept FMEA should be conducted to ensure robustness of the design. This technique enables the design team to uncover potential shortcomings in the conceptual design by identifying and evaluating

1. Design functions and requirements
2. Foreseeable sequence of events both in design and production
3. Failure modes, effects, and potential hazards
4. Potential controls to minimize the impact of the end effects

This early FMEA document may then become the basis of a more formal and rigorous design failure modes and effects analysis (DFMEA) and process failure modes and effects analysis (PFMEA). It should be noted that both DFMEA and PFMEA are living (version-controlled) documents that are initiated before design requirements have been fully established, prior to the completion of the design, and updated throughout the life cycle of the product including design changes, where appropriate. As the design matures, so does the FMEA with more details and complexities. For a more in-depth review of this topic, refer to Jamnia (2017).

It may easily be argued that a rigorous FMEA effort would inevitably lead to a more robust design by identifying and mitigating design flaws. At the same time, the same activity would identify potential causes of failure due to reliability and wear out, thus enabling reliability engineers to design proper reliability tests to examine and challenge specific identified components.

The first step in creating the DFMEA charts is to ask, What are the requirements and/or functions? The next step is to ask, How will these functions fail? Let it suffice to say that functions fail in one or a few of the following ways (though rarely more than three ways):

1. No function—For example, consider that the function of a vacuum cleaner is to produce suction to remove dirt. No function means that you turn a vacuum cleaner on and nothing happens!
2. Excessive function—By way of the vacuum cleaner example, the suction is so strong that you cannot move the cleaner around as it is stuck to the surface being cleaned.
3. Weak function—There is hardly any suction present.
4. Intermittent function—Vacuum cleaner works fine but unexpectedly it loses power; shortly after, its power is restored.
5. Decaying function—Vacuum cleaner starts working fine, but over time (either short or long term) loses its suction, and it does not recover.

Example

An electromechanical product called the Learning Station has been developed to aid educating of children with cognitive delays and physical impairments. Three of the functions (or requirements) associated with the Learning Station are shown in Table 7.14. The general functions are (1) protect the device from environmental factors such as fluid spills or other factors, (2) enable the user to develop social skills, and (3) enable the user to run educational software. Table 7.14 provides a flow down of how these general functions are satisfied and what the failure modes associated with each function might be.

Table 7.15 contains the failure modes and effects associated with the Learning Station. In a typical FMEA table, the module to be studied is mentioned. By suggesting that the

TABLE 7.14

Learning Station Functions and Failure Modes

General Function	Specific Function	Failure Mode
Protect from environment	Stop liquids from entering internal compartments	Liquid leaks inside
	Resist electrostatic discharge from reaching sensitive internal components	Electrostatic discharge reaches sensitive components
	Stop users from reaching (or touching) internal components	Users reach (or touch) internal components
Develop social skills	Devise a turntable for turn-taking between users	Turntable fails to turn over time
	Devise system operation by tokens to enforce social skills development	Tokens fail to work
		Token operation fails to work over time
Run educational software	Execute commercially available software	Software does not run
		Software begins to run but stops

TABLE 7.15

Learning Station Initial Design Failure Modes and Effects Highlighted Boxes Show Potential Areas for Reliability Testing

Item	Module	Function	Potential Failure Mode	Potential Effects of Failure	Severity	Potential Causes of Failure
1	Learning Station	Protect from Environment	Liquid leaks inside	Unit stops working temporarily		Lack of gasket and/or drain paths in the design
				Unit stops working permanently		
2			Electrostatic discharge reaches sensitive components	Unit stops working temporarily		Lack of ESD/EMI barriers in the design
				Unit stops working permanently		
3			Users reach (or touch) internal components	Electric shock		Excessively large openings in the enclosure
				Bodily injury		
4		Enable developing social skills	Turntable fails to turn over time	Unit does not rotate on its axis		Turntable deflects under unit weight
						Turntable bearings have corroded
5			Tokens fail to work	Unit stops functioning		Wrong tokens were used
6			Token operation fails to work over time	Unit stops functioning		**Component life is too short**
7		Run educational software	Software does not run	Software does not accept commands		Incompatible software with operating system
			Software begins to run but stops	Software does not accept commands		System is overheated

module is the system, the design team will be concerned with the system-level failures. Also, the specific function under observation is mentioned. Finally, a potential cause of failure is also provided. At this point, the severity column is left blank, though, it can just as easily be filled out.

Note we are using the term "failure modes and effects"; however, at the early stages of design, we have more interest in potential causes. Why? Because once we can identify the root causes of failures, we can attempt to make changes to the design to either remove the failure mode(s) or attempt to reduce their frequency, or mitigate them in such a way as to reduce their impact. It should be noted that *effects* impact product risk, whereas *causes* influence product design configuration and possibly reliability. Hence, it may easily be argued that a rigorous FMEA effort would inevitably lead to a more robust design by identifying and mitigating design flaws. At the same time, the same activity would identify potential causes of failure due to reliability and wear out, thus enabling reliability engineers to design proper reliability tests to examine and challenge specific identified components. In Chapter 8, we will discuss component-level reliability testing in more detail.

In the context of the Learning Station design, the first item in the initial DFMEA reminds the design team of the possible need for either a gasket or a drain path in the housing (enclosure) to guide liquids away from the inner cavity. The second item ensures that both the electromagnetic emission interference (EMI) and electrostatic discharge (ESD) have been properly considered and mitigated.

Another point worth noting is the potential conflict between mitigating overheating (item 7) by having large openings in the unit's housing and mitigating users reaching/touching internal components by having no openings (item 3) at all. Here is an example of a conflict that can be resolved. Openings are needed to remove heat generated by the electronics, and at the same time, any opening provides an opportunity for users to reach the inside and hurt themselves. Clearly, this conflict should be resolved satisfactorily.

Finally, the initial DFMEA provides other design insights to be considered. First, before specifying the turntable willy-nilly, features such as its weight-bearing capacity and its resistance to corrosion should be considered. Another insight is that products do not last forever (item 6). Product and component life expectancy and their impact on the service organization should not be overlooked. One last point that is brought to light is that certain failures may not be avoided altogether and require mitigations. For instance, an identified cause in item 7 is software incompatibility with the operating system. A sensible solution and mitigation is through labeling. Typically, the point-of-purchase packaging of many application software provides information on the appropriate class of operating systems the intended software can run on. To be clear, labeling is not just a sticker that is applied to a product. In general, any printed material either attached to the product or included within the packaging, or even the packaging material is considered as labeling.

Suppose that through this exercise, the Learning Station design team decided to design a gasket at the interfaces of the housing. The DFMEA matrix is modified as shown in Table 7.16. One may notice that this DFMEA matrix has an extra column compared to Table 7.15. This column called "Prevention" is a direct response to potential causes and provides a plan of what needs to be done to either eliminate the cause of the failure or to mitigate it.

Generally speaking, once potential causes of failure are recognized, a control plan should be developed to minimize their impact. At the concept level, this control plan would dictate the elements of detailed design. For instance, for the Learning Station as shown in Table 7.16 these elements may be:

TABLE 7.16

A Populated Initial Design Failure Modes and Effects Matrix for the Learning Station

Item	Module	Function	Potential Failure Mode	Potential Effects of Failure	Severity	Potential Causes of Failure	Occurrence	Prevention
1	Learning Station	Protect from environment	Liquid leaks inside	Unit stops working temporarily / Unit stops working permanently		Excessive warping of gasket		Finite element analysis
2			Electrostatic discharge reaches sensitive components	Unit stops working temporarily / Unit stops working permanently		Inadequate ESD/EMI barriers in the design		Design review
3		Develop social skills	Users reach (or touch) internal components	Electric shock / Bodily injury		Excessively large openings in the enclosure		Human factors study
4			Turntable fails to turn over time	Unit does not rotate on its axis		Turntable deflects under unit weight		Load analysis
						Turntable bearings have corroded		Material compatibility
5			Tokens fail to work	Unit stops functioning		Wrong tokens were used		Labeling
6			Token operation fails to work over time	Unit stops functioning		Component life is too short		Reliability analysis
7		Run educational software	Software does not run	Software does not accept commands		Incompatible software with operating system		Labeling
			Software begins to run but stops	Software does not accept commands		System is overheated		Thermal analysis

Notes: The only item that requires reliability testing has been highlighted. Severity and occurrence are ignored for the purposes of this example.

1. Once a gasket is designed, a finite element analysis (FEA) should be conducted to ensure that the gasket does not warp under the clamping load of the enclosure in such a way as to create openings for penetration of liquids inside.

2. The size and shape of any openings should be evaluated by human factors engineers to ensure that any potential harm or injury is minimized and mitigated.

3. Proper load and material analysis should be conducted to ensure the proper workings of the turntable. Note that in Table 7.13, the turntable was identified as potentially needing reliability testing to prevent corrosion, however, in the final analysis, it was ruled out in favor of conducting material compatibility studies.

4. Through the judicious use of labeling, any product misuse is mitigated and reduced.

5. Finally, both thermal and reliability analyses of the product (in the design phase) should be done to ensure any thermal or aging failures are either resolved or remediated.

As design engineers we do all we can to remove the causes of failures; however, in practice may of the identified failure modes are not completely eliminated. We just mange to reduce their failure rate. Associated with these remaining failures—albeit at a reduced rate—are remaining risks to the user or property. These remaining risks are called the overall residual risk (ORR) and should be quantified. In Chapters 12 and 13, we will review how risk may be calculated and its association with reliability.

Life-Expectancy Calculations

Earlier in this chapter, we discussed use profiles and conditions. For certain mechanical elements such as cams or springs, it is sufficient to know how many times a member needs to be activated. For other members that may undergo cyclic loading, life calculations may not be as easy and straightforward. In the following pages, we will be reviewing a few life-expectancy models.

Life Expectancy for Pure Fatigue Conditions

Miner's index is used to predict the lifetime of a member of a system that experiences different levels of stress and the corresponding frequencies. This equation is particularly useful in vibration environments:

$$D_f = \sum_i \frac{n_i}{N_i} \tag{7.11}$$

N_i is the cycles to failure at stress level S_i and n_i is the actual cycles of vibration at that stress level. Clearly (and theoretically), at end of life $D_f = 1$.

Example

Suppose that a vibratory member made of 300M alloy undergoes a stress level of 82,000 psi at a frequency of 60 Hz. Considering that the S–N curve (Figure 7.18) gives the number of cycles it can endure at this stress level to be 1×10^9 cycles, determine the time to failure.

Frequency is defined as the number of cycles per unit of time (usually seconds). Thus, to calculate the number of cycles for a given frequency and time, one may multiply frequency and time to obtain the number of cycles:

$$D_f = \sum_i \frac{n_i}{N_i}, i = 1$$

$$D_f = \frac{n}{N} = \frac{\text{number of cycles}}{N} = \frac{\text{time} \times \text{frequency}}{N} = \frac{tf}{N}$$

$$1 = \frac{t \times 60}{1,000,000,000}$$

$$t = 16,666,666.67 \text{ seconds or } 4629.63 \text{ hours}$$

Does this mean that the part will fail after 4629.63 hours? The correct answer is that if the used S–N curve provided data for a 50% probability of failure, then it is expected that by $t = 4629.63$, 50% of the samples have failed. A different way of expressing this number is to say

$$L_{50} = 4629.63 \text{ hours}$$

This annotation is often referred to L_x life, where the subscript x refers to percent of failed samples.

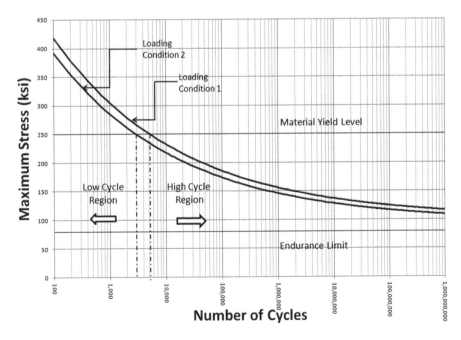

FIGURE 7.18
Fatigue S–N curve for 300M alloy.

Life Expectancy for Random Vibration Conditions

Random excitations cause all the natural frequencies (and mode shapes) of a continuous system to be present. However, the notable response of the system takes place at what is termed as *apparent* frequency, which is mainly influenced by the first natural frequency and contributions from other higher ones. In a random vibration environment, the number of positive zero crossings that occur per unit of time significantly influences the apparent resonant frequency. The number of positive zero crossings refers to the number of times that the stress (or displacement) crosses the zero line with a positive slope. Based on this definition, one apparent cycle is between two positive zero crossings (Sloan 1985; Steinberg 1988).

The lifetime of a piece of equipment under vibration may be estimated using the following relationship (Sloan 1985):

$$L_D = \frac{B}{4\,fS^b\left[.2718\left(\dfrac{2}{3}\right)^b + .0428\right]}\,\text{Sec} \tag{7.12}$$

where f is the response frequency, S is stress, and B and b are constants and depend on the material used. For example, $b = 11.36$, $B = 7.253 \times 10^{55}$ psi/cycle for G10 glass epoxy,[11] and for aluminum, $b = 9.10$, $B = 2.307 \times 10^{47}$ psi/cycle.

This equation is based on a skewed Gaussian distribution and incorporates that number of zero crossings. The drawback of this formulation is the relatively limited information on constants B and b for a large class of materials.

Example

Suppose that in a vibration environment three stress values for a 300M alloy have been calculated to be 60 ksi (68% of the time), 120 ksi (27% of the time), and 180 ksi (4% of the time). What would be the expected life of the part should the system be approximated as a single degree of freedom with a frequency of vibration of 60 Hz?

It is possible that a relationship may be developed between the stress values and the number of cycles to failure. For the 300M, the following relationship has been proposed (MIL-HDBK-5J 2003):

$$\log N = 14.8 - 5.2\log\left[\frac{S_{max}}{(1-R)^{0.38}} - 94.2\right]$$

where R is the stress ratio, which is the ratio of minimum stress to maximum stress in a fatigue cycle, and S_{max} is in ksi. For a fully reversible load, $R = -1$.

For $R = -1$, we have:

At $S_{max} = 60$ ksi, $N_1 \approx \infty$ (Endurance limit for 300M is about 80 ksi),

$n_1 = $ Frequency $\times 68\%$ of times at this stress level)

At $S_{max} = 120$ ksi, $N_2 = 3.0 \times 10^5$,

$n_2 = $ Frequency $\times 27\%$ of times at this stress level

At $S_{max} = 180$ ksi, $N_3 = 4175$,

$n_3 = $ Frequency $\times 4\%$ of times at this stress level

or

$N_1 \approx \infty, \qquad n_1 = (60)(0.68t)$

$N_2 = 3.0 \times 10^5, \; n_2 = (60)(0.27t)$

$N_3 = 4175, \qquad n_3 = (60)(0.04t)$

Now apply Miner's index:

$$MI = \frac{n_1}{N_1} + \frac{n_2}{N_2} + \frac{n_3}{N_3} = 1$$

$$MI = \frac{60 \times .68t}{\infty} + \frac{60 \times 0.27t}{3 \times 10^5} + \frac{60 \times 0.04t}{4175} = 1$$

$$t = 1590 \text{ seconds}$$

That is 1590 seconds or 26.5 minutes. If a number of these systems were being tested in a laboratory, this number means that about 50% of the samples would have failed within nearly half an hour (or $L_{50} = 26.5$).

Life Expectancy for Pure Creep Conditions

For pure creep damage, Robinson's index may be used to predict the lifetime of a member of a system that experiences different levels of temperature under a static load:

$$D_c = \sum_i \frac{t_i}{t_{ri}}$$

where t_i is the hold time spent at temperature T_i. t_{ri} is the rupture time at the same temperature.

On the one hand, the premise of Robinson's equation is that the part is under a static load. On the other hand, the impact of creep becomes significant when local temperatures approach about 50% of the melting point of metals and some plastics and the heat-deflection point of other plastics. In electromechanical products, only solder and some encapsulants have melt or glass transition temperatures low enough that creep may play an important role in their behavior.

Life Expectancy for Creep–Fatigue Interactions

In a static sense, the only stresses found in electromechanical products are residual stresses in solders or encapsulants. Relaxation of these stresses because of creep does not cause any damage and yet tends to reduce stresses in soldered leads in certain instances involving low frequency vibrations (Steinberg 1991). Thermal cycling, however, creates a more severe environment for encapsulants and particularly for solder. This is because, in addition to creep, fatigue begins to play a role because of induced cyclic stresses caused by the thermal coefficient mismatch between various components and materials on the board.

A significant amount of research has been conducted to develop a reliable model for predicting solder life in a variety of environments. A very popular model is the Coffin–Manson or modified Coffin–Manson relationship. These models are constructed based on empirical data and as such are only accurate for the specific materials and/or conditions on which they were based (Tasooji et al. 2007). A basic form of the Coffin–Manson equation is as follows:

$$N_f = \frac{A}{(\Delta T)^B} \tag{7.13}$$

In this equation, N_f is the number of temperature cycles to failure, ΔT is the temperature range of thermal cycling, and A and B are dependent constants. Although A depends mainly on material type and behavior, it has been shown that B has a range of one to three for solders (Jeon et al. 2007), with a typical value of two. Considering the empirical nature of this relationship, a number of improvements have been proposed in order to improve the applicability and accuracy of the life predictions. These improvements are collectively referred to as modified Coffin–Manson equations and are generally accurate within the criteria and conditions for which they were developed (Tasooji et al. 2007).

One such modification has been proposed by Ross and Wen (1993) and has the following ramification:

$$N_1^{\beta_1} = N_2^{\beta_2}$$

$$\beta = 0.442 + 6.0 \times 10^{-4}\, T_m - 0.0174 \ln\left(1 + \frac{360}{\tau}\right)$$

where T_m is the mean solder temperature (°C), and τ is the half-cycle dwell time (min), and the subscripts 1 and 2 refer to two different dwell times when all other conditions are kept the same. In other words, if the number of cycles to failure at a given temperature and dwell time is known, it may be possible to calculate the number of cycles to failure at a different temperature and a different dwell time.

Example

Suppose that a system fails after 1500 cycles at a temperature of 80°C and a dwell time of 30 min. What would the number of cycles to failure be if the dwell time is reduced to 15 and 5 min at the same temperature?

$$\beta = 0.422 + 6.0 \times 10^{-4}\, T_m - 0.0174 \ln\left(1 + \frac{360}{\tau}\right)$$

$$\beta_1 = 0.422 + 6.0 \times 10^{-4}(80) - 0.0174 \ln\left(1 + \frac{360}{30}\right)$$

$$\beta_1 = 0.445$$

$$\beta_2 = 0.422 + 6.0 \times 10^{-4}(80) - 0.0174 \ln\left(1 + \frac{360}{15}\right)$$

$$\beta_2 = 0.434$$

$$\beta_3 = 0.422 + 6.0 \times 10^{-4}(80) - 0.0174\ln\left(1 + \frac{360}{5}\right)$$

$$\beta_3 = 0.415$$

Now that βs are calculated, we can calculate the number of cycles to failure:

$$N_1^{\beta_1} = N_2^{\beta_2}$$

or

$$N_2 = N_1^{\beta_1/\beta_2}$$

$$N_2 = 1500^{(0.445/0.434)}$$

$$N_2 = 1817 \text{ (Life at 15 minutes dwell time)}$$

$$N_3 = 1500^{(0.445/0.415)}$$

$$N_3 = 2545 \text{ (Life at 5 minutes dwell time)}$$

These results indicate that the life of the product increases dramatically when it is subjected to high temperatures for lower dwell times.

Example

Suppose that a system is tested in a chamber that varies the temperature between −10°C and 80°C. The dwell time in each chamber is 30 min. The unit is operational during the testing and it is known that its critical components have a temperature rise of 45°C above the environment. Fifty percent of the units tested fail after 825 cycles.

We need to determine the life of this product if it is being operated at room temperature for nearly 2 hours per day, 200 days a year:

$$\beta = 0.422 + 6.0 \times 10^{-4}T_m - 0.0174\ln\left(1 + \frac{360}{\tau}\right)$$

$$\beta_1 = 0.422 + 6.0 \times 10^{-4}(80 + 45) - 0.0174\ln\left(1 + \frac{360}{30}\right)$$

$$\beta_1 = 0.472$$

Room temperature is assumed to be 25°C:

$$\beta_2 = 0.422 + 6.0 \times 10^{-4}(25 + 45) - 0.0174 \ln\left(1 + \frac{360}{120}\right)$$

$$\beta_2 = 0.460$$

Now that β's are calculated, we can calculate the number of cycles to failure:

$$N_1^{\beta_1} = N_2^{\beta_2}$$

or

$$N_2 = N_1^{\beta_1/\beta_2}$$

$$N_2 = 825^{(.472/.460)}$$

$$N_2 = 990$$

Thus the units will survive 990 cycles. Since each cycle is one day, the data indicates that 50% of the units will fail after 990/200 = 4.95 years or $L_{50} = 4.95$.

Design Life, Reliability, and Failure Rate

In the previous section, we developed formulae for calculating the life expectancy of mechanical components undergoing either fatigue or creep. It was also mentioned that there may be interaction with other factors from environmental conditions impacting component's life. Furthermore, it was pointed out that the calculated expected life should not be viewed in an absolute sense, that it is a probabilistic value.

Often, life expectancy is equated with the design life of a component. Design life is defined as the length of time during which the component is expected to operate within its specifications. However, this definition appears to suggest that failures do not happen at all, although that is not the case. For this reason, *operating life expectancy*, or *operating life*, has been used in place of design life. Furthermore, operating life expectancy has been defined as the length of time that a component operates within its *expected failure* rate.

The question is how component failure rates may be calculated based on their life expectancy values. Shortly, we will demonstrate how this information at a component level may be used to calculate the overall failure rate of the entire unit.

Chapter 6 provides more details and other possible ways of studying reliability. For now, the purpose of this section is to bring the reader's attention to the fact that ultimately, any life calculations should be tied to the concept of reliability and, from a practical point of view, to an expected failure rate.

Device Failure Rate Prediction

In this chapter, the focus has been on understanding potential causes of failures and design flaws, and how they impact the behavior of a product either in its manufacture stage or when it launched in the field. The final element of reliability in design is to calculate the failure rate (or the reliability) of the device (or system) that is under development. This knowledge will enable the design team to develop a baseline for product reliability and identify areas where improvements may be required. As discussed in Chapter 1, the field of reliability began with the attempt to make this calculation. To do this, we need to have a knowledge of the failure rates associated with the components and then a knowledge of the system itself. There are two approaches: the first is using component failure rates from a database and the second is to use our knowledge of how components fail as well as an understanding of product use profile (i.e., the physics of failure).

Failure Prediction Using Databases

Often, databases (such as MIL-HDBK-217F or Telecordia) provide generic reliability or failure rates for electronic components. Mechanical devices are generally associated with a life expectancy such L_{10} or L_{50} and maybe found in databases such as NPRD.[12]

These generic rates should be adapted to the specific conditions that may exist in the particular device of interest. To do so, we need to develop an understanding of the impact of temperature, and electrical and environmental stresses on the failures rates. Mechanical components are not typically affected by these stressors, but electric and electronic components are. Temperature and electrical stresses (even humidity) impact the life of electronic components. In Chapter 8, we will review these stressors, but for now, let it suffice to examine how the hazard rates of these components are affected.

Temperature Effects

The failure rates provided in reliability databases are based on a reference temperature (such as 40°C). If the operating temperature of a given component is different than the reference temperature, then the corresponding hazard rate should be calculated. Klinger et al. (1990) provides the following conversion factor:

$$\lambda_u = \pi_T \, \lambda_r$$

where λ is the hazard (or failure) rate and the subscript r denotes the referenced value. The subscript u denotes the use conditions. For our discussion here, π_T may be considered as the conversion factor, but as we will see in Chapter 8, it is also called an acceleration factor. It is calculated as follows:

$$\pi_T = \exp\left[\frac{E_a}{K_b}\left(\frac{1}{T_r} - \frac{1}{T_u}\right)\right]$$

E_a is the activation energy. K_b is the Boltzmann constant. T_r is the reference temperature and T_u is the use temperature, i.e., actual component temperature in the device. Temperatures are in absolute values.

Electrical Stress Effects

Other factors such as electrical stress may have adverse effects on the device reliability as well. A component is generally rated for a specific voltage, current, power, etc. Reliability studies have shown that once a particular percentage of this rating is exceeded, the hazard rate is increased. In MIL-HDBK-217F (1991), this percentage is assumed to be 25% and is denoted by P_0.

For electrical stress (Klinger et al. 1990), MIL-HDBK-217F (1991) provides the following relationship:

$$\lambda_{u} = \pi_S \, \lambda_r$$

where

$$\pi_S = \exp\left[m(P_1 - P_0)\right]$$

P_0 and P_1 are percentages of maximum-rated electrical stress. P_0 is a reference value and is usually set to 25% for critical applications and 50% for other noncritical applications. P_1 is the percentage of the maximum rated electrical stress. m is provided by MIL-HDBK-217F (Klinger et al. 1990) and other standards such as Bellcore TR-332 (1995). Generally, each component is rated at a particular electrical stress level. At that prescribed value, A_E is equal to one. Now, should the component be operated at stress levels below the set value, its reliability does not change; however, should the operating stress go beyond this limit, its reliability decreases.

Environmental Factors

It stands to reason that an identical product would have a different reliability value if it is used in a benign environment as opposed to a harsh environment. These effects are accounted in a factor generally denoted as π_E. MIL-HDBK-217F (1991) provides a table for π_E, which varies between 0.5 for benign environments and 220 for extremely severe conditions.

Calculating System Failure Rate

To calculate a system's reliability and its associated hazard rate, we need to have knowledge of the hazard rates associated with the components and then knowledge of the system itself. As it will soon become clear, this approach is also called a "parts count" method. Because of its simplicity it is frequently used to estimate product reliability at relatively early stages of design.

Component Hazard Rate

The component hazard rate is simply a product of a hazard rate multiplied by various stress factors, namely, electrical, temperature, and quality (another factor undefined as yet). The quality factor takes into account manufacturing issues. Generally, this factor is equal to 3 for commercial items, and 2 or 0.9 for parts that are to be used in military applications.

$$\lambda_{SS_i} = \lambda_{G_i} \pi_{Q_i} \pi_{S_i} \pi_{T_i}$$

λ is the hazard rate and π denotes a stress multiplier. λ_{G_i} denotes the generic value of the hazard rate, which is generally provided in handbooks and supplier published data; π_{Q_i} is the quality factor as previously defined; and π_{S_i} is the electrical stress factor. Its value is unity for the reference stress value. π_{T_i} is the thermal stress factor. Similarly, its value is unity at a database reference temperature (such as 40°C in MIL-HDBK-217).

System Hazard Rate

A system may be composed of a group of subsystems, each made up of many circuit boards. The circuit boards in a subsystem and, indeed, the subsystems within a system may be arranged in series, parallel, or a complex arrangement. A discussion of parallel and complex arrangements is beyond the scope of this work, and the reader is referred to other sources such as Klinger et al. (1990) and MIL-HDBK-338B (1988).

For a series network, the hazard rate of the system may be calculated as follows:

$$\lambda_{SS} = \pi_E \sum_{i=1}^{n} N_i \lambda_{SS_i}$$

where π_E is the environment factor and N_i is the number of each component in the system. This approach provides the strictest sense of reliability calculation in which the failure of any component will flag the failure of the entire system.

To calculate the hazard rate for a system, tabulate the following items for each component:

1. Generic hazard rate (λ_G)
2. Thermal stress factor (π_T)
3. Electric stress factor (π_S)
4. Quality factor (π_Q)

The steady-state hazard rate is obtained by multiplying the above numbers ($\lambda_{SS_i} = \lambda_{G_i} \pi_{Q_i} \pi_{S_i} \pi_{T_i}$). To obtain the failure rate of the entire system, add component hazard rates and multiply by an environmental factor

$$\left(\lambda_{SS} = \pi_E \sum_{i=1}^{n} N_i \lambda_{SS_i} \right).$$

Values of various stress and environmental factors are available from various sources such as Bellcore (1995) or other standards.

Case Study

Renal failure is a major cause of death among children with spina bifida or spinal-cord injured individuals. A possible method to diagnose kidney failure is to measure bladder pressure. For this purpose, a handheld device to measure this pressure has been developed for home use. This device measures the bladder pressure on a regular basis and then reports it to a central database for clinical monitoring of the patient condition.

Our role in the design of this unit is to determine the failure (hazard) rate and the mean time between failures of this device. This system is designed to be used only 1 hour a day in a benign environment. Next, we need to evaluate the first-year repair volume should there be an annual production of 12,500 units. Finally, we should determine whether these units should be subjected to a postproduction environmental stress screening prior to shipment to reduce any defects due to manufacturing. The bill of materials (BOM) along with generic reliability data are provided in Table 7.17.

TABLE 7.17

Failure (Hazard) Rate Calculation Data

Item	Quantity	Description	Generic	Quality	Electric	Thermal	Multiply All
1	13	0.1 uF Cap.	1.00×10^{-9}	3	1	0.9	3.51×10^{-9}
2	1	10 uF Cap.	1.00×10^{-9}	3	1	0.9	2.70×10^{-9}
3	2	22 pF Cap.	1.00×10^{-9}	3	1	0.9	5.40×10^{-9}
4	1	Diode	3.00×10^{-9}	3	1	0.9	8.10×10^{-9}
5	1	Net. Res.	5.00×10^{-10}	3	1	0.9	1.35×10^{-9}
6	4	1.0 M Res.	1.00×10^{-8}	3	1	0.9	1.08×10^{-8}
7	1	39 K Res.	1.00×10^{-9}	3	1	0.9	2.70×10^{-9}
8	2	470 K Res.	1.00×10^{-9}	3	1	0.9	5.40×10^{-9}
9	3	Potentiometer	1.70×10^{-7}	3	1	0.9	1.38×10^{-6}
10	2	10 K Res.	1.00×10^{-9}	3	1	0.9	5.40×10^{-9}
11	1	IC	1.00×10^{-5}	3	1	0.9	2.70×10^{-5}
12	1	LCD	3.00×10^{-9}	3	1	0.9	8.10×10^{-9}
13	1	Sensor	2.50×10^{-8}	3	1	0.9	6.75×10^{-8}
14	1	Battery	1.00×10^{-7}	3	1	0.9	2.70×10^{-7}
15	3	Connector	2.00×10^{-10}	3	1	0.9	1.62×10^{-9}
16	3	Switches	1.00×10^{-8}	3	1	0.9	8.10×10^{-8}
						Total	2.89×10^{-5}

Let's assume that in the long term, the failure distribution is exponential and the failure (hazard) rate is constant. On this basis, the system failure rate may be based on the sum of the component hazard rates:

$$\lambda_{\text{system}} = \pi_E \sum_{i=1}^{n} N_i \, \lambda_{SS_i}$$

Based on the data provided in Table 7.15, we have

$$\lambda_{\text{system}} = \pi_E \times 2.89 \times 10^{-5}$$

Since the environment is benign, the environmental factor π_E is 0.5 (MIL-HDBK-217F). Thus, the system hazard rate is

$$\lambda_{\text{system}} = 1.45 \times 10^{-5}$$

Using this rate, the number of failed units in the first year and the associated mean time between failures (MTBF) is calculated as follows:

$$R(t) = e^{-\lambda t}$$

$$R(8760) = e^{-8760(1.45 \times 10^{-5})}$$

$$R = 0.881$$

$$F = 1 - R = 1 - 0.881$$

$$F = 0.119$$

$$\text{MTBF} = \frac{1}{\lambda_{\text{system}}} = \frac{1}{1.45 \times 10^{-5}} = 69204 \text{ hours}$$

This figure indicates that an 11.9% ($F = 0.119$) first-year population failure may exist, which is equivalent to 1488 ($= 0.119 \times 12,500$ first-year production) failed units. This would be true only if the system was operated continuously. However, the duty cycle is only 1 hour per day. Thus, a new adjusted MTBF must be found based on the duty cycle. In this particular example, each hour of operation is equivalent to 1 day. Thus, the adjusted MTBF must be calculated by multiplying the calculated MTBF by 24:

$$\text{MTBF}_{\text{adjusted}} = 24 \times \text{MTBF}_{\text{calculated}}$$

$$\text{MTBF}_{\text{adjusted}} = 1660900 \text{ hours}$$

$$\lambda_{\text{adjusted}} = \frac{1}{\text{MTBF}_{\text{adjusted}}} = \frac{1}{1660900}$$

$$\lambda_{\text{adjusted}} = 6.02 \times 10^{-7}$$

$$R(t) = e^{-\lambda t}$$

$$R(8760) = e^{-8760\left(6.02 \times 10^{-7}\right)}$$

$$R = 0.9948$$

$$F = 1 - R = 1 - 0.9948$$

$$F = 0.0052$$

This indicates that only 0.5% of production or 65 units will likely fail in the first year. It should be noted that this percent failure (even though calculated for the first year) is assuming a "steady-state" condition. In practice, there are a number of process deficiencies associated with a newly launched product that lead to higher than expected failures (i.e., infant mortality period of the bathtub curve). Once these deficiencies are addressed, failures of the product drop leading to increasing product reliabilities. One way of identifying defective products prior to shipping is to establish postproduction stress screening, which leads us to the last question that we wanted to answer. It was whether a postproduction stress screening would be needed.

To answer this question, we take the following approach. First, if there is no stress screening in place, we need to take the infant mortality into account. A good model for approximating the infant mortality is the AT&T Weibull equation (see Chapter 6). Then, based on prior data and experience, we select a shape factor (a). Finally, the scale parameter (λ_0) is calculated based on the steady-state failure rate.

Assume the shape factor $a = 0.75$. The AT&T model assumes that the steady state (or constant hazard rate) begins at 10,000 hours. Hence, we have

$$\lambda = \lambda_0 t^{1-a}$$

$$6.02 \times 10^{-7} = \lambda_0 (10,000)^{1-0.75}$$

$$\lambda_0 = 6.037 \times 10^{-4}$$

Now the number of failed units in one year (8760 hours) may be calculated:

$$R(t) = \exp\left(-\frac{\lambda_0 t^{1-a}}{1-a}\right)$$

$$R(8760) = \exp\left(-\frac{(6.037 \times 10^{-4}) \times 8760^{(1-0.75)}}{1-.75}\right)$$

$$R = 0.9769$$

$$F(t) = 1 - R(t)$$

$$F = 0.0231$$

The number of first-year failures in a year is 281 (=0.0231×12500) units, should they be shipped with the infant failures not screened out. This number is expected to drop to 65 should such a stress screening be put into place. This data should provide the information needed to make a decision on instituting stress screening, which is generally a management decision.

Failure Rates Based on Physics of Failure

In Chapter 1, we discussed that the failure rate information provided in various databases were in fact tailored to a large extent to the industry that supported them. For example, MIL-HDBK-217 was best used for predicting military equipment. Later, as Bellcore and Telecordia developed their own information based on MIL-HDBK-217, it was no surprise that equipment in these industries was modeled more accurately than equipment that was used in other industries. Nevertheless, use of the database approach in reliability predictions fell out of favor in some circles. Instead, the emphasis was placed on understanding the probability of component(s) failure in the context of the probability of the component use profile (or the system within which the component resides). This approach requires the reliability engineer to be able to calculate the expected life of various components in an assembly or system based on the applied physics. For instance, we may be interested in the time it would take for a truck wheel rim to develop microcracks, which could eventually lead to a catastrophic failure of the truck wheel. In other words, we are interested in calculating the life of the wheel. To do so, we first recognize that the wheel is subjected to static loads when the truck has stopped, and to alternating compressive and tensile loads when the truck moves. Hence, the primary physics of failure in this scenario is fatigue. To calculate the life of a truck wheel, we proceed to calculate the level of alternating stresses first, and then, with the aid of S–N curves for the wheel material, the number of cycles to failure is estimated. This number may be translated into a "life" value by understanding the product use profile, i.e., how may cycles in a day, etc. For more information on this subject, see Jamnia (2016). To illustrate this approach better, a case study is provided.

Case Study

Consider the handheld bladder pressure gage discussed in the previous case study. A clinical version of this device has been developed with minor modifications. First, it has been modified to operate with a power supply as well as a rechargeable battery. Second, a supercapacitor has been incorporated in the unit as a way of driving an alarm circuit in case of a depleted battery and an indicator that the unit is not drawing power from an AC source. Finally, this gage is incorporated with a larger medical device monitor that has a relatively large temperature rise from the ambient condition. We need to calculate the failure rate of this supercapacitor and understand whether it will be able to support a 10-year design life. It should be noted that a database instantaneous hazard rate for a supercapacitor has been estimated to be less than 0.1×10^{-6} per hour.

Let us first calculate the failure rate of supercapacitors in a 10-year span based on this data:

$$\text{Failure} = 1 - e^{-\lambda t}$$

For $\lambda = 0.1 \times 10^{-6}$ and $t = 87600$ hours (or 10 years), we have

$$\text{Failure} = 1 - e^{-0.1 \times 10^{-6}(87600)} = 8.72 \times 10^{-3}$$

Now, let us take a more realistic approach. To predict the failure rate of this component over time, we first need to understand the factors that affect the life of the component and their relationships. For a supercapacitor, temperature is a major contributor along with the driving voltage. By consulting the technical datasheets provided by manufacturers, these relationships may be established. For example, see Williard et al. (2015). In this scenario, the following relationship is known:

$$\text{Life} = e^{(aT+b)} e^{-2\left(\frac{V}{5.5}-1\right)} \tag{7.7}$$

where T and V are temperature and usage (or drive) voltage, respectively. Furthermore, $a = -0.069$ and $b = 13.94$ for temperatures below 70°C. For temperature ranges above this limit but below 85°C, we have $a = -0.099$ and $a = 16.00$.

The next element needed in the life calculation is the use profile of the device. In this case, the device is used an average of 10 hours a day with a 3-hour standard deviation. Additionally, the device is used 4.5 days a week; i.e., 65% of the time with a standard deviation of 15%.

Finally, we need to have the operational conditions of the device. For this bladder pressure gage, the unit operates at three different temperature conditions. When the unit is not being operated, the temperature range is between 20°C and 25°C (ambient conditions). Under normal operating conditions, the internal temperature rise is between 24°C and 30°C above ambient conditions. However, during a daily 3-hour battery charging period, the internal temperature rise is between 30°C and 36°C. The use voltage of is 4.8 volts.

Table 7.18 summarizes these input values as well as results of the sample calculations shown next:

Product life = 10 years = 10 × 365 = 3650 days = 3650 × 24 = 87600 hours

Use case – days per year = 65%

Hours per use days = 10 hours

Use profile – charging = Hours charging (3) × Use case (65%)/24 (hours per day)

Use profile – normal = 1 – Use profile (charging) – Use case (unit off)

Use profile – unit off = 1 – (Use case × Hours per use days)/24 (hours per day)

TABLE 7.18

A Summary of Sample Calculations for Supercapacitor Life and Failure Rates

Factors	Value	Mean	Std. Dev.
Hours per day used	10	10	3
Duty use cycle	0.65	0.65	0.15
Ambient temperature condition	22.5	22.5	0.83
Charging temperature rise	33	33	1
Normal temperature rise	27	27	1
Product life requirement	10	years	
	3650	days	
	87600	hours	
Use case duty cycle (days per year)	65%		
Hours used per day (only days in use)	10	hours	
Use case (charging)	8.1%	(3-hour charge time per day)	
Use case (normal operation)	19.0%		
Use case (unit off)	72.9%		
Use voltage	4.8	V	
Usage at ambient temperature	63875	hours	
Usage at charging temperature	7117.5	hours	
Usage at normal operation	16607.5	hours	
Voltage (V)	4.8		
Component life at ambient temperature	307077.14	hours	
Component life at charging temperature	31077.502	hours	
Component life at normal operation	47132.349	hours	
Minor's index	0.7893927		

Use voltage = 4.8 volts

Usage at ambient temperature = (Product life hours) × (Use profile – unit off)

Usage at charging temperature = (Product life hours) × (Use profile – charging)

Usage at normal operation = (Product life hours) × (Use profile – normal)

With this information, the component life at the three usage temperatures may be calculated using Equation 7.7:

$$\text{Life (at ambient temperature)} = e^{\left(-0.069(22.5)+13.94\right)} e^{-2\left(\frac{4.8}{5.5}-1\right)}$$

$$\text{Life (at charging temperature)} = e^{\left(-0.069(55.5)+13.94\right)} e^{-2\left(\frac{4.8}{5.5}-1\right)}$$

$$\text{Life (at normal operation)} = e^{\left(-0.069(49.5)+13.94\right)} e^{-2\left(\frac{4.8}{5.5}-1\right)}$$

Now, we can calculate Minor's index:

$$R = \frac{\text{Usage at ambient temperature}}{\text{Life (at ambient temperature)}} + \frac{\text{Usage at charging temperature}}{\text{Life (at charging temperature)}}$$

$$+ \frac{\text{Usage at normal operation}}{\text{Life (at normal operation)}}$$

Failures occur when $R \geq 1.0$.

Under the conditions specified in Table 7.16, Minor's index is calculated at about 79%. This indicates that under these conditions the supercapacitor does not have a significant failure rate during the first 10 years of its life. However, the input values to this calculation (such as use profile or even temperature values) are not single values but come from different distributions. We can employ the Monte Carlo approach once again and evaluate the distribution of Minor's index as depicted in Figure 7.19. An output of this analysis is the percent of the values about the threshold of 1.0. In this calculation, this excess value (therefore failure rate) is 6.0% at 10 years.

By rerunning the same analysis for different product life values (i.e., 1 to 9 years), the failure rate at various ages may be calculated. Figure 7.20 provides these results.

Now, we have arrived at the solution that is predicting the failure of the clinical model of the bladder pressure gage. Based on this prediction, the first 6 years may be considered as failure-free when any failure may have random causes. However, starting with the sixth year, the wear-out period of the supercapacitor begins and by the tenth year, the failure rate due to wear-out may be as high as 6%. This knowledge gives the foresight to the design team to take the following actions:

1. Locate the PCBA containing the super capacitor such that it may be easily replaced
2. Develop a service schedule to replace the PCBA prior to its failure around the sixth year of operation
3. Develop a self-monitoring mechanism to identify when the component life nears its end and provide a warning

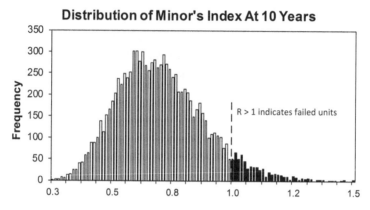

FIGURE 7.19
A distribution of Minor's index (*R*) values. Failure occurs when *R* reaches 1. Therefore, a calculated R value greater than 1.0 indicates that the component is already in a failed state.

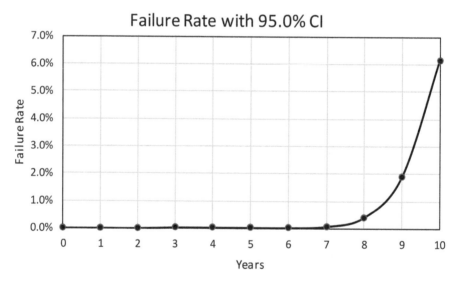

FIGURE 7.20
Failure rate of the supercapacitor and its growth in the 10-year life.

Notes

1. 2017 PDA/FDA Joint Regulatory Conference.
2. In our opinion, there is a difference between testing and designing experiments. We have witnessed too many instances where design engineers have no firm view of how their design may function. They run tests only to develop an understanding of how their design may operate.
3. For simplicity sake, we are assuming that the yield strength is a constant. However, in the previous example, we demonstrated that it is a distribution and not a single point.
4. In other words, supplier A is 99.8% (=100% – 0.2%) reliable; whereas, supplier B is 97.14% (=100% – 2.86%) reliable.

5. This is not to be confused with *mechanical* wear, which is material removed due to friction.
6. We can also use the Microsoft Excel command NORMDIST for calculation.
7. Otherwise, a normal distribution is assumed.
8. For this example, the actual units (i.e., mm or in) is irrelevant.
9. We are not providing this table due to space constraints. Reader are encouraged to conduct numerical experiments of their own.
10. We used NORMDIST command of Microsoft Excel.
11. Some printed circuit boards are made with G10 glass epoxy.
12. At times, the B_x nomenclature is used in place of L_x.

8

Component and Subsystem Reliability Testing

Introduction

In Chapter 3, we reviewed the stages of the design for reliability (DfR) process within the V-model. This process is depicted in Figure 8.1. Chapter 7 corresponds to steps 4 through 7 of the V-model where we discussed and reviewed how reliability and robustness in design may be established. In fact, by keeping the principles of reliability in mind during the design process, a number of failure modes may be avoided leading to more robust and reliable assemblies and subsystem designs. However, regardless of how meticulous the paper design and reliability analyses and predications are, we need to build what we have designed and examine whether we have achieved the reliability that we anticipated. The V-model calls this step *verification*. Though, we need to keep in mind that due to the iterative (and integrative) nature of the design process—particularly in an Agile product development environment—there may be a number of verification stages.

Reliability verification may be based on legacy data. For instance, an assembly or module design has been replicated from a previously fielded or tested product. This information may be used to justify why additional verification activities may not be necessary. Similarly, empirical formulae, or physics of failure (PoF) analysis, may be used for the same justification. This being said, the most robust and accurate method for verifying and demonstrating reliability of a design is limit or life testing.

In this chapter, we will review the basics, concepts, and the applications of reliability testing. We will answer questions such as *what to test* and *how to test* in terms of reliability. In conjunction with these two questions, we will review various reliability test design methodologies and discuss sample size and test duration.

Robustness versus Reliability Testing

Before diving into the what and the how of reliability testing, it is important to know about the interrelationship between robustness and reliability. In Chapter 1, we provided the classic definition of reliability as being the probability of a design performing its intended function under defined operating conditions for a specified period of time. This definition implies that reliability is a function of time, and by merit, reliability of a product is expected to degrade and decline as the product or device ages. Eventually, the product loses the capability to support the applied stresses and it fails.

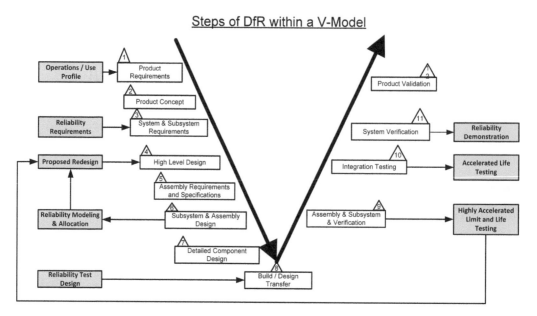

FIGURE 8.1
Steps of the DfR within the V-model.

Robustness is different than reliability in that it is not impacted by the element of time. A Google search for *robustness* gives the following definition: "the ability to withstand or overcome adverse conditions." In the context of product development, robustness is a product's ability to perform as intended without eliminating the causes of variations that influences it. In simple engineering terms, it is having adequate design margins. *Design margins* are the threshold settings that define the design and yet the design remains impervious to them. Also called *noise factors*, they include factors such as piece-to-piece manufacturing variations, customer usage, and interaction with other systems. A design is not robust if it is inherently sensitive to variations and uncertainties. In essence, robustness is *the ability of the design to perform its intended function consistently in the presence of noise factors*.

Figure 8.2 depicts reliability and robustness graphically and introduces a related concept called design capability, which is subset of quality. *Design capability* is the totality of a product's characteristics to meet its intended use and requirements. The relationship between capability and stress defines robustness. As time goes by, due to a variety of factors, a design's capability degrades and eventually the stress distribution overlaps a product's capability distribution. The common area between the two defines the probability of failure or unreliability. As the degradation increases, so does this probability.

As product developers and designers, our goal is to bring to market a product that is both robust and reliable. Uncovering robustness issues during the design process requires that relatively large samples be evaluated under a variety of use and environmental conditions. Discovering reliability issues necessitates testing the product over a period of time—in the presence of noise factors—to allow for increasing the chances of lowering a product's strength and its interference with stresses. In essence, reliability testing is testing robustness over time.

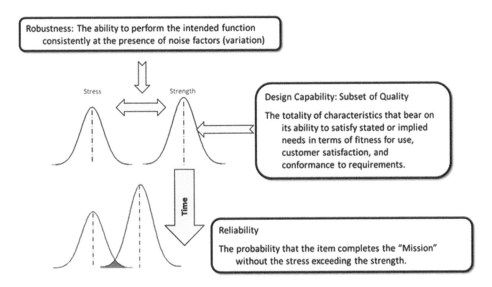

FIGURE 8.2
Relationship of design capability, robustness, and reliability in terms of stress, strength, and time.

The What and How of Reliability Testing

In reliability test design, the first question that begs an answer is what should be tested. The general answer to this question begins by suggesting that one should focus on what is new in the design, or what is unique or different. Here is the rationale for this suggestion. Should one really spend time and resources on testing a component that is well understood? We believe not. The reliability and failure rates of commonly used parts with known failure modes may be extended to new products and their models. Now, it may be possible that a common part may be designed for a unique or different environment or use conditions. Should this be the case, the component should be tested.

Beyond this general rule of thumb, the first step in the test design of component or subassemblies is to define the *reliability-critical items* list along with their critical design parameters outputs. There are different means of creating this list; however, to do this systematically, one of the following approaches should be used.

1. Reliability feasibility analysis—As explained in Chapter 4, a gap analysis between reliability allocation and feasibility at the concept phase highlights subassembly or modules that may not meet their intended targets or correspond with the highest failure rates. These components are subject to scrutiny by the design process as well as reliability testing in order to improve the design and meet its reliability goals.

2. Parameter diagram (P-diagram)—P-diagrams provide a holistic and bird's-eye view of factors impacting specific functions of a particular system. It helps to fully define expected functions of a product along with its associated failure modes. Although this diagram is commonly used to complete design failure modes and effects analyses (DFMEAs), it is a strong deductive tool that helps with a systematic

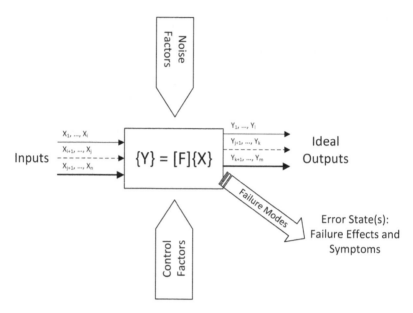

FIGURE 8.3
A general overview of a parameter diagram.

approach for test design. The P-diagram helps to identify noise and control factors that may cause deviation of the output function of the design. This would lead to the identification of the critical parameters that need to be monitored or controlled during reliability testing. Figure 8.3 provides a concept chart of a P-diagram; however, for an in-depth description of P-diagrams, see Jamnia (2018).

3. Design failure modes and effects analysis (DFMEA)—In Chapter 7, we provided an example of a DFMEA. We demonstrated the use of this tool to identify and prioritize areas of concern. DFMEAs highlight how a component may fail and the impact of the failure on the behavior of the product and its effect on the environment or people. This tool helps the design team to create a prioritized list based on the perceived risk of failure. It is often used in conjunction with new features, technology, or applications.

Types of Reliability Testing

Once we identify what needs to be tested, we will focus on how it should be tested. The lion's share of time and resources allocated to reliability belongs to this stage. How we test an assembly depends very much on our test objectives. Having said this, there are four general categories of reliability testing based on test objectives. They are

1. Reliability design margin development and characterization
2. Reliability demonstration test
3. Reliability growth (development) test
4. Production reliability stress screening

In this chapter, we will discuss the first two in some detail. We will leave a more detailed treatment of the latter two topics to Chapter 9. For now, let us provide a brief definition of each.

Reliability Design Margin Development and Characterization

During the detailed design stage of the V-model, engineers make certain assumptions about how their components work within a larger assembly. Once an assembly is built, these assumptions need to be tested. Reliability design margin development and characterization is a series of activities to challenge these assumptions. For instance, an assumption is that the subassembly operates under certain environmental conditions, such as, say, temperature and humidity ranges. To challenge this assumption, we will subject the subassembly to these conditions and monitor whether it is operational. Should it be proven that the unit works as intended, we will expand the test condition envelope and continue monitoring the operation of the unit until it ceases to function.

Another approach is called *testing the corners*. This implies that testing is designed so that the corner cases where a unit is expected to operate is examined. For instance, a subassembly is fabricated with its components at the maximum material conditions and then is placed in an environment that represents the maximum operation temperature. If the unit functions as expected, it is then said that it has passed the corner test.

We should note that these tests do not provide any quantitative metrics of reliability; however, they are essential tools for providing feedback and an understanding of reliability to the design team. Examples of these types of tests are design of experiment (DOE) and highly accelerated limit test (HALT).

Reliability Demonstration Test

Traditionally, this test is conducted to demonstrates the reliability of the final engineering build of a subassembly or even the product. It is done by degrading the inherent strength by subjecting the unit under test to realistic environments as well as prolonged test duration. In this test, design capabilities or strength is worn out due to repeated use (over time) and elevated stresses. In a way, reliability demonstration testing (RDT) and reliability growth testing (RGT) are very similar. However, there are two major differences. The first difference is that in RDT redesign of components are not acceptable. However, depending on the acceptance criteria, repair of the system and replacement of a failed module may be possible. The second is that in RGT, the test duration is driven by what has been discovered as well as the integration timelines. In RDT, the test duration is driven by the expected service life for repairable systems, and design life for nonrepairable units. RDT is considered to be a verification study; therefore, acceptance criteria, that is, verification of reliability requirements, should be set *a priori*.

Production Reliability Stress Screening

Production reliability stress screening, also known as environmental stress screening, attempts to identify and remove products with manufacturing defects. This test is typically deployed at the start of production and is applied to the entire production when feasible, or to a selected sample should the entire production prove impractical. The test utilizes information learned from design margin testing to precipitate manufacturing defects in a very short time, usually in hours. The objective of this test is not to improve or to demonstrate the reliability of the design, rather it is to ensure that only good products are shipped out.

Highly Accelerated Limit Testing

HALT is a test methodology that involves application of environmental conditions that exceed product specification limits. This method seeks to identify the environmental conditions that limit the operation of the design and, ultimately, those that cause its destruction.

> The actual specification limits at which a design can operate as intended during its design life are probabilistic values defined by the inherent variation in characteristics and capabilities of the design materials and quality. Similarly, the operating limits of a design as intended are also probabilistic even though its life is substantially reduced.

HALT finds the weakest design elements that lead to either recoverable or permanent failures. By exposing these weak links, design improvements can be implemented leading to improved reliability and durability of the design. The stresses utilized in this methodology to precipitate failures are hot and cold temperatures, thermal cycling, vibration, power margining, and power cycling. HALT does not *simulate* a typical field environment, rather it *stimulates* damages accrued and precipitates failure modes.

In HALT, there are three widely used terms. They are *specification limits, operating or operational limits*, and *destruct limits*. Figure 8.4 provides a pictorial view of the relationship between these limits at the lower and upper bounds.

The dashed line in Figure 8.4 represents the nominal value of the normal operating conditions along with a distribution as a reminder that variations are present in the operation

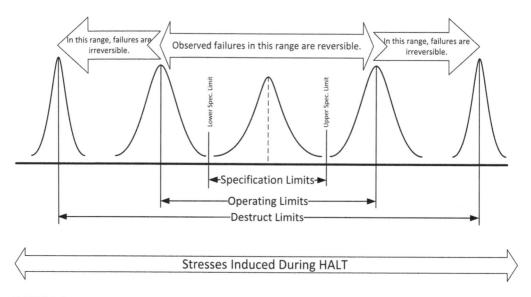

FIGURE 8.4
Operating and destruct limits defined from highly accelerated limit test.

conditions. A product is (and should be) designed to function in a range of operating conditions. For instance, it is common for a product to have a temperature specification, such as, say, from 10°C to 50°C, or a humidity specification from 30% RH to 75% RH.

Typically, specification limits are based on user needs or customer expectations. These limits are not the same as a product's capabilities. In other words, a product that has a 10°C to 50°C temperature specification will in fact operate at temperatures below 10°C or above 50°C. It may just not be as reliable under those extended conditions.

The reason is that in the course of the design, off-the-shelf or custom-designed components are selected with higher specification limits than the product being designed. Since all components are selected to operate beyond the specification limits, the product should be able to operate beyond the same limits. The unknown is that we do not typically design for these upper and lower *operating limits*. As we said earlier, just because the product operates beyond its specification limits, it does not mean that it can last as long as expected. Or, it may have a higher frequency of intermittent failures or malfunctions.

There is another aspect of Figure 8.4 that we would like to emphasize. One may notice that all the limits are expressed as a distribution and not a single limit. One reason for having a distribution as a "limit" is because of variations inherent in the material properties and design characteristics of components. Another reason is due to the interactions between a system's components and the uncertainties in these interactions.

During HALT, we come to learn of the inherent operational limits of the design. This is done by identifying the stress levels at which the design will fail to perform as intended. The signature of an operational limit is that once the applied stress is removed, the product begins to function as intended again. So, we begin to increase (or decrease) the stress level and monitor functioning of the product until we reach a stress level that causes a malfunction. We will then begin to lower the stress level for the function to resume. These two levels of stress mark the bounds of the operational limit.

Note that we are using the word *malfunction*. We are implying that there may be aspects of the product that is still functioning. For instance, say we are testing a printed circuit board assembly (PCBA) with a small liquid crystal display (LCD). As we increase or decrease the temperature of the PCBA in the HALT chamber, we may observe that the LCD may lose focus or turn off completely. At the same time, we may monitor other aspects of the board and see that other components are still operating as expected. If we were to back off from the applied temperature levels and observe that the LCD gains its functioning, we have identified an operational limit.

Once we have identified and verified the first operational limit, we can once again increase the applied stresses. The outcome of this increase may be twofold. First, it may be that we identify other points on the operational limit distribution. Second, once we go beyond the operational limit, we will discover a malfunction that we will not be able to recover once we back off the stress levels. In other words, the product has turned into a brick! This limit is called the *destruct limit*. As one can imagine, this limit may have an upper bound and—depending on the stresses applied—a lower bound as well.

The failures that we discover during HALT should be examined and evaluated by the design team. They make the decision on whether and how to eliminate or mitigate these failures. Any design improvements will undoubtedly increase the range of the operating and destruct limits. Improvement will tend to increase design robustness and reliability as well.

Having said this, we need to keep in mind that a major advantage of running HALT is to find design weaknesses in a very short period of time, maybe even in minutes. This information is then looped back into the design process for improvements. This lends itself

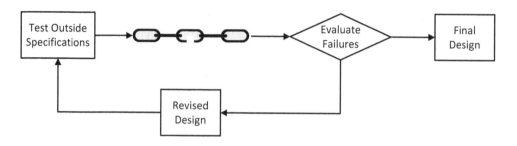

A) Use of HALT During the Design Process

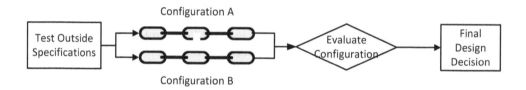

B) Use of HALT to Compare Product Configurations

FIGURE 8.5
(A) Using HALT in the design process or (B) in comparing product configurations from, say, different manufacturers.

to the realization that HALT is an iterative approach performed to uncover latent defects in the product design process, component selection, and/or manufacturing that would not otherwise be found through conventional qualification methods (Figure 8.5A). HALT can also be used in different test scenarios to compare alternative designs, or different suppliers and/or manufacturing processes of the same design (Figure 8.5B).

HALT Chamber and Test Setup

Up to this point, we have talked about HALT methodology and the outcomes that we can expect. These are the whats of HALT. Now, we need to talk about the hows. Figure 8.6 shows a simplistic view of a HALT chamber. It is comprised of two segments, namely, the test chamber and the drive mechanism, which controls both the heat input and the shake table. Typically, both hot and cold air enter into the chamber via air ducts and are directed to the units under test using flexible hoses. At the bottom of the test chamber sits the shake table that is connected to and is driven by the drive mechanism.

We are often asked whether the unit under test should be mounted in the chamber with the same orientation as it would have in the product. The answer is that in HALT, we aim to stimulate and not to simulate. Therefore, we recommend that the unit under test be mounted directly to the shake table to the extent possible, and to use flexible hoses to direct the flow of air directly over the sample. An example of this setup is shown in Figure 8.7. This approach will enable a direct transfer of both heat and vibrational energies to the unit under test (UUT). In all cases, the unit under test is securely mounted in the chamber.

Figure 8.7 shows that the unit under test (here a PCBA) is instrumented. It is set up in such a way that it can be operated while the test is in progress. Considering that our

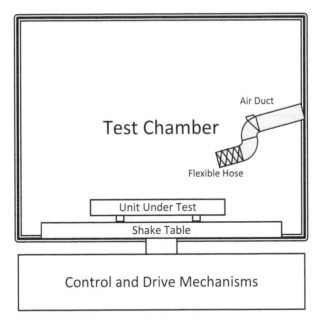

FIGURE 8.6
A typical setup of a HALT chamber.

A) An Instrumented Printed Circuit
Board in the HALT Chamber

B) Data Acquisition Equipment to
Monitor Input and Output Signals.

FIGURE 8.7
An example of printed circuit board assembly setup in HALT chamber along with the data logging equipment for input and output signals.

intentions are to identify both the operational and destruct limits, this step in the setup is crucial. Once we have ensured that we can adequately operate and monitor the unit under test, we can begin operating the chamber.

Figure 8.8 illustrates a typical HALT profile for temperature and vibration, and how to establish upper and lower operating and destruct limits. For each profile, UUT is monitored for functionality and failures. When the first failure occurs, the stress (such as temperature) level is returned to the previous level. Once the stress stabilizes, the functional test is repeated to examine whether an operating or destruct limit has been encountered.

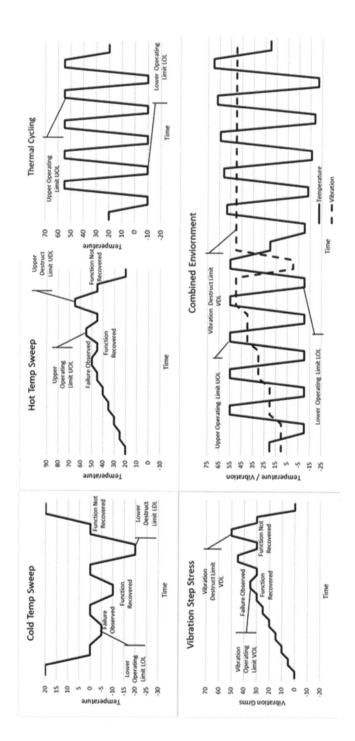

FIGURE 8.8
HALT typical temperature (hot/cold) and vibration stress profiles.

The HALT profiles include the following test cases. As mentioned earlier, in all cases, the UUT is securely mounted in the chamber.

- Cold temperature step stress test—Start at room temperature and decrease temperature with a dwell period at each temperature level. The dwell period allows the UUT to acclimate to 80% of chamber conditions and may last 5, 10, or 15 minutes. The UUT is monitored for functionality and failures.

- Hot temperature step stress test—Start at room temperature and increase temperature with a dwell period at each temperature level. The dwell period allows the UUT to acclimate to 80% of chamber conditions and may last 5, 10, or 15 minutes. The UUT is monitored for functionality and failures.

- Rapid thermal cycling test—Starting at the identified temperature operating limits determined during the cold and hot thermal step stress processes, subject the UUT to 5 to 10 rapid thermal cycles (typically at rates above 50°C per minute). Dwell time at each cycle may be 5 to 10 minutes again to allow the UUT to acclimate to the chamber conditions. Monitor the UUT throughout the rapid thermal transition process.

- Vibration step stress test—This test case requires accelerometers to be attached to the UUT to measure the vibration response on the product itself. Vibration starts at 5 G_{rms} and increases in steps of 5 G_{rms} until failure is observed or limits of the chamber itself are reached. Dwell time is typically 10 minutes at each step. Monitor the UUT throughout the vibration sweep.

- Combined vibration and thermal cycling 1—In this test case, the UUT will be subjected simultaneously to thermal and vibrational stresses combined. In this first test case, temperature cycles between the lower and upper operational limits and dwells for a period of time to allow acclamation of the UUT. At the same time, vibration increases from benign levels (5 G_{rms}) and reach the vibration operating limit at steps of 20% of the vibration operating limit. If no failures are observed, we move to the second and final combined test.

- Combined vibration and thermal cycling 2—In this pass, vibration levels are maintained at 80% of the vibration operating limit. Simultaneously, the thermal cycle begins between lower and upper operational limits; however, the bound widens in order to search for a combination of temperature and vibration levels where failures may be observed.

- Finally, once all these profiles are executed, it is recommended that a tear down of the UUT be completed in order to better understand the impact of the test on the UUT.

HALT is most useful at the subsystem (subassembly) level, as it can be used to identify design weaknesses or robustness concerns when more than several components are interacting to perform an intended set of functions. Another aspect of HALT is that the identified lower and upper operating and destruct limits are used in other reliability tests. For instance, by identifying the operating limit, we would know the maximum stress that may be applied in an accelerated life test (ALT), or what boundaries and limits need to be used in product screening tests such as environmental stress screening (ESS).

Examples

Once a profile is completed, the UUT is inspected for any signs of damage. Figure 8.9 provides pictorial examples of the type of damage that may be sustained by a PCBA during HALT. Figure 8.9A depicts damages on two different PCBAs that occurred during the thermal cycle. In one, a component is missing and in the other, a component has moved but it has not been dislodged. The root cause of both incidents may be due to a manufacturing defect and cold solder flow. Figure 8.9B shows the broken leads of a component. This damage took place during the vibration cycle. The root cause is likely the large mass of the component relative to the lead size.

HALT and Realistic Failures

Design engineers who do not have previous experience with the value that HALT brings to the table usually object to results observed through this methodology. They argue that their product or design will never experience these levels of stress or that the product is not labeled to operate under these conditions. When we hear these types of objections, our response is that HALT stimulates the inherent failure modes and mechanisms of a design. It does not simulate operating conditions under which failures may occur in the field. We recognize that, ultimately, design engineering owns the decision of whether to remediate or mitigate any or all of the reported failure modes and mechanisms.

Let us explore the design engineer's objections a bit further. The objection is that the product will never experience this level of stress, particularly under vibration. It is true

Component has
fallen off.

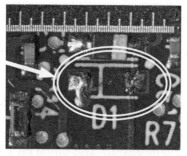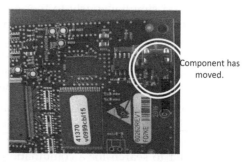

Component has
moved.

A) Observed Damaged During Thermal Cycling

Component has
fallen off during
vibration sweep.

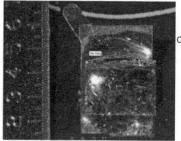

Close up view of
fallen
component

B) Observed Damaged During Vibration Sweep

FIGURE 8.9
Missing components at thermal cycling or vibration test during HALT.

that under the random vibration of HALT, more weaknesses are precipitated and failures observed than in thermal cycling. This is to be expected since on the one hand, the UUT undergoes hundreds of thousands of vibration cycles leading to fatigue fractures compared to tens of thermal cycles. On the other hand, due to the mechanical fatigue damage, time is accelerated exponentially during HALT (Hobbs 2005).

Design engineers are also correct in the observation that, with a few notable exceptions,[1] the majority of fielded products experience low-cycle thermal fatigue more readily than high-cycle vibration-induced fatigue. So why is that we still insist that HALT produces realistic failure modes?

Hobbs (2005) defined a phenomenon called the *crossover* effect. He argued that materials experience stress but are "unaware" of the source of the stress. It is the stress levels that dictate how the failure occurs and what the mechanism of stress is. At certain stress levels, which he called crossover, the same failure mechanism can be produced by either temperature or vibration. This is depicted in Figure 8.10. Based on the physics, Hobbs recommends that the focus in HALT be on the failures detected, not on the stress type or the stress levels beyond field stresses. In other words, if we experience a failure in HALT, we will experience the same failure in the field sooner or later.

Another common question that we have been asked is whether design engineering *has* to fix the reported failure. At times, this seems like a trick question. If we are aware of a weak point in our design, wouldn't we want to remove it by improving the design? But, if we were to fix the weak point and run HALT again, wouldn't we discover the next weak spot? When do we stop? The truth is that these types of questions do not have a standard answer, or the answer may be that it depends. In new product development, there are competing constraints. On the one hand, we need to develop a product that the customer accepts and gladly uses. On the other hand, there are market pressures that force us to launch the product as soon as possible. In short, we need to seek our answer or decision matrix in a cost–trade-off analysis. The elements of this analysis may include the time and resources needed to identify the root causes, implementing design changes, and establishing a screening test such as HASS[2] will provide the rationale needed to make an informed decision.

We have also been asked how do we learn of other potentially catastrophic failures if we do not fix the first failure that stopped the test. Ironically, the answer to this question is also related to an objection that we hear to running HALT. This objection is this: we know

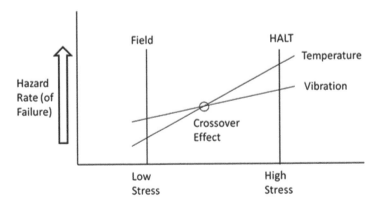

FIGURE 8.10
Crossover effect of instantaneous failure rate observed in HALT and in the field.

we have a weak design, we shouldn't run HALT. A technique that was recommended by Hobbs (2005) and has been widely adopted is to mask known weak areas and components. For instance, room-temperature vulcanizing (RTV) silicone is used to secure relatively large components on a printed circuit board. This will protect these components from damage due to vibration so that other weak areas may be exposed.

Reliability Demonstration Testing

In the V-model of product development, and in particular within an Agile project management environment, subassemblies are designed and developed with an eye on a later integration step. As a part of the development cycle, the requirements of these subassemblies have to be verified prior to the integration. In reliability, this verification is called reliability demonstration testing.

As mentioned earlier, RDT is done by degrading the inherent strength of the product by subjecting the unit under test to elevated environmental stresses[3] as well as prolonged test durations. In this test, design capabilities or strength is worn out due to repeated use (over time) and elevated stresses. IEC[4] Standard 62506 provides an extensive overview of accelerating stresses.

In RDT, we are interested in answering two questions. First, does the design meet its design life, and, second, what is the failure rate within this period. By knowing these two parameters, we can in fact calculate other reliability metrics. We are not sure if there is a commonly accepted definition of design life; however, a general definition may be the period of time during which an acceptable percentage of the units fail. We need to keep in mind that in general, we deal with either a constant failure rate, which is indicative of random failures, or with increasing rates, which indicate that some components are wearing out. Mechanical components usually exhibit wear. Bearings, valves, linkages, and the like are usually the first examples that we think about when asked about mechanical components. However, corrosion and wear on electrical contacts, cracking and crazing of encapsulants (of integrated circuits), or even cracking of solder joints due to thermal cycling are also examples of the wear-out failure mechanism. As we learned in Chapter 6, we can use the Weibull distribution to model both a constant as well as an increasing (wear-out) failure rate.

> Unless a certain failure mode and mechanism is identified as wear out, it is always recommended to consider a constant failure rate and the use of exponential distribution for reliability test designs and/or reliability calculations.

The main task in the test design for reliability demonstration is to determine the sample size and the required duration of the test that can adequately demonstrate and verify the reliability requirement at an acceptable statistical confidence. This may not be as simple a task as it may appear. Here are two reasons why. For obvious reasons, the test duration may not be the actual design life because design life is often expressed in years.

As a result, we need to accelerate the test duration in a such a way so that when we are done, the results would be representative of how a product would behave during its design life. This brings us to the second reason why RDT may not be as simple as it may sound. To conduct an accelerated life test, we need to understand the underlying physics of various failures that we would anticipate. For instance, in some cases, moisture absorption may be the cause of failure, and in others, plastic creep. The mechanisms for time acceleration will be different in these two cases. In the next section, we will discuss some of the models that have been developed over the last few decades that are commonly used to understand this.

Accelerated Life Testing

Accelerated life testing (ALT) is a methodology in which a test is executed under elevated stress levels to shorten its duration. Although the applied stresses are beyond specification limits, they have to be bounded by the operating limits, as illustrated in Figure 8.11, to ensure that the test is producing the same failure modes as in real-life stress levels. The operating limits can be identified using HALT as previously explained (and depicted in Figure 8.4). These limits can also be identified using engineering judgment based on knowledge of material properties and failure mechanisms. Based on IEC standard 62506 there are two distinct objectives in conducting accelerated life testing: the first focuses on identifying potential failures and the second on demonstrating dependability (reliability) of the design.

Needless to say, we need to select the level of stress in such a way so that it produces relevant (or realistic) failure modes, similar to those that would occur in real life. In designing an accelerated life test, there are two distinct scenarios. In the first scenario, we have a priori knowledge of failure mechanisms and a good basis of the physics of failure and stress–life relationships. We can use this knowledge along with mathematical predictive models to calculate an accelerated factor. The acceleration factor is the ratio of life expectancy under normal stresses to life expectancy under elevated stresses. The second scenario is that we do not have a priori knowledge. In this case, there is more work and exploration that needs to take place.

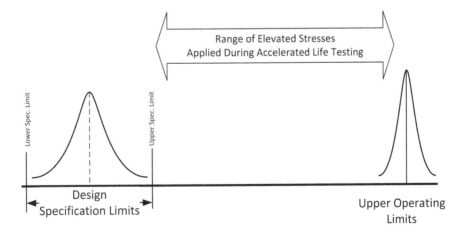

FIGURE 8.11
Relationship of design specification limits, operating limits, and selected elevated stress for ALT.

In the two next sections, we will explain methods and examples of conducting ALT for both known and unknown stress–life relationships on the component (module or subsystem) level.

Known Stress–Life Relationships

In general, the life expectancy of electronic assemblies depends on temperature, electrical, mechanical, and environmental factors such as moisture, dust, and even radiation. As many as 20 different life models have been proposed (Viswanadham and Singh 1988; Nelson 2004). Herein, the more commonly known Arrhenius, Eyring, Black, Q10, inverse power, and Arrhenius–Peck models are reviewed.

Arrhenius Model

The Arrhenius model is based on empirical data and is generally used to describe thermally or chemically activated mechanisms (Condra 2001). We find application of this model in accelerated aging of electric insulations, dielectrics, solid-state semiconductors, intermetallic diffusion, and battery cell degradations. Additionally, this model is applicable to lubricants and greases as well as plastics.

The Arrhenius model is a rate equation that describes temperature dependency as follows:

$$\text{Rate} = C \, \text{Exp}\left(-\frac{E_a}{K_b T} \right)$$

where C is a constant, E_a is the activation energy, K_b is the Boltzmann constant (8.617×10^{-5} eV/K), and T is temperature in absolute value.

This relationship clearly does not provide a life expectancy period. The implication of this equation is that at a given temperature, the rate of a given mechanism multiplied by the time required for completion of the mechanism is constant. In other words

$$\text{Rate}_{@T_2} \times \text{Time to failure}_{@T_2} = \text{Rate}_{@T_1} \times \text{Time to failure}_{@T_1} = \text{Constant}$$

where T_1 and T_2 are two different temperature values.

$$t_{@T_2} \, \text{Exp}\left(-\frac{E_a}{K_b T_2} \right) = t_{@T_1} \text{Exp}\left(-\frac{E_a}{K_b T_1} \right)$$

$$t_{@T_2} = t_{@T_1} \text{Exp}\left(\frac{E_a}{K_b} \left(\frac{1}{T_2} - \frac{1}{T_1} \right) \right) \tag{8.6}$$

or

$$AF = \frac{t_{@T_2}}{t_{@T_1}} = \text{Exp}\left(\frac{E_a}{K_b} \left(\frac{1}{T_2} - \frac{1}{T_1} \right) \right) \tag{8.7}$$

Therefore, by running an experiment at two different temperatures, failure times t_1 and t_2 may be measured leading to a determination of E_a for a particular device:

$$E_a = \frac{T_2 T_1}{T_1 - T_2} K_b \ln AF = \frac{T_2 T_1}{T_1 - T_2} \ln \frac{t_{@T_2}}{t_{@T_1}} \tag{8.8}$$

Then, the life expectancy of that device at a particular temperature may be calculated using Equation 8.6 or 8.7. The question is whether we need to run an experiment to determine the activation energy for the type of product that we are dealing with. While, in general, this is true, this energy level has been studied and typical values are readily available. For instance, values of 0.6 to 0.8 eV have been used for electronic components such as integrated circuits, diodes, and transistors. Some use values as high as 1.0 for systems and larger assemblies to include potential interactions.

Example

A manufacturer selected a new component for use in an engineering system under development. The activation energy (E_a) of this component has been determined experimentally and it is 0.68 eV. This component is rated at 125°C and will be used in an environment of 60°C. We are asked to demonstrate that the component can survive 10 years in the field. We need to design an accelerated test and calculate how long this test will take. To be conservative, we decide to test the component at 110°C. To calculate the test duration needed to demonstrate a 10-year life, we first need to calculate the acceleration factor. Then, we divide the needed life (i.e., 10 years) by this factor.

We start by applying the Arrhenius model:

$$AF = \exp\left(-\frac{E_a}{K_b}\left(\frac{1}{T_{\text{Accelerated}}} - \frac{1}{T_{\text{Normal}}}\right)\right)$$

$$AF = \exp\left(-\frac{0.68}{0.00008617385}\left(\frac{1}{(110+273)} - \frac{1}{(60+273)}\right)\right)$$

$$AF = 22.1$$

$$\text{test duration} = \frac{10 \text{ years}}{22.1} = 0.45 \text{ year or } 165 \text{ days}$$

This means that it will take 165 days to demonstrate the 10-year life at an elevated temperature of 110°C. If we decide that this duration is too long, we may push the test temperature to 120°C (just shy of its limit). Under this condition, the test duration would reduce to 98 days.

In this example, the activation energy was readily available. However, more realistically, the activation energy of a module of interest or its components is not known and experiments should be designed to calculate this energy based on time to failure. In practice, most reliability engineers use their engineering judgment and experience to choose values from 0.6 to 1.2.

An Example of How to Calculate Activation Energy

A new electronic product has been developed. In order to determine its expected life, two different experiments were conducted. In the first, a sample of units was placed in a 100°C chamber and operated. The average failure time of this population was 148 hr. A second set of samples was tested at 75°C, and their average failure time was 876 hr. What should the expected life be if these units are operated at 40°C?

From Equation 8.7 we have

$$AF = \frac{t_{@T_2}}{t_{@T_1}} = \frac{876}{148} = 5.919$$

And from Equation 8.8, we can obtain the activation energy:

$$E_a = \frac{T_2 T_1}{T_1 - T_2} K_b \ln AF = \frac{(75+273)(100+273)}{100-75}\left(8.617\times10^{-5}\right)\ln\left(5.919\right)$$

$$E_a = 0.796 \text{ eV}$$

From Equation 8.6:

$$t_{@40} = t_{@100}\text{Exp}\left(\frac{E_a}{K_b}\left(\frac{1}{T_2}-\frac{1}{T_1}\right)\right) = 148\text{Exp}\left(\frac{0.796}{8.617\text{x}10^{-5}}\left(\frac{1}{(40+273)}-\frac{1}{(100+273)}\right)\right)$$

$$t_{@40} = 17017 \text{ hr}$$

Hence, this device will have an average life expectancy of nearly 17,000 hours (or nearly 2 years) at 100% duty cycle.

Eyring Model

While the Arrhenius equation includes only temperature as the stressor, a similar equation may be developed that includes a second stressor in addition to temperature. The general form of this equation is called the Eyring model (Condra 2001; Viswanadham and Singh 1988):

$$t_f = \frac{A}{B}S^{-n}\text{Exp}\left(\frac{E_a}{K_b T}\right)$$

where t_f is the expected life, A and B are the constants, S is the applied stress (such as percent relative humidity or electrical current), n is a parameter related to the stress, E_a is the activation energy, K_b is the Boltzmann constant, and T is temperature in absolute value. The Eyring model can be used to represent failures due to elevated currents and electric fields.

Black's Equation for Electrical Stress Effects

Other factors such as electrical stress (current, voltage, and power) may have adverse effects on an electronic device life as well. A component is generally rated for a specific voltage, current, power, etc. Reliability studies have shown that once this rating is exceeded, failure rates increase.

Black applied the Eyring model to relate electrical stress to the component's life as follows (Condra 2001; Viswanadham and Singh 1988):

$$t_f = \frac{a_m b_m}{A}J^{-2}\text{Exp}\left(\frac{E_a}{K_b T}\right)$$

where t_f is the expected life, a_m and b_m are the thickness and width of metallization (conductor), A is a geometry- and material-related constant, J is the current density, E_a is the activation energy, K_b is the Boltzmann constant, and T is temperature in absolute value. The Black model is suitable for capacitor failures and electromigration in aluminum conductors.

Q_{10} Equation for Polymer Aging

A polymer is a compound formed by a chemical reaction that combines particles into groups of repeating large molecules. We determine that the acceleration factor, AF, of the chemical reaction that leads to the breakdown of polymers follows a well-known and commonly used formula:

$$AF = Q_{10}^{(T_2 - T_1)/10}$$

where T_1 and T_2 are the normal usage and elevated test temperatures, respectively. Q_{10} is the reaction rate coefficient. In most polymer-based material applications, it is often assumed to be 2.0, though ASTM F1980 standard does provide a guideline for selecting a more accurate estimate.

Inverse Power Model

Another model commonly used for the application of nonthermal stressors such as fatigue, voltage, pressure, humidity, and mechanical loading is called the inverse power model. The acceleration factor in this model is calculated by the following relationship:

$$AF = \left(\frac{S_{\text{Normal}}}{S_{\text{Accelerated}}} \right)^{-n}$$

where n, a positive number, is a model parameter. It is an indicator of how applied stresses impact life; the higher the value of n, the lower the life expectancy. Similar to the Q_{10} equation, n is often assumed to be 2.0 for electric power and 2.66 for humidity, though it can be determined empirically.

Arrhenius–Peck Model

The equation that governs failures from humidity is called the Arrhenius–Peck model. It is a product of the Arrhenius and inverse power law equations because humidity and temperature are interdependent:

$$AF = \left(\frac{RH_{\text{Normal}}}{RH_{\text{Accelerated}}} \right)^{-n} \exp\left(-\frac{E_a}{K_b} \left(\frac{1}{T_{\text{Accelerated}}} - \frac{1}{T_{\text{Normal}}} \right) \right)$$

where $RH_{\text{Accelerated}}$ and RH_{Normal} are accelerated and normal-use relative humidity values, respectively. E_a is the activation energy, and K_b is the Boltzmann constant (8.617×10^{-5} eV/K). T is temperature in absolute value. The typical failure mechanisms demonstrated by this model include:

- Degradation or hermetically sealed products
- Moisture absorption of plastic packaging materials used in semiconductors
- Loss of material physical strength
- Oxidation and galvanic corrosion of metals
- Changes in electrical and thermal insulating characteristics

- Degradation of electronic components
- Delamination of composite material
- Swelling of materials due to water absorption
- Degradation of optical element transmission quality

Unknown Stress–Life Relationships

When a mathematical model of the stress–life relationship is not available, we have to create this relationship by first executing experiments to collect needed data and then by developing the model that we would need. The experimental approach begins by conducting a series of tests at three different elevated stress levels (S_{Low}, S_{Medium}, and S_{High}). These tests must continue until a number of failures are obtained. Once a timeline of failures are created, we create a life distribution at each stress level. Then, we develop a mathematical relationship between different values of life and stresses.

All elevated stress levels (i.e., S_{Low}, S_{Medium}, and S_{High}) should be located above the upper design or specification limit but below the operating limits (Figure 8.12). These levels of elevated stresses will ensure that unit under test will not experience "quick" destruct failures and that it will demonstrate certain "life" before failure. The operating limits of the design are to be defined either by an engineering assessment of the design or using results from HALT.

Even though we will treat the evaluation of sample size later in this chapter, it is important to discuss it here within the context of developing a life–stress model. We need to keep in mind that while our task is to develop such a model, we would not know when and under what stresses our units under test would fail. A rule of thumb is that we place a smaller number of samples under higher stress levels and place a higher number at lower stresses. The reason is that the probability of failure is higher at the higher stress level as failure mechanisms are stimulated at a faster rate. Furthermore, it is more likely that *all* samples at the highest stress level will fail, while most of the samples at the lowest stress level may survive the test. A common practice is to allocate 3 times more samples at S_{Low} compared to 2 times in S_{medium} than the number of samples used in S_{High}. For instance, if we wanted to place 5 units at the highest stress levels, we would place 10 units at medium stress levels and 15 units at the lowest stress levels for a total of 30 samples.

Figure 8.13 illustrates the graphical analysis used to plot the test results at elevated stresses. It illustrates how to extrapolate the probability of failure distribution at normal

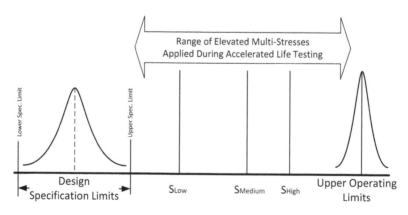

FIGURE 8.12
Selection of elevated stress levels for multilevel ALT.

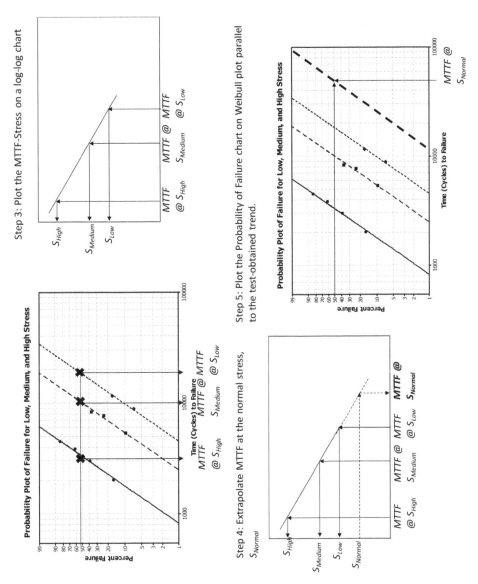

FIGURE 8.13
Graphical data analysis and extrapolation of probability of failure at normal-use stress level using Weibull plot.

stress levels based on a graphical solution approach. This is summarized in the following four steps (Morris, et.al. 1993):

1. Testing is executed at three elevated stress levels and continues until failures are obtained at each level.
2. Next, the time-to-failure distribution of each stress level is plotted to obtain the mean time to failure (MTTF) at each level.
3. The third step is to create the stress–life relationship graphically by using MTTF and stress values.
4. Finally, by using this empirically developed relationship, MTTF at normal stress levels are extrapolated and calculated.

Case Study: Expected Design Life of a Linear Pneumatic Pump

An original equipment manufacturer (OEM) design team has selected a pneumatic linear reciprocating pump to provide pressurized air to operate a fluidic circuit. This OEM did not previously use the pump, and the failure rate or modes of that pump was unknown. The pump supplier had determined that the linear pump was prone to wear out of the piston Teflon lining. This failure was manifested by an increase in the gap between the piston and the cylinder walls, as shown in Figure 8.14, leading to a play or rattling of the piston.

This failure mode is expected to produce chatter between the cylinder walls and piston leading to increased noise. The OEM did not have any information on the Teflon lining stress–life relationship. The only available information was that lining material broke down with temperature. Additionally, the pump was rated to operate at or below 120°C. The design and operations teams wanted to understand life expectancy of the pump to establish a preventive maintenance strategy. This was to be a proactive measure to avoid high rates of field failures.

In this scenario, it is not clear which stress–life relationship may be applicable. As a result, we designed the experiment at three different temperature levels, all within the 120°C operating level but above the 45°C design specification. There were 20 samples available for testing and they were allocated to the temperature levels as shown in Table 8.1. Failure was defined as any sample exhibiting noise levels above 35 dB.

The test was conducted and noise failures were observed as noise level exceeded the design specification level set at 35 dB. The time to failure in hours was recorded as shown

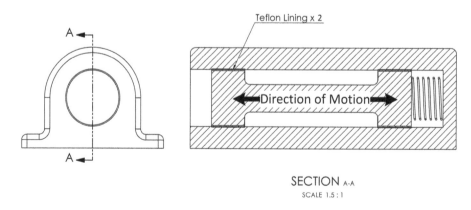

FIGURE 8.14
A linear pneumatic pump with a reciprocating piston expected to wear out the Teflon lining.

in Table 8.2. All three samples at 105°C failed before reaching 3800 hours. For 85°C, we terminated the experiment at 10,000 hours after we observed four failures, and at 12,134 hours at 65°C after two failures.

Figure 8.15 shows the Weibull plot of the time to failure and the probability of failure of the devices tested at three elevated temperatures. As shown in Figure 8.15, all failures

TABLE 8.1

Multitemperature Test Design

Temperature	Sample Size
65°C	10
85°C	7
105°C	3

TABLE 8.2

Time to Failure (Hours) for Multitemperature Test Results

105°C	F = Fail S = Survived	85°C	F = Fail S = Survived	65°C	F = Fail S = Survived
3928	F	5342	F	8760	F
1923	F	7473	F	11986	F
3776	F	8385	F	12134	S
		9800	F	12134	S
		10000	S	12134	S
		10000	S	12134	S
		10000	S	12134	S
				12134	S
				12134	S
				12134	S

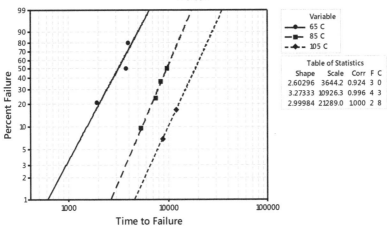

Probability Plot of Time to Failure at 65 C, 85 C, 105 C
Weibull

FIGURE 8.15

Weibull plot of the time to failure and the probability of failure at three elevated temperature levels.

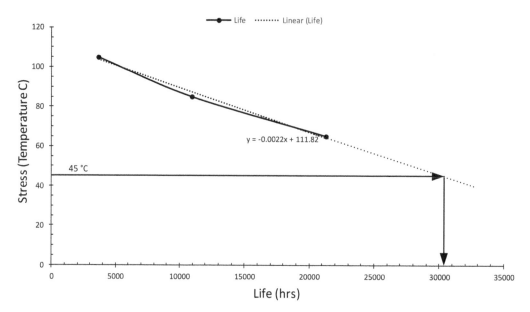

FIGURE 8.16
Graphical analysis at three different temperature levels and the projection of use temperature.

are occurring at an increasing rate, i.e., wear out. The Weibull shape factor observed at all temperature levels ranges between 2.6 and 3.2. The samples were all inspected and the failure mode was confirmed to be identical on all samples at all temperatures. This is an important check that confirms the validity of the test results to identify the acceleration factors of this failure mechanism. As shown in Figure 8.15, the MTTFs are 21,289, 10,926, and 3,644 hours at 65°C, 85°C, and 105°C, respectively.

To calculate the life at the 45°C temperature, we use a graphical approach as explained earlier and in Figure 8.13. As shown in Figure 8.16, we plot the MTTFs calculated from the Weibull analysis against the applied stresses (i.e., temperature). Then, by using the least-squares method, we fit a straight line to this graph. The equation of this line is

$$\text{Temperature} = -0.0022\,(\text{Life}) + 111.82$$

We can solve for life as a function of temperature:

$$\text{Life} = 50827.27 - 454.54\,(\text{Temperature})$$

At 45°C, we obtain a life of 30,373 hours.

Reliability Test Duration and Sample Size

An aspect of reliability demonstration testing is to determine the length of time to run reliability testing and how many samples to use. Earlier, under the section title "Accelerated Life Testing," we learned how to accelerate a test to shorten the life of a device. For instance, by using a known life model and applied elevated stresses, we may reduce a 10-year life to a few months. Now the question remains whether we need to test our units to these calculated times. Additionally, we would want to determine how many samples we should use and what level of confidence would we have in our results.

To answer these questions, we will start with single-use components to see an application of the binomial equation and how sample size may be calculated. Then, we will extend the argument for sample size to nonrepairable components of repairable systems.

Single-Use or Nonrepairable Components

Failures of single-use items are a nuisance to end users causing interruption in their work. Excessive failures lead to unsatisfied customers who may then look for a more reliable source. Single-use items have a limited mission time that could range between seconds and hours; however, its function is executed once and only once. Hence, the reliability of a single-use item cannot degrade over time because the element of time does not play a role. Also, the probability of success of each item is statistically independent in each trial.

In single-use items, reliability concerns itself with piece-to-piece variations due to manufacturing defects. The question it tries to answer is not *when the failure will happen*, but *whether failure will happen*. As a result, the reliability of a single-use item is defined as *the probability of success* of each new item (disposable) to achieve the design function. Considering that success or failure is totally independent of that of the previous or the subsequent items, the proper model to express this probability is binomial distribution.

Considering that in single-use items defects are typically less than one in a few thousand, to detect a defected population we may need to examine as large of a sample size as possible. We like to note that reliability testing of single-use items during the design and development phase is not the same as an *acceptance sampling* test, a topic beyond the scope of this book.[5]

To design a reliability test, we begin by making use of the *binomial distribution* formula as described in Chapter 6. $P(x)$ is the probability of finding x defected items in n samples. It is calculated based on the probability of success (q) and failure (p) of each single item:

$$P(x) = \sum_{x=0}^{r} \binom{n}{x} p^x q^{(n-x)}$$

If we want to demonstrate that each single item has an average reliability of R, then the term q denotes reliability, and the term p is the unreliability of each single item. Then the preceding formula can be rewritten as

$$P(x) = \sum_{x=0}^{x=r} \binom{n}{x} (1-R)^x R^{(n-x)}$$

An intriguing question to ask is this: If we were to repeat the same test n times, in what percent of times will the target reliability R will be demonstrated? A different way of asking this question is, What is the confidence level (CL) of repeating the same results? If $P(x)$ is indicative of failure and CL of success, then we can say

$$P(x) = 1 - CL$$

Then, the binomial expression may be written as

$$1 - CL = \sum_{x=0}^{x=r} \binom{n}{0} (1-R)^x R^{(n-0)}$$

So for a test of n samples with "zero" failures, this equation further reduces to

$$1 - CL = \binom{n}{0}(1-R)^0 R^{(n-0)}$$

$$1 - CL = \frac{n!}{0!(n-0)!} 1 R^n$$

$$1 - CL = R^n$$

$$\ln(1-CL) = n \ln(R)$$

We can now calculate the number of samples required to demonstrate a certain reliability (or probability of success) at a given statistical confidence level as follows for "zero" failures:

$$n = \frac{\ln(1-CL)}{\ln(R)} \tag{8.9}$$

This equation may be rearranged to calculate the reliability metric based on sample size and confidence level:

$$R = \exp\left(\frac{\ln(1-CL)}{n}\right)$$

Table 8.3 illustrates the typical sample size for reliability test plans based on set reliability and confidence levels using Equation 8.9. It is clear that as reliability approaches 1, the sample size increases rapidly.

We need to keep in mind that testing the probability of survival (reliability) of a single-use item is different from testing the time between failures of a device that repeatedly uses the disposables to conduct a function. When testing the reliability of a single-use item , we are looking for the probability that one of the items is defective, independent of all other items in the population. In contrast when we test for the reliability of the device that uses these items, we are looking for the average time between failure to process any of those single-use items. Failure of a device to handle or process single-use items or repeated functions will be addressed in more detail in Chapter 9.

TABLE 8.3

Sample Size Required for Demonstrating Attribute Reliability at Different Statistical Confidence for Single-Use Items

Target Reliability		Sample Size		
		90%	95%	99%
Confidence Level	90%	22	45	229
	95%	28	58	298
	99%	44	90	458

Success-Run Reliability Test Duration and Sample Size

At times, Equation 8.9 is called the Bayes formula. Lipson and Sheth (1973) suggested that by combining this equation with that of the Weibull reliability expression

$$R(t) = \exp\left[-\left(\frac{t}{MTTF}\right)^{\beta}\right]$$

we will arrive at an expression that some may call Wei–Bayes and we call the Bayes–Lipson equation. It is expressed as (Abernethy and Fulton 1992)

$$R(L_D) = \exp\left[\frac{\ln(1-CL)}{n\left(\dfrac{T}{L_D}\right)^{\beta}}\right] \tag{8.10}$$

It is prudent to keep in mind that the derivation of this equation is based on the assumption that at the end of the test period, there are no observed failures. In other words, this is a *success-run* test.

In this equation, there are interrelated factors of which if four are known, the fifth may be calculated using the Bayes–Lipson equation. These four factors are the desired reliability at the design life [$R(L_D)$], sample size (n), test time to design life ratio (T/L_D),[6] confidence level (CL), and the Weibull shape actor (β). We recall from Chapter 6 that β is equal to 1.0 for constant failure rates. It is greater than 1.0 for wear-out failure trends when the failure rate increases with time. Additionally, β is less than 1.0 for early life failures. Based on Equation 8.10, the sample size required for demonstrating a reliability $R(L_D)$ based on a given shape parameter is

$$n = \frac{\ln(1-CL)}{\ln(R(L_D))\left(\dfrac{T}{L_D}\right)^{\beta}}$$

Alternatively, the ratio of test duration, T, to design life, L_D, on sample size n for a given confidence limit of (CL) is expressed by

$$\frac{T}{L_D} = \left[\frac{\ln(1-CL)}{n\,\ln(R(L_D))}\right]^{1/\beta}$$

This is a curious equation; one that we need to discuss in some detail. We have calculated this ratio for a reliability of 95% with 90% confidence, a sample size of 1 but for a variety of Weibull shape factors (Figure 8.16). We note that the test duration is substantially shorter for larger shape factors. In other words, the more severe the wear-out phase, the shorter the test duration. The implication is that we need to have a firm understanding of the failure trend and use an appropriate Weibull shape parameter; otherwise, we are either wasting time by spending too much time under test or missing important failures because we have not tested long enough. Assuming a lower wear trend or rate would lead to an overestimation of test duration (Figure 8.17).

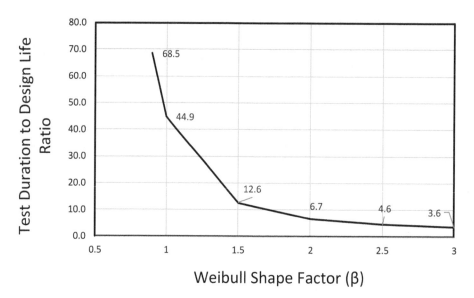

FIGURE 8.17

Ratio of test duration to design life (L_D) for different failure trends to demonstrate reliability.

TABLE 8.4

Test Duration to Design Life Ratio for a Reliability of 95% with 90% Confidence for Different Sample Sizes (n) and Shape Factors (β)

	N			
β	0.5	1	2	3
1	2015.2	44.9	6.7	3.6
2	503.8	22.4	4.7	2.8
5	80.6	9.0	3.0	2.1
10	20.2	4.5	2.1	1.6
20	5.0	2.2	1.5	1.3
30	2.2	1.5	1.2	1.1
50	0.8	0.9	0.9	1.0
100	0.20	0.45	0.67	0.77
1000	0.002	0.045	0.212	0.355

The next point of focus is the impact of sample size on the test duration. Consider the information contained in Table 8.4. This table depicts test duration to design life ratios for different values of shape factors and sample sizes.

Table 8.4 indicates that the larger the number of samples, the shorter the test duration needs to be. We have taken this to some ridiculous numbers to make a point here. Note that this table suggests that for extremely large sample sizes, we need to test for a very small fraction of design life. This is equivalent to saying that if the MTTF of a device is 1000 hours, we can either test a single unit to 1000 hours or 1000 units to only one hour; neither of which may produce realistic results. While mathematics provides guidelines for selecting the test duration and sample size, we need to exercise engineering judgment to ensure proper test design.

Having pointed out the potential pitfalls of this equation, this equation remains quite useful provided that we keep our engineering mindset in place. This should be discussed in the context of the expected failure mechanism. If we are testing a part or a subsystem for wear-out mechanisms, it does not make sense to test many samples for short period of time each. In Chapter 9, we will discuss the optimum sample size for system reliability testing.

The Risk of Success-Run Reliability Testing

When the component design is finalized with either the mitigation of its failure modes or assurance that they occur beyond the design life, a common practice by many manufacturers is to run a reliability demonstration test plan with "zero" failures. The test plan is based on rewriting Equation 8.10 with the following assumptions:

1. Test time, T, is predetermined to be equal to the design life, L_D.
2. Assume all samples will survive to the design life, L_D.

$$R(L_D) = \exp\left[\frac{\ln(1-CL)}{n\left(\frac{L_D}{L_D}\right)^\beta}\right]$$

This will reduce Equation 8.10 into the single-use item reliability equation given earlier as

$$R = \exp\left(\frac{\ln(1-CL)}{n}\right)$$

Hence, the sample size for each component running the full length of the design life, L_D, will be given by Equation 8.9.

This approach is simple and straightforward, however, there is a risk that should a failure be observed in the units under test that the stated reliability of the design cannot be verified. Consider this: Assume that our goal is to demonstrate a 90% reliability of a component or module at 90% statistical confidence. Furthermore, the component design life is specified as 10 years. Based on Equation 8.9, 22 samples must successfully complete an accelerated test run equivalent to 10 years without a single failure. If 21 samples survive and 1 sample fails even after 9.5 years of testing, the test plan has failed and the 90% reliability cannot be verified. Furthermore, the reliability of the units may not be determined either. In a way, we need to make sure that the samples would survive and the test succeeds before embarking on this test plan. The alternative test plan is based on testing n samples for the test time, T, and monitoring time to failure (if any) and calculating the true reliability using means that we have discussed.

Two Special Cases

In reliability testing, there are often two scenarios that need to be reconciled. First, test time (or duration) is often dictated by the business due to market pressures for releasing a product. The second scenario is that sample size may be limited due to the cost of manufacture. As a result, either we can have as many samples as needed to demonstrate a certain reliability within a given time, or we can have a limited number of samples but ample time to demonstrate the same reliability.

By taking advantage of the Weibull form of the reliability equation, we can reformulate Equation 8.10 for use in either of these two different scenarios. Recall that reliability at the design life, L_D, is given by

$$R(L_D) = \exp-\left(\frac{L_D}{\text{MTTF}}\right)^{\beta}$$

By equating this to reliability obtained from Equation 8.9 we obtain

$$\exp\left\{-\left(\frac{L_D}{\text{MTTF}}\right)^{\beta}\right\} = R(L_D) = \exp\left[\frac{\ln(1-CL)}{n\left(\frac{T}{L_D}\right)^{\beta}}\right]$$

from which

$$-\frac{L_D^{\beta}}{\text{MTTF}^{\beta}} = \frac{\ln(1-CL)}{n\dfrac{T^{\beta}}{L_D^{\beta}}}$$

$$-\frac{L_D^{\beta}}{\text{MTTF}^{\beta}} = \frac{\ln(1-CL)L_D^{\beta}}{nT^{\beta}}$$

This reduces to

$$nT^{\beta} = -\text{MTTF}^{\beta}\ln(1-CL) \tag{8.11}$$

Now, depending on whether we have control over test time (T) or sample size (n), we can reformulate Equation 8.11 either in terms of required number of samples where test duration (T) is identified or dictated:

$$n = \frac{\text{MTTF}^{\beta}}{T^{\beta}}\left[-\ln(1-CL)\right]$$

or in terms of test time on each sample, if the sample size is limited due to cost or program constraints:

$$T = \text{MTTF}\left[\frac{-\ln(1-CL)}{n}\right]^{1/\beta}$$

This particular form of the equation may be used to calculate a test duration to demonstrate a needed MTTF using a sample size of n. We need to keep in mind that the fundamental assumption of this equation is that none of the samples fail, i.e., zero failures or success run.

Case Study: Field Corrective Action Testing for Urgent Launch

Analysis of field failures of a particular product has identified the root cause to be the wear-out mode of a component with a Weibull shape factor (β) of 2.5. The design team has updated the design with a material change. Even though this redesign will not eliminate the failure mechanism, it is intended to decrease its rate. Our goal as the reliability team is to design a test to implement this change as part of a field corrective action (FCA). The action is urgently needed to contain the failure in the field that has created a high rate of complaints and customer dissatisfaction.

A minimum reliability target to release the change is an MTTF of 36 at a 90% statistical confidence level. Our time frame for this release (i.e., test time) is 2 months. We need to calculate the number of samples needed to properly demonstrate this requirement. The test engineering team determined that an acceleration factor of 12 can be obtained by increasing speed and eliminating idle time.

To find the minimum sample size, we apply the equation

$$n = \frac{\text{MTTF}^{\beta}}{T^{\beta}}\left[-\ln\left(1-CL\right)\right]$$

where MTTF is 36, β is 2.5, and CL is 0.9. Test time (T) is determined by multiplying the time frame for release (i.e., 2 months) by an acceleration factor (i.e., 12):

$$T = 2 \times 12 = 24 \text{ months}$$

We can now calculate the sample size

$$n = \frac{36^{2.5}}{(24)^{2.5}}\left[-\ln\left(1-0.9\right)\right] = 6.34 \text{ Samples}$$

We will select seven samples. This will be the minimum test required to release this design change into the field based on a 36-month MTTF at a 90% confidence.

Once we complete this test, we were tasked to demonstrate a higher reliability of 95% at 60 months of operation. This implies that after 60 months of operation no more than 5% of the population would fail.

We can continue the initial test but now we need to determine the additional time needed to demonstrate the ultimate reliability target of 95% at 60 months of usage at a 90% statistical confidence. Recall that

$$\frac{T}{L_D} = \left[\frac{\ln\left(1-CL\right)}{n\ln\left(R(t)\right)}\right]^{1/\beta}$$

where R(60 months) is 0.95

$$\frac{T}{60} = \left[\frac{\ln\left(1-0.9\right)}{(7)\ln\left(0.95\right)}\right]^{1/2.5} = 60\left[\frac{-2.303}{-0.359}\right]^{0.4}$$

$$T = 126.77 \text{ months}$$

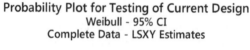

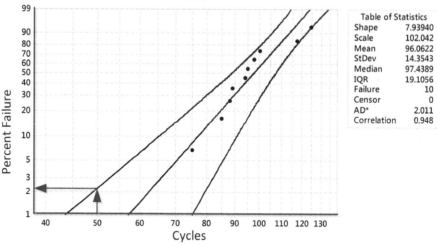

FIGURE 8.18

Probability of failure plot for the current design prior to supplier improvements. The current design has a 2% probability of failure with 95% confidence at 50 cycles.

To calculate the actual test time, we need to take the acceleration factor into account. To calculate the test duration, we need to divide this calculated time by the acceleration factor:

$$\text{Total Test Duration} = \frac{126.77}{12} = 10.56 \text{ months}$$

Since, in the first section, the units were tested for 2 months, the remaining time is 8.56 (= 10.56 − 2) months.

Case Study: Design Reliability Improvement Demonstration

In a new product development process, the design team was not satisfied with the reliability of a selected component. The supplier was required to make modification to the component design to improve reliability. The supplier claims that the change of material should reduce the probability of failure by 50% at the quoted design life of 50 cycles with 95% confidence. We need to design a reliability test to substantiate the new component design and the supplier's claim.

Our first action prior to designing a test is to collect test data related to the existing component design and analyze any failure trends. Here, the use cycles of ten reported failures are 75, 85, 88, 89, 94, 95, 97, 100, 117, and 124.

The Weibull plot for this test data is depicted in Figure 8.18. It shows that the probability of failure is increasing over time with a shape factor of 7.9. In other words, a severe wear-out mode exists in this design. Figure 8.18 also shows that the probability of failure at the specified mission time of 50 cycles is about 2% at the lower 95% confidence level (conservative estimate).

Based on the supplier's claim that the redesign reduces the failure rate by 50% with 95% confidence, our reliability requirements should be set at a failure rate of 1% or lower at a confidence level equal to or greater than 95%.

In a nutshell, we can summarize what we know and what needs to be done as follows:

- Probability of failure of the current design is 2% at 50 cycles at a 95% statistical confidence. This is gained by studying Figure 8.18.
- The reliability of the current design is 98% (= 100% − 2%) at 50 cycles with a 95% statistical confidence.
- Design team claims that the new design probability of failure at this design life is reduced by half, i.e., it is 1%.
- We need to demonstrate that the reliability of the new design at 50 cycles with a 95% statistical confidence is 99% (100% − 1%) or higher.
- Our task is to determine the most economical sample size and an acceptable test duration.

We will explore the reliability test design for different sample sizes to define the most economical plan within the time constraints of the program and the resources available. Recall Equation 8.10 and apply L_D to be 50 cycles, β to be 7.939, CL to be 0.95, and $R(50)$ to be 0.99. We now have a relationship between test duration (T) and sample size (n):

$$R\left(50 \text{ Cycles}\right) = 0.99 = \exp\left[\frac{\ln\left(1 - 0.95\right)}{n\left(\dfrac{T}{50}\right)^{7.939}}\right]$$

Table 8.5 provides a tabulation of test durations needed based on available sample sizes.

Notice that increasing the number of sample sizes does not have a major impact on reducing the test duration required on each sample. For example, increasing the sample size from one to ten, reduced the test duration by only 34% of the design life (from 2.05x to 1.53x). If the failure mode is confirmed to be wear out, we can reduce the financial cost of this test by using a smaller sample size. The risk of this decision is that the units under test must survive a test duration much longer than the 50 cycles (design life) to demonstrate the target reliability of 99% at 50 cycles.

The major takeaway from this example is this. During a development cycle, engineering builds are often expensive and are not plentiful. Should we be in a position to test for a wear-out failure mechanism and we are not under a strict timeline, we can in fact test and demonstrate with a few sample sizes, provided that each sample is run beyond the target

TABLE 8.5

Test Time to Demonstrate Reliability for Wear-Out Failure Mode (Numbers Are Rounded to the Nearest Integer)

Samples, n	Test Duration (Cycles) on Each Sample	Test Duration to Design Life Ratio (T/50)
1	102	2.05
2	94	1.88
3	89	1.78
4	86	1.72
5	84	1.67
10	77	1.53

design life. A major contributor to the test duration to design life ratio is the severity of the wear-out mechanism as depicted in the magnitude of the Weibull shape parameter. The next example will explore this relationship further.

Case Study: Failure Distribution Model and Test Duration

During an audit of an OEM's design records, the auditor noticed that two different subsystems with similar reliability targets had different test durations. The target reliability to be demonstrated was 99% at 50 cycles at a confidence level of 95%. This was similar to the previous case study that was reviewed. The only difference was the subsystem was an electronic subsystem. The auditor noted that the test protocol required had determined the test protocol to be 1490 cycles to demonstrate the target reliability on a sample size of ten. The auditor was puzzled why the previous case required only 77 cycles on the same sample size to demonstrate the same reliability target.

We explained that failures associated with the electronics were random in nature and were not related to aging or usage over time. In other words, the failure distribution could be modeled by the exponential distribution and a constant failure rate. Recall Equation 8.10 and apply L_D to be 50 cycles, β to be 1.0, CL to be 0.95, and $R(50)$ to be 0.99. We now have a relationship between test duration (T) and the sample size (n):

$$R(50 \text{ Cycles}) = 0.99 = \exp\left[\frac{\ln(1-0.95)}{n\left(\dfrac{T}{50}\right)^{1.0}}\right]$$

Table 8.6 provides a tabulation of test durations needed based on available sample sizes based on the assumption of a constant failure rate, i.e., Weibull shape factor, $\delta = 1.0$.

Unlike the previous scenario, increasing sample size from one to ten, reduced the test duration by nearly a factor of 10 (from 298x to 30x). This shows that we can shorten the test duration by increasing the sample size we know as a fact that the failure rate is a constant over time.

In this project, there were the double constraints of project schedule and lack of sample availability. Ultimately, only ten samples were available for testing and the schedule was adjusted to allow for the 1490 cycles. The assumption was that none of the samples would fail during this period.

Now, we need to convince ourselves that the probability of failures is low. To do so, we calculate the MTTF for this subsystem. Based on an exponential distribution, we have

$$R(t) = e^{-t/\text{MTTF}}$$

TABLE 8.6

Test Time to Demonstrate Reliability for a Constant Failure Trend Over Time (Numbers Are Rounded to the Nearest Integer)

Samples, n	Test Duration (Cycles) on Each Sample	Test Duration to Design Life Ratio ($T/50$)
1	14904	298
2	7452	149
3	4968	99
4	3726	75
5	2981	60
10	1490	30

$$\ln\left(R(t)\right) = -t/\mathrm{MTTF}$$

$$\mathrm{MTTF} = -t/\ln\left(R(t)\right)$$

Thus

$$\mathrm{MTTF} = -50/\ln\left(0.99\right) = 4975$$

Recall that by the time we arrive at the MTTF, 63.2% of all our samples are expected to fail. Hence, it may be reasonable that we would not experience any failures during the 1490 cycles.

It would be curious to examine the same parameters for the previous case where there was a severe wear-out mode present:

$$R(t) = e^{-(t/\mathrm{MTTF})^{\beta}}$$

$$\ln\left(R(t)\right) = \left(-t/\mathrm{MTTF}\right)^{\beta}$$

$$\ln\ln\left(R(t)\right) = \beta\ln\left(\mathrm{MTTF}\right) - \beta\ln\left(t\right)$$

$$\ln\ln\left(0.99\right) = 7.939\ln\left(\mathrm{MTTF}\right) - 7.939\ln\left(50\right)$$

$$\mathrm{MTTF} = 95 \text{ Cycles}$$

Note the substantial difference in the two MTTF values (95 vs. 4975). A subsystem with a severe wear-out trend will have a high reliability at 50 cycles, but as it approaches end of life the wear-out mechanism takes over and the reliability declines rapidly, where 63.2% of the population will fail by 95 cycles. A constant failure trend component will have a constant failure rate over time, meaning that some portion of the population fails at any point of time, even during early life, until the total population failure accumulates to 63.2% at 4975 cycles.

Limited Sample Availability

These last two examples point to an important element of reliability test design: the inter-relatedness of test duration, sample size, and presence of wear-out mode (if any). One of the issues facing reliability engineers is having enough samples to test. There are a number of reasons for this. One reason is that engineering prototypes and builds are often expensive.[7] This combined with lack of proper planning by the development team often leads to sample shortages. We have also planned the right number of samples for testing only to give them up in the last minute because of some last-minute design flaw discoveries that require more units for root-cause investigation. In a way, we are suggesting that we need to be flexible to know how to adjust to testing using smaller sample sizes than we initially planned. We can adjust the test duration based on sample size and presence or lack of any wear-out modes.

There is another reason to know how to adjust test parameters to accommodate smaller sample sizes. Our test equipment, such as test chambers and vibration tables, may not accommodate large sample sizes. The alternative would be to either test them sequentially or reduce the sample size. The former may not be a practical solution because of program timelines. Hence, we need to understand how to adjust test parameters to reduce sample size.

Kececioglu (2002) suggested that an exponential distribution for reliability be assumed (i.e., $\beta = 1$) and then accept the experimental data at face value (i.e., $CL = 0.5$ or 50%). Under these conditions Equation 8.10 may be manipulated to give us the test duration as a multiple of the required design life:

$$T = \sqrt[\beta]{\left[\frac{\ln(0.5)}{n\ln(R(L_D))}\right]} L_D \qquad\qquad (8.12)$$

For example, for a sample size of one, i.e., $n = 1$, and a constant failure rate, the test duration is for a 95% reliability (i.e., $R(L_D) = .95$), $T_{\text{test}} = 13.5\ L_D$, meaning that the test duration should be 13.5 times longer than the design life of the product. For instance, if during the design life of, say, a switch, it is designed to be actuated 1000 times to reach a reliability of 95%, with a sample size of one unit, it should be actuated 13,500 times without any failures.

One may suggest that a test duration time to design life ratio of 13.5 seems to be high. Physically, chances are that within this period, components may begin to wear out. Should that be the case, for a 95% reliability, $T_{\text{test}} = \left[\sqrt[\beta]{13.5}\right] L_D$. Now, for a typical wear-out shape factor $\beta = 2$. Now, $T_{\text{test}} = \{3.67\} L_D$. In other words—as in the previous example—should 1000 actuations be needed in a design life, to test to a 95% reliability, the test units should be actuated only 3670 times (compared to 13,500 calculated earlier). Again, as demonstrated earlier in the case studies and here, knowing the wear-out mode can help significantly reduce test durations.

Cumulative Damage Reliability Testing

Thus far, the focus of this chapter has been on identifying means to accelerate aging of units under test along with sample size and test duration calculations for component reliability testing. We have also reminded ourselves that we cannot do this without understanding and working through practical limitations that exist in terms of resource planning, sample size availability, and test duration limitations. Up to now, we had assumed that the test is designed for a single stress or load with sample availability for each test.

In practice, a subassembly or component of a system is often subjected to multiple stressors often at the same time. For instance, consider the engine control module (ECM) in today's modern cars. This module, which controls various functions of the engine, is subjected to high temperatures caused by the running of the engine. At the same time as the engine starts and shuts down, the ECM heats up and cools down with it. In addition, as the seasons change, so does the temperature of the environment, which impacts the overall temperature and temperature cycling experienced by this unit. Now in addition to this variety of thermal loadings, add the impact of humidity or vibration caused not only by the engine but also by the road and the shock absorber design.

Considering all these test conditions, the needed sample size would be prohibitive. However, we need to keep in mind that should we apply each of these loads to separate groups samples, we would not be replicating physics correctly.

In engineering, the *principle of superposition* suggests that if a linear homogeneous system is subjected to a number of independent loads, the resulting stress may be calculated by the sum of the stresses induced by each load separately. In other words, to replicate physics of multiple loads, the same samples should be subjected to them.

Edson (2018) suggests that before we begin the test regiment, first we need to understand and quantify the damage that a unit under test may experience as a part of its design life. Damage may be due to high (or low) temperatures, or mechanical and thermal shocks as well as other factors such a power and temperature cycling. In certain cases, other physical factors such as humidity or dust may be influential as well. Once the nature of the damage is understood, we can use one of the governing equations such as the Arrhenius or power law models to calculate an acceleration factor so that a design life may be accelerated as we have previously defined.

Damage Caused by Use and Environmental Profiles

In previous sections, we based our foundation of reliability on the study of failures, failure modes, and their timings based on using a number of samples under different stress conditions. We discussed how to develop accelerated testing based on increased stress levels in order to shorten test times. The question is how we apply stresses associated with various use and environmental profiles and how we develop accelerated conditions.

Edson (2018) suggests that in order to develop an accelerated reliability test, first one needs to understand and calculate the damage caused by a given stress condition. Next, the damages caused by all stress conditions need to be calculated. Finally, the total accumulated damage is the sum total of damages done by individual stresses. Once this accumulated damage is calculated, an acceleration factor may be determined based on increased stress values. In other words, for a given stress (e.g., fatigue or thermal aging)

$$\text{Total Damage} = \sum_i \text{Damage}_{\text{stress level } i}$$

and

$$\text{Total Damage}_{\text{Normal Stresses}} = \text{Damage}_{\text{Accelerated Conditions}}$$

In short, the use and environmental profiles are used to calculate the cumulative damage to a component, module, or a system. For most systems, the use and environmental profiles exert one of the following types of stress:

1. Thermal aging, which is modeled by the Arrhenius equation, i.e., $\text{Life} = A e^{\frac{B}{T}}$, where A and B are constants.

2. Thermal or mechanical/vibration fatigue, which is modeled by the inverse power law, i.e., $\text{Life} = A(S^{-m})$, where A and m are constants and S is the applied stress level. This model also applies to power and voltage as well as pressure levels.

3. Humidity, which is modeled by the Arrhenius–Peck equation—a combination of the inverse power law and the Arrhenius equation.

Once we identify what stresses are applied during the life of a product, we can calculate the cumulative damage caused by each type of stress. Then, we can accelerate the test time

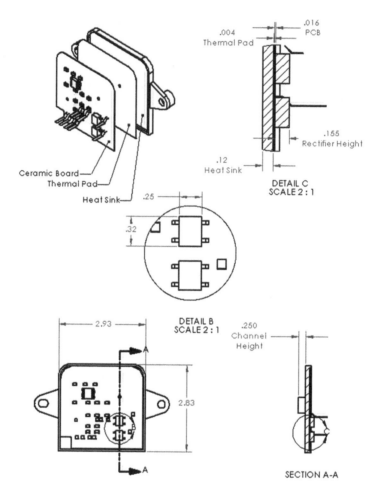

FIGURE 8.19
A depiction of an engine control module.

by increasing the applied stress of the same type to create an equivalent level of damage during the design life of the product under study. To illustrate what we mean here, let us consider the following case study. We need to keep in mind that in this approach, the major assumption is that sequence of application of stresses will not impact the final cumulative damage.

Case Study: Cumulative Damage Testing

An ECM as shown in Figure 8.19 has been designed for a 10-year life. We need to design a reliability test to demonstrate a reliability of 95% after the 10-year design life. The use conditions are as follows:

- Engine run-time = 7 hours per day
- Engine running temperature rise above ambient = 50°C
- Summer daytime temperature = 40°C
- Summer nighttime temperature = 20°C

- Winter daytime temperature = 15°C
- Winter night-time temperature = −5°C

For simplicity, we assume that there are 360 days in a year, equally divided between summer and winter. Furthermore, we assume that daylight in summer lasts 16 hours and in winter 10 hours.[8]

To design a reliability demonstration test, first, we need to identify the stresses that the unit undergoes. For the ECM, these stresses are

1. Temperature aging due to high environmental temperature conditions when the car is not being operated.
2. Thermal cycling and associated thermal fatigue. This is due to the unit's exposure to the engine's high temperature conditions while it is running, and its return to ambient, cooler temperatures when the car is not running.
3. Humidity variations between summer and winter months.
4. Vibrations transmitted to the ECM caused by engine vibration and road conditions.

Now, let us consider each type of stress and its associated damage separately. The reliability test will comprise of testing each stress sequentially.

Temperature Aging

Temperature aging may be modeled using the Arrhenius equation. The acceleration factor associated with this equation is

$$AF = \exp\left[\left(\frac{E_a}{K}\right)\left(\frac{1}{T_u} - \frac{1}{T_a}\right)\right] \tag{8.13}$$

where E_a is the activation energy and K is Boltzmann constant (8.61739×10^{-5}). T_u and T_a and are the use temperature and accelerated temperature, respectively, in degrees Kelvin.

To design a cumulative damage reliability test for temperature aging, we first need to understand the system's exposure time to the expected (or use) temperature. We can start by tabulating the daily temperature variations. For instance, in this problem, the engine runs for 7 hours daily. suggests that in summertime, we have 9 hours where the ECM is at daytime temperatures and 8 hours at nighttime temperatures. In wintertime, these exposures are 3 and 14 hours, respectively. Given that there are 24 hours in a day, the percent Exposures for any temperature may be calculated. However, we need to keep in mind that in this case study, we are assuming that summer hours apply for only half of the year. For instance, the percent exposure apportionment for 90°C is calculated as

$$\%\text{Exposure Apportionment} = \frac{7}{24 \times 2} = 14.58\%$$

Now, for this device, the design life is 10 years. Based on a 364-day year, there are 8736 hours in a year and 87,360 hours in 10 years. Thus, in 10 years, the ECM's exposure to 90°C is

$$\text{Life Time Exposure to } 90°\text{C} = 14.58\% \times 87360 = 12740 \text{ hr}$$

We intend to run the aging acceleration in environmental conditions set to 150°C. The next step is to calculate the acceleration factor when the thermal aging is conducted in this chamber. Using Equation 8.14 and assuming an activation energy equal to 0.8, we have

$$AF = \exp\left[\left(\frac{0.8}{8.61739 \times 10^{-5}}\right)\left(\frac{1}{90 + 273} - \frac{1}{150 + 273}\right)\right] = 37.622$$

The acceleration factor is the ratio of lifetime exposure to the test duration; hence, the test duration may be calculated as

$$\text{Test Duration} = \frac{\text{Life time Exposure}}{AF}$$

In this scenario, the test duration for aging the ECM to 90°C conditions for a period of 12,740 hours is

$$\text{Test Duration} = \frac{12740}{37.622} = 338.6 \text{ hours}$$

This means that under 150°C, the ECM receives an equivalent damage to the 90°C in only 338.6 hours or 2.01 weeks. Table 8.7 provides the calculation for the other temperature conditions.

Notice that acceleration factors for temperatures (particularly winter nights) are quite large. This really suggests that under cold and low temperatures, electronic parts do not really age; hence, their accumulated damage in a design life is relatively small and may be replicated quickly under high temperatures. Table 8.7 indicates that we can effectively replicate (or accelerate) damage due to 10 years of aging in only 2.37 weeks, if we have large enough samples.

TABLE 8.7

Accelerated Life Calculation for a Design Life of 10 Years Under 150°C Test Conditions

	Daily Exposure (hours)	Temp in °C	% Exposure Apportionment	Life Time Exposure (Hours)	Temp in °K	Acceleration Factor	Test Duration (hours)
Summer							
Day time - not running	9	40	18.75%	16380	313	2.24E+03	7.322
Night time - not running	8	20	16.67%	14560	293	1.69E+04	0.86
Running time	7	90	14.58%	12740	363	37.622	338.6
Winter							
Daytime, not running	3	15	6.25%	5460	288	2.94E+04	0.19
Nighttime, not running	14	−5	29.17%	25480	268	3.26E+05	0.08
Running time	7	65	14.58%	12740	338	249.436	51.1
		Total =	100.00%	87360			
			Accelerated Design Life (hours) =				398.2
			Accelerated Design Life (weeks) =				2.37

Should we have only one sample to test, we need to adjust the test duration based on Equation 8.12:

$$T_{test} = \left[\sqrt[\beta]{\frac{\ln(0.5)}{n \ln\left[R(T_D)\right]}} \right] L_D$$

$\beta = 1.8$ is a typical value for aging studies. For a minimum reliability of 95%, we can calculate the test duration as follows:

$$T_{test} = \left[\sqrt[1.8]{\frac{\ln(0.5)}{\ln[0.95]}} \right](2.37) = 10 \text{ weeks}$$

So, to demonstrate that our ECM has a minimum of 95% reliability under aging conditions, one unit has to survive 10 weeks in an environmental chamber at 150°C. Once this test is completed, the same unit should also pass the thermal fatigue test.

Thermal Fatigue

Now that we have designed a test for thermal aging, we need to focus on the thermal cycling that the ECM experiences. Thermal fatigue may be modeled using the inverse power law. Earlier, we mentioned that damage may be calculated as a product of life and stress. By "life" what we really mean is exposure time. So,

$$\text{Damage} = C(S)^m$$

where C is the number of exposure cycles, S is the applied stress (such as temperature range of cycling), and m is the power law coefficient. Edson (2018) suggests that for cyclic fatigue of solder, $m = 2.5$. With this in mind, we can calculate the damage due to cyclic temperature in summer daytime when the engine runs and the ECM heats to 90°C, then shuts off 7 hours later, and the engine cools to the environment temperature of 40°C. Note that we have assumed that the 364-day year is divided equally between summer and winter. Hence there are 182 summer days when this heating and cooling cycle is repeated. In real life, we need to take more granularity into account for monthly and daily variations.

$$S = 90°C - 40°C = 50°C$$

$$\text{Damage} = 182(50)^{2.5} = 3.2 \times 10^6$$

Note that the units of damage are not relevant at this time. Table 8.8 provides calculation of damage caused by other thermal conditions and the cumulative damage over a design life of 10 years. Additionally, it provides the number of thermal cycles needed to create an equivalent damage under a thermal cycle test that varies the chamber temperature between 150°C and −40°C.[9] To calculate the equivalent number of thermal cycles in a chamber, we need to set the damage caused by the test conditions to the cumulative damaged caused during the 10-year design life:

$$\text{Damage} = C(S)^{2.5} = 7.1 \times 10^7$$

TABLE 8.8

Accumulated Damage Due to Thermal Cycling and Fatigue

	Air Low Temp	Air High Temp	S Delta-T	Exposure Cycles	Damage
Summer, running	40	90	50	182	3.2E+06
Summer, not running	20	40	20	182	3.3E+05
Winter, running	15	65	50	182	3.2E+06
Winter, not running	−5	15	20	182	3.3E+05
				Damage in One Year =	7.1E+06
Design Life in Years =		10	Total Damage In Design Life =		7.1E+07
	Hi Temp	Low Temp	Delta-T	Number Of Thermal Cycles	Accumulated Damage
Equivalent Test	150	−40	190	143.0	7.1E+07

$$S = 150°C - (-40°C) = 190°C$$

$$C(190)^{2.5} = 7.1 \times 10^7$$

$$C = 143 \text{ Cycles}$$

Similar to thermal aging, this number of cycles may be used if we have an adequate number of samples (~30) available. For a sample size of one, we need to adjust this number based on Equation 8.12. The assumption of $\beta = 1.8$ holds here as well:

$$C_{test} = \left[\sqrt[1.8]{\frac{\ln(0.5)}{\ln[0.95]}} \right] (143) = 608 \text{ cycles}$$

This number indicates that should one sample of the ECM survive 608 thermal cycles, it has demonstrated a minimum reliability of 95%.

Humidity Aging

Humidity aging may be modeled using the Arrhenius–Peck equation:

$$AF = \left(\frac{RH_u}{RH_a} \right)^m \exp\left[\left(\frac{E_a}{K_b} \right) \left(\frac{1}{T_u} - \frac{1}{T_a} \right) \right] \tag{8.14}$$

where RH is relative humidity, and u and a are subscripts referring to use and accelerated conditions. Peck (1986) suggests that appropriate values are $m = -2.66$ and $E_a = 0.8$.

Based on available test chambers, it is our decision to set the test conditions for humidity to be 85°C and 85% RH (relative humidity). We have determined that the ECM is subjected to 40°C and 50% RH during summer daytime. Based on these values, the acceleration factor due to humidity is

$$AF = \left(\frac{50}{85} \right)^{-2.66} \exp\left[\left(\frac{0.8}{8.61739 \times 10^{-5}} \right) \left(\frac{1}{40+273} - \frac{1}{85+273} \right) \right] = 170.42$$

Now, we need to calculate the exposure hours under the test conditions. Assuming that humidity is only a factor during the 16 hours of daytime during the summertime, we calculate the design life exposure to be 29,120 hours (= 16 hours × 182 days a year). By dividing this number by the acceleration factor, we obtain the test duration for one design life:

$$\text{One design life for Humidity} = \frac{29120}{170.42} = 171 \text{ hours or } 7.1 \text{ days}$$

Similarly, we need to calculate the test time based on one sample:

$$T_{\text{test}} = \left[1.8 \sqrt{\frac{\ln(0.5)}{\ln[0.95]}} \right] (171) = 727 \text{ hours or } 30 \text{ days}$$

Random Vibration

Similar to thermal cycling, the damage caused by random vibration is also fatigue and may be modeled using the power law. Earlier, we mentioned that damage may be calculated as a product of life and stress. By "life" what we really mean is exposure time. So,

$$\text{Damage} = T(S)^n$$

where T is the exposure time, S is the applied stress (i.e., acceleration), and n is the power law coefficient.[10] Edson (2018) suggests that for random vibration fatigue of solder, $n = 4$. We also need to keep in mind that random vibrations are applied only when the engine runs. For a primer on random vibration, see Jamnia (2016). In this test case, our assumption for random vibrations input is given in Table 8.9; however, data may need to be extracted either from other tests or standards. Briefly, we are suggesting that the ECM experiences four different levels of root mean square accelerations (G_{rms}) during its 7-hour operation. For instance, in the travel sequence 1, the ECM is subjected to one hour of random vibration at $G_{\text{rms}} = 0.02$. For a 364-day year, this means that the exposure to this stress level is 364 hours, and 3540 hours during a 10-year design life. Knowing this information, we can calculate the damage caused by this stress level during the design life:

$$\text{Damage} = 3640(0.02)^4 = 0.001$$

Table 8.9 provides the damage caused by the other segments of the travel sequence.

TABLE 8.9

Accumulated Damage Due to Random Vibration

Travel Sequence	Daily Hours	G_{rms}	Cumulated Hours in 1 Year	Cumulated Hours in Design Life	Design Life Damage
1	1	0.02	364	3640	0.001
2	3	0.1	1092	10920	1.092
3	2	0.05	728	7280	0.046
4	1	0.02	364	3640	0.001
Total	7		Sum	25480	1.139

As in prior calculations, in order to find the accelerated design life, we need to first sum the damage caused by all segments of the travel sequence. Then, we can compute what exposure under test conditions is needed to generate equivalent damage. We have

$$T(0.5)^4 = 1.139$$

$$T = 18.2 \text{ hours}$$

Similar to thermal aging, this exposure time may be used if we have an adequate number of samples (~30) available. For single-sample testing, we need to adjust this number based on Equation 8.12 with an assumption of $\beta = 2$, which is more reflective of failures due to vibration:

$$T_{\text{test}} = \left[\sqrt[2]{\frac{\ln(0.5)}{\ln[0.95]}} \right](18.2) = 67 \text{ hours}$$

This number indicates that should one sample of the ECM survive 67 hours of random vibration at a $G_{\text{rms}} = 0.5$ on each axis, it has demonstrated a minimum reliability of 95%.

In summary, the reliability testing of the ECM consists of the successful outcome of exposing a single unit to the following conditions:

1. 10 weeks in an environmental chamber at 150°C
2. 608 thermal cycles between 150°C and −40°C
3. 30 days of exposure to 85°C and 85% RH
4. 67 hours of random vibration at a $G_{\text{rms}} = 0.5$ on each axis

Test Duration with Anticipated Failures

There is a different approach to calculating the test duration when we anticipate the failure rate is constant. Furthermore, we may observe failures during the test. In these types of scenarios, our test design is to demonstrate an expected value of the MTTF or to calculate this metric based on the accumulated test time on the samples and the total number of observed failures.

The appropriate model is the chi-square distribution, which is used for reliability demonstration test design when an exponential distribution is applicable:

$$T = \frac{\text{MTTF} \times \chi^2_{Cl,\, 2(r+1)}}{2} \tag{8.15}$$

Herein, T is an accumulated test time on all samples. r is the acceptable number of failures observed. Cl is the statistical confidence at which the results are to be demonstrated. Finally, χ^2 is the chi-square distribution factor.

Example: Test Design with Anticipated Failures

An original equipment manufacturer of an electronic component wants to design a test to demonstrate an MTTF of at least 20 years using accelerated life testing. Samples in this test are expected to fail due to multiple failure mechanisms. We designed the test to run 20

TABLE 8.10

Tabulation of Total Test Time with Number of Observed Failures

Observed No. of Failures	Required Total Test Time	Required Test Time per Sample
0	46.1	2.3
1	77.8	3.9
2	106.4	5.3
3	133.6	6.7
4	159.9	8.0

samples and expect to demonstrate a MTTF of 20 years with a 90% statistical confidence. However, we cannot anticipate when we would observe failures. As result, the test duration depends on when failures are observed.

By applying Equation 8.15 and our test parameters we get the following formula:

$$T = \frac{(20)\chi^2_{0.90,\, 2(r+1)}}{2}$$

Next, we can tabulate the test time (T) associated with the number of observed failures. This data is provided in Table 8.10. We should note that the test time is the accumulated time on all samples. So, based on the data in this table, should we not observe any failures for 46.1 (accelerated) years, we can claim that we have demonstrated a MTTF of 20 years with 90% confidence. This accumulated time for 20 samples implies that each sample needed to be tested for 2.3 years (accelerated). However, should we observe one failure, we need to increase our test duration to 3.9 years per sample for a total of 77.8 years and so on.

Degradation Testing

So far we have defined failure as the inability of a unit under test to perform as intended. We have attempted to design reliability tests in order to obtain sufficient information about these failures to define the life characteristics of the design.

A different approach to identify design reliability is to design a test to monitor degradation of functional or design specifications even before any failures are observed. This data can then be utilized to project the trend over time and predict when the design may fail. This methodology can be used to shorten the duration of the test.

Key design parameters and loading conditions need to be defined, monitored, and tracked during the test and plotted against time (or repeated cycles). The rate of degradation or decline of key design parameters can be projected following the pattern or trend of the obtained data. Knowledge of the failure mechanisms and contributing factors helps us predict the trajectory of the predicted trend. This methodology can be used for measuring and tracking direct design parameters or key indicators of other physical properties. Examples of such parameters are remaining capacity in batteries, pump flow rate, gasket leak rate, or noise levels in case of worn bearings. We should note that the accuracy of this approach depends to a large extent to the linearity of the relationship between various input and output parameters. The more nonlinear the trend of decline, the less accurate the projection of the degradation rate will be.

Having said this, we create and develop trend lines through statistical analysis of design parameters or key indicators of the units under test on regular intervals. Once we gain confidence that we have collected enough data and in the calculated trend, we can project the trend line to the desired time span and predict whether failure may occur or that design specification levels or thresholds are violated.

Case Study: Lead–Acid Battery Remaining Capacity

In a new product, a rechargeable battery is used as backup power source. The user requirements for the battery is to be able to be recharged, perform its function, and discharge at least 100 times over the expected service life of the product. The design specification of the battery is to have a remaining capacity (RC) of at least 70% of the initial capacity after 300 cycles.

We conducted a test where 20 batteries have been charged and discharged only 100 times. The remaining capacity of the batteries was measured at charge–discharge cycles along the test, and the following RC values were recorded. Table 8.11 shows the remaining capacity of the 20 samples over 50 cycles at an interval of 10 cycles. The mean value and standard deviation of all 20 batteries were calculated at each cycle interval and trended over the cycle count as shown in this table and Figure 8.20.

TABLE 8.11

Batteries Remaining Capacity Testing Results

Battery	Cycles (Charge/Discharge)					
	0	10	20	30	40	50
B1	0.99	0.921	0.911	0.902	0.892	0.877
B2	0.95	0.893	0.872	0.857	0.832	0.804
B3	0.96	0.914	0.894	0.876	0.855	0.833
B4	0.96	0.914	0.894	0.874	0.838	0.802
B5	0.96	0.922	0.857	0.838	0.819	0.797
B6	0.97	0.935	0.884	0.862	0.829	0.79
B7	0.97	0.935	0.894	0.872	0.84	0.804
B8	0.97	0.935	0.894	0.872	0.834	0.796
B9	0.98	0.951	0.913	0.885	0.848	0.807
B10	0.98	0.951	0.919	0.889	0.85	0.809
B11	0.98	0.951	0.919	0.889	0.848	0.805
B12	0.98	0.953	0.921	0.891	0.846	0.798
B13	0.98	0.956	0.933	0.902	0.863	0.823
B14	0.98	0.956	0.933	0.895	0.855	0.814
B15	0.98	0.949	0.901	0.862	0.814	0.763
B16	0.98	0.921	0.903	0.863	0.823	0.781
B17	0.98	0.96	0.944	0.897	0.837	0.778
B18	0.98	0.966	0.937	0.886	0.831	0.779
B19	0.99	0.984	0.925	0.913	0.882	0.845
B20	0.99	0.984	0.952	0.885	0.822	0.756
Mean Value	0.976	0.934	0.920	0.900	0.830	0.803
Standard Deviation	0.011	0.024	0.024	0.018	0.020	0.028

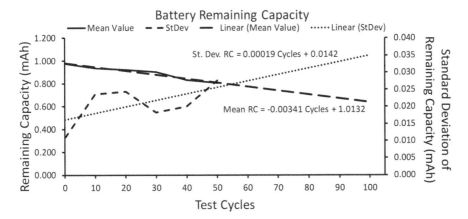

FIGURE 8.20
Battery remaining capacity (RC) degradation and projection over testing cycles.

As can be seen in Table 8.11, the average value of the remaining capacities of all 20 batteries drops over test cycles, as expected, due to the aging and grid corrosion inside the batteries. However, it is also noticeable that due to piece-to-piece variation between batteries, the standard deviation of the remaining capacity degradation increases over test cycles.

As shown in Figure 8.20, the projected mean value and standard deviation of the remaining capacity at 100 cycles can be calculated from the linear trend line plotted on the chart as follows:

$$\text{Mean } RC = \mu_{RC} = -0.0341 \text{ Cycles} + 1.0132$$

$$\text{St. Dev. of } RC = \sigma_{RC} = 0.0019 \text{ Cycles} + 0.0142$$

$$\mu_{RC} = -0.00341 \times 100 + 1.0132 = 0.6722 \text{ mAh}$$

$$\sigma_{RC} = 0.00019 \times 100 + 0.0142 = 0.0332 \text{ mAh}$$

Based on the projected value, and the failure criteria of RC <70%, the reliability of the batteries at 100 cycles can be calculated using the standard normal distribution as explained in Chapter 6:

$$Z = \frac{0.7 - \mu_{RC}}{\sigma_{RC}}$$

$$Z = \frac{0.7 - 0.6722}{0.0322} = 0.863$$

$$Z = 0.863$$

As explained in Chapter 6, we can calculate the probability that the mean RC is less than 70% using published tables (NIST 2012) or standard function in Microsoft Excel:

$$P(z \geq 0.863) = 0.8059$$

This means that the probability that the battery can perform its function after 100 charge–discharge cycles is 80.6%. This is a relatively low reliability. The implication is that either the design team has to select a more reliable battery or for the business to manage the risk of battery failures and field returns.

Notes

1. There are exceptions that prove this rule, particularly in industries such as avionics.
2. HASS stands for highly accelerated stress screening. We will discuss this methodology in Chapter 10.
3. Stressors are often temperature and humidity, but others such as vibration and electrical loads may also be used.
4. IEC stands for International Electrotechnical Commission. It is a regulatory body to develop a number of standards.
5. This test is conducted to accept or reject a production lot based on using a small sample size pulled from a production batch. For more on acceptance sampling Montgomery (2013) and Kenett and Zacks (2014) are recommended.
6. Now that we know about accelerated life testing, it should be clear that all times and durations are calculated with the acceleration factor in mind.
7. Typically, engineering build samples are an order of magnitude more expensive that the production units.
8. We have simplified this case study for the sake of this book; however, in practice, we may need to obtain daily temperature variations.
9. As test designers, we set these parameters ourselves based on what the component may be able to withstand and what our test equipment may be capable of.
10. We are using different nomenclature between the equation use for thermal cycling and random vibration to emphasize that even though the equation may be the same, the numerical value of the factors are not the same.

9

System Reliability Testing

Introduction

In Chapter 8, we described component- and subsystem-level reliability testing. We defined the test design aspects in terms of duration, sample size, and the factors that may influence test execution. Although the techniques described in that chapter are equally applicable to system-level testing, there are subtle differences that we need to be aware of. For instance, when we test components or subassemblies a goal may be to determine whether wear-out modes exist in the component and if so how severe they are. Armed with that information, we can then demonstrate either reliability at certain times or the unit's mean time to failure (MTTF). The reason is that most components and subassemblies are designed to be non-repairable. Additionally, they usually exhibit a prominent failure mode. Hence, by understanding the failure-in-time behavior of various components, modules, and subassemblies, we can understand a system's behavior and offer design improvements and suggestions.

It is no surprise, then, that most engineering systems' reliability behavior is at least the sum total and the cumulative opportunities for the failure of its components. We say "at least the sum total." The reason is that at the system level, various components interact and this interaction may have adverse influence on one or a few of the interacting subassemblies. As mentioned, components and modules are not repaired once they fail. However, it is expected that systems will have failures during their service life and that we can restore functionality by replacing the failed component(s). It is, therefore, more appropriate to use the term mean time between failures (MTBF), or availability, to describe system reliability during its expected service life while maintaining an acceptable failure rate.

In Chapter 8, we reviewed the techniques to understand module-level design margins as well as how to demonstrate subassembly level reliability. In this chapter, we will focus on bringing these components together and understanding reliability at the system level. There are two aspects to this activity. First, as components are assembled into systems, their interactions are unknown to some degree. This would lead to unexpected failures that need to be resolved. This set of activities is called reliability growth testing (RGT). Once we are satisfied with the growth, we are then set to demonstrate reliability at the system level. From our point of view, reliability testing does not stop at system-level testing. It does continue to production through reliability stress screening (RSS). In this chapter, we will review reliability growth as well as demonstration testing; however, stress screening will be discussed in Chapter 10.

In this chapter, we will explain the most common strategies and methods used to explore, grow, and demonstrate system reliability. These methods include tools used to determine *the what and the how to test a system* such as the parameter diagram (P-diagram) and failure modes and effects analysis (FMEA). They are used to identify critical items to be tested, and all the noise and control factors to be included in reliability testing.

We will explain how reliability testing can start as early as possible in the design and development process, even before the final design is mature. Reliability growth analysis, management, and projection are powerful approaches, especially when software is immature and not fully developed and we have limited engineering builds and prototypes prior to launch. Reliability growth predictions may be used to project the final reliability of the system and its timing based on design improvements across various builds and configurations.[1]

Next, we will provide guidelines on different techniques of testing of mature or final design. This includes testing for including minimum MTBF and system reliability demonstration over service life. Finally, we will discuss the optimum sample size for system reliability testing. Many reliability standards present guidelines on system testing duration, but rarely do they suggest the means to calculate sample size based on statistical confidence levels.

The What and How of System-Level Reliability Testing

Planning reliability testing of a system requires a different mindset than that if testing its components. Within the design and development process, it is natural to design and test subassemblies first. The expectation is that by the time of system integration, we have high confidence in the reliability of its modules.[2] It is commonly suggested that system-level failures occur at the interface of its modules and subassemblies. This is influenced by the operating conditions and use profiles of the system. Hence to design a robust test, we need to understand the stresses that may potentially flush out and precipitate all possible failures of the system enough to produce. In other words, we need to ensure the reliability of the design under as realistic usage scenarios and conditions as possible. There are two tools to help us with this task: one is P-diagram and the second is FMEA.

Parameter Diagram

Any system operates on the basis of receiving inputs in order to operate and function as expected. These inputs may be material, energy, or data. For instance, for a car to operate, it would require gasoline as one of its inputs. The ideal output of the function of a system may also be material, energy, or data. Again, in the case of the car as an example, the system output is its movement, which may be expressed in terms of kinetic energy. A system operates in an environment that it shares with other systems. Changes in this environment may influence its ideal operation and output, and lead to malfunctions and failures. There are other external influences that may adversely impact the ideal output of a system. These may be the inherent variation in the manufacture of the product or even how different users use (or misuse) a product. These influences are called *noise factors*. They are defined as factors that are beyond the ability of the design team to specify. We mentioned environment; however, others are manufacturing, degradation over time, usage, and system interactions. Example noise factors are

- External environment, such as during operation, shipping, and storage
- Piece-to-piece variation due to manufacturing defects and process control

- Change over time due to aging of material, or degradation due to usage and wear out
- Customer usage in terms of misuse scenarios, handling, user error, cleaning
- System interaction with disposable sets, data entry into the device, connectivity

Finally, there are elements—called *control factors*—that are within the reach of the design team to undo the influence of noise factors. Optimum design of the control factors will make a system more robust; i.e., less sensitive to the variation in the noise factors. Examples of control factors may be troubleshooting training, power-source redundancy, and fail-safe functionality, among others.

Figure 9.1 shows a generic P-diagram. From a reliability test design point of view, a P-diagram helps us understand the relationship between all conditions, variables, and inputs involved in the reliability performance of the system before designing the test. In particular, we use the noise factors to "perturb" the system to examine whether it is stable or results in failures.

Figure 9.2 shows an example of a P-diagram for an electromechanical system. At a brainstorming session, we invited a cross-functional team of subject matter experts to populate all the noise and control factors affecting the performance and failure of the system. This diagram helped us to identify the test setup and the main elements to be considered when conducting effective system reliability tests. Our goal was to demonstrate the actual reliability of the system in terms of what may go wrong and how the system would fail.

Developing a system-level reliability test and incorporating a P-diagram may seem to be a daunting task. However, it can be easily completed if the following activities are adopted.

Keep Records

One of the most important rules of tests and experiments is to keep detailed records of all activities and observations. This includes a list of activities to be performed on a regular basis and records that these activities were in fact completed. These records become an

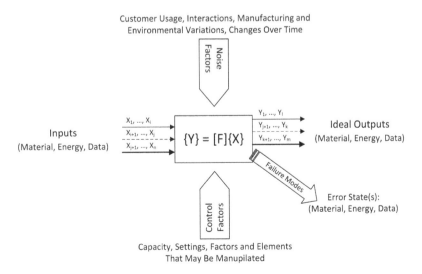

FIGURE 9.1
A depiction of the main elements of a parameter (P-) diagram.

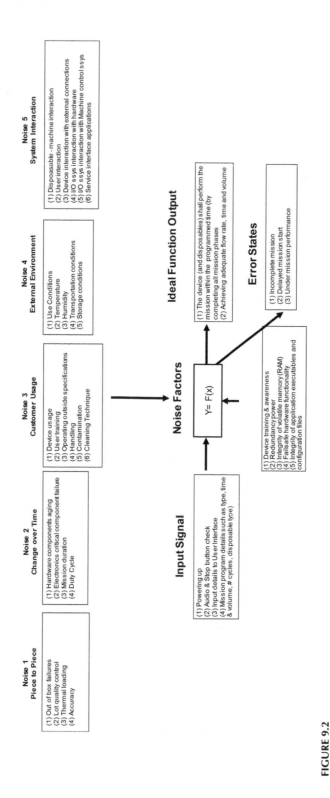

FIGURE 9.2

An example P-diagram for the operation and errors of an electromechanical system.

essential part of root cause investigations when failures occur. In addition to what we mentioned, records to keep include the hardware and software configurations, the unit's history and records prior to the start of the test,[3] records of all calibrated values on the unit under test (UUT), out-of-box inspection, and other measurements and inspections of vital parameters of the UUT.

Develop Measurement Baselines

Another set of activities prior to the start of the test is to measure various elements of the product to be tested. For instance, if we have moving parts, we should know their dimensions, particularly around interface areas. Or, if we have sensors, we need to know their output voltages (or their signal-to-noise ratios). These pretest measurements of critical parameters are important so that they can be compared to posttest values to determine any potential degradation or wear-out.

Test Parameter Settings

The test parameter settings are the control and noise factors that affect the performance of the system. Some of these factors are related to the product's inherent design, while others are affected by the user's error or usage profile and human factors. As an example (Figure 9.2), factors used to design the test setting are temperature ranges, duty cycle of the system missions, actuation of moving parts such as doors, power cycling per expected use frequency in the field, connecting and disconnecting of USB data storage, stowing of keyboard and display, and cleaning, among others.

Test Data Generation and Outputs

In a way, the reason to do all the previous activities is to collect data on our system under test. This data includes monitoring of output functions, inspection, and operational checkouts. Once a failure is observed, teardown and inspection of physical degradation and wear-out would add to the data collected. In the example as depicted in Figure 9.2, the output data includes mission time, observed failures and alarms, measurements of material output (flow, volume, pressure, temperature) of injected medication, and visual inspection of any physical damages.

Automation of Functions in the Test

We digress here for a moment to discuss the impact of test automation on system-level reliability testing. Unlike component-level testing where test automation is naturally assumed, we need to exercise caution in automating system-level reliability testing. Considering that manual operation of systems is time consuming and labor intensive, a full or a high degree of automation could provide savings in terms of time and finances as well as degree of repeatability. Yet, automation may hide much of users' errors. It may minimize any evaluation on design-related human errors. While we are not advocating against automation, we are suggesting that test cases exercised in system-level testing should be as comprehensive as possible to simulate real-life usage representing the profiles of the entire population. In essence, system reliability testing should be designed so that it demonstrates the target reliability via producing all types of failures or issues to minimize error in overestimating system MTBF.

Failure Modes and Effects Analysis

Failure modes and effect analysis is another tool that is usually constructed after the P-diagram is developed to identify the consequences and causes of all that "could go wrong" in the design. This bottom-up approach is a useful tool to identify what to test and what failure modes and mechanisms to test for. On this basis, a test may be designed to either identify the end of life of the system or, in the case of design improvement, confirm elimination of a failure mechanism. FMEA and its contribution to what to test and how to test is explained in detail in Chapter 7.

As an example, Table 9.1 depicts a portion of a system-level FMEA for a product called the Learning Station. It includes critical functionality failures, causes, and their effects. Typically, an FMEA matrix shows prevention and elimination methods for the anticipated failures and the severity rating of their effects—as shown in Table 9.1. These prevention and elimination methods are reviewed and examined as part of reliability test design. As shown in Table 9.1, when reliability is one of the mitigations, demonstration testing should be conducted to demonstrate the effectiveness of these mitigation measures. Failure modes and causes listed in the system FMEA should be examined thoroughly in the reliability testing and analysis of this system. As shown in Table 9.1, items 4 and 6 would require

TABLE 9.1

FMEA for Learning System Machine Reliability Growth Testing

	Module	Function	Potential Failure Mode	Potential Effects of Failure	Potential Causes of Failure	Prevention
1	Learning Station	Protect from environment	Liquid leaks inside	Stops working temporarily	Excessive warping of gasket	Finite element analysis
				Stops working permanently		
2			Electrostatic discharge reaches sensitive components	Stops working temporarily	Inadequate ESD/EMI barriers in the design	Design review
				Stops working permanently		
3			Users reach (or touch) internal components	Electric shock	Excessively large openings in the enclosure	Human factors study
				Bodily injury		
4		Develop social skills	Turntable fails to turn over time	Unit does not rotate on its axis	Turntable mechanism wears out	Reliability analysis/test
					Turntable bearings have corroded	Material compatibility/durability
5			Tokens fail to work	Unit stops functioning	Wrong tokens were used	Labeling
6			Token operation fails to work over time	Unit stops functioning	Component life is too short	Reliability analysis/test
7		Run educational software	Software does not run	Software does not accept commands	Incompatible software with operating system	Labeling
			Software begins to run but stops	Software does not accept commands	System is overheated	Thermal analysis

reliability testing to demonstrate durability of mechanisms to survive the expected service life of the product. Items 1, 2, and 7 do not require reliability testing; however, they need to be examined for robustness. Items 3 and 5 require executing studies with cooperation from the human factors team to substantiate the design.

Reliability growth testing (RGT) is a process by which we repeatedly test a system, identify its failures, repair or replace failed components—at times with design improvements—and test again. The purpose of RGT is to challenge and, in a way, attempt to demonstrate the reliability of a system during the design and development phase. As it stands, RGT does not require a fully mature system design.

At the early stages of the development process, testing subassemblies (or the entire system if available) reveals design flaws relative to robustness as opposed to reliability. By discovering failure modes during RGT and to implement design fixes, we grow the reliability of our design. An outcome of this test is to plan and manage reliability improvement tasks as well as the duration of tests needed to verify reliability requirements. Additionally, it determines whether extra efforts or additional resources will be needed.

A secondary outcome of RGT is to project the final reliability of the product based on its growth at early stages by using prototypes and engineering builds. Reliability growth is tracked by monitoring the number of failures, timing of observed new failure modes relative to the start of the test and to each other, and the increase in the time between failures as design improvements are implemented. In the last few paragraphs, we have mentioned implementing design fixes. We want to be clear that there is not a single approach to doing so. Depending on the program needs and design team resources, failed modules may be replaced with a new design almost immediately as it fails, or the failed component may be replaced with an identical one. In case of the latter option, other failure modes may be discovered while waiting for the next engineering build to be developed. While, in the former approach, the design improvements may be tested along with their interactions.

We need to be mindful not to ignore the impact of software on reliability. A real problem in testing is that software may mask hardware flaws or vice versa. When software plays a significant role in the functioning of a product, our reliability growth test strategy should combine physical design improvements with software updates as they provide an increasing number of functionalities. In other words, with each new engineering build, our test plan should be flexible to examine not only the hardware improvements but also software functionalities as well.

We have selected three common reliability growth models mentioned in reliability standards (IEC 61164 2004) for the following sections, namely, the Duane model (Duane 1964), the Crow–AAMSA NHPP model, and grouped data reliability growth model. For more information on these models, one may consult O'Connor and Kleyner (2012); and Modarres, Kaminskiy, and Krivtsov (2017), as well as MIL-HDBK-189 and ISO 61164 standard.

Duane Model

The Duane model is based on the assumption that design improvement of multiple failure modes is implemented during the test, and that the test continues with an improved component or module. A second assumption of the Duane model is that reliability growth is linear,[4] i.e., growth has a constant rate. As a result, this model is best suited for mature designs and the latest design configurations.

Based on a linear model for reliability growth, we can say

$$\text{Log} \, (\text{MTBF}_c) = \text{Log} \, (\text{MTBF}_0) + a \left(\text{Log} \, T_c - \text{Log} \, T_0 \right)$$

where MTBF is the mean time between failure and T is the test time. Subscripts 0 and c represent initial and cumulative values, respectively.

$$\text{MTBF}_0 = \frac{T_0}{N_0}$$

$$\text{MTBF}_c = \frac{T_c}{N_c}$$

Here, N represents the cumulative—meaning from the very beginning of the test—number of failures found either at the beginning of the test or cumulatively by the end of the test.

The reliability growth index, α, reflects the effectiveness of the reliability improvement program. It is determined empirically based on data from previous product development efforts of similar designs or from early rounds of reliability improvements and growth. It is worth mentioning that though the Duane model is based on a linearity assumption, in reality, it is piecewise linear. In other words, α may not be constant over the entire test duration, rather it has to be evaluated periodically from each test–fix–test cycle.

The reliability growth index, α, can also be estimated based on the manufacturer's previous new product development experiences and based on the known availability of design team resources. What we mean is that the project management team may plan resource allocations for a desired reliability growth rate. Let us review this aspect of planning in conjunction with α values. Typically, values of α range from 0.0 to about 0.6.

1. For α between 0.4 and 0.6, eliminating failure modes is a top priority. We need to apply stress testing and conduct failure analysis and corrective actions for all observed failures. This requires extensive design team engagement and resource allocation.

2. For α between 0.3 and 0.4, we need to give priority to addressing top failure modes only and applying corrective actions for them. Our test methodology should include stress testing. While resource needs are not as extensive as the previous criterion, design team engagement and dedicated resources are needed.

3. For α equal to 0.2, no stress testing is needed. We need to conduct corrective action for important failure modes. Under these conditions, the majority of the design team may be allocated to other projects, while a few are still involved in making corrective actions.

4. For α between 0.0 and 0.2, we do not expect significant growth any longer. We may apply corrective actions for important failure modes. As additional reliability growth is no longer expected, corrective actions require resources on a part-time basis.

The main objective of applying the Duane model is to calculate the reliability metrics—MTBF—at any instant during the test, which will be referred to as the instantaneous MTBF_i. We can rewrite the log form of the Duane model as follows:

$$\text{MTBF}_c = \text{MTBF}_0 \left[\frac{T}{T_0} \right]^\alpha$$

Finally, since the cumulative MTBF_c is calculated as T_C/N_C the model identifies the instantaneous, achieved MTBF_i after reliability growth, T_i, through the differentiation of the cumulative MTBF equation as follows:

$$\text{MTBF}_i = \frac{dT}{dN}$$

Since

$$N = \frac{T}{\text{MTBF}_c}$$

and

$$\text{MTBF}_c = \text{MTBF}_0 \left[\frac{T}{T_0} \right]^a$$

then

$$N = \frac{T}{\text{MTBF}_0 \left(T / T_0 \right)^a}$$

$$N = T(1-a)\left(\frac{T_0^a}{\text{MTBF}_0} \right)$$

Since $\dfrac{T_0^a}{\text{MTBF}_0}$ is a constant value, then

$$\frac{dN}{dT} = (1-a)T^{-a}\left(\frac{T_0^a}{\text{MTBF}_0} \right)$$

$$\frac{dN}{dT} = (1-a)\left(\frac{T_0}{T} \right)\frac{1}{\text{MTBF}_0}$$

$$\frac{dN}{dT} = \frac{(1-a)}{\text{MTBF}_c}$$

Hence, the instantaneous MTBF is calculated as

$$\text{MTBF}_i = \frac{dT}{dN} = \frac{\text{MTBF}_c}{(1-a)}$$

Duane Model Example

In a new product development project, the systems engineering team conducted verification testing and accumulated 1578 hours of operations. During this time, it observed ten occurrences of seven different failure modes. The design team decided to continue testing while root causes were identified and design solutions were implemented for all failure modes.

Project management allocated all engineering resources to the project to improve reliability. Reliability requirements at launch was targeted to be 600 hours MTBF. The question that the team needed to answer was the test duration needed to demonstrate this target MTBF.

To calculate the test duration, the Duane model was applied. The initial MTBF accumulated over 1578 hours of testing was given by

$$MTBF_0 = \frac{1578}{10} = 157.8 \text{ hours}$$

Since the reliability requirement target was 600 hours, we considered this value to be the instantaneous MTBF at the launch time. The cumulative $MTBF_c$ required to achieve an $MTBF_i$ of 600 hours was given by

$$MTBF_i = \frac{MTBF_c}{1-a}$$

where

$$MTBF_c = (1-a)MTBF_i$$

Project management had assigned rigorous testing and eliminating all failure modes as a top priority, and all available engineering resources were allocated to this project. Hence, we assumed $\alpha = 0.5$.

$$MTBF_c = (1-0.5)600 = 300 \text{ hours}$$

Using the relationship between $MTBF_c$ and $MTBF_o$

$$MTBF_c = MTBF_0 \left[\frac{T}{T_0} \right]^{\alpha}$$

$$T = T_0 \left[\frac{MTBF_c}{MTBF_0} \right]^{1/\alpha}$$

Thus

$$T = 1578 \left[\frac{300}{157.8} \right]^{1/0.5}$$

$$T = 5703 \text{ hours}$$

This is the total number of test hours needed. The team had already tested the product for a total of 1578 hours. Therefore, project management had to plan for continuing testing for another 4125 hours (= 5703 – 1578), which is nearly 6 months.

As previously explained, project management may control reliability growth rate by increasing or reducing allocated engineering resources. Or, project management can reset the scope of reliability improvement of the design. Suppose that project management decided to reduce resources and focus on top failure modes identified as "high" risk to the end users. Under this assumption, the reliability growth index will be reduced to 0.3. The result is

$$\text{MTBF}_c = (1 - 0.3)600 = 420 \text{ hours}$$

$$T = 1578 \left[\frac{420}{157.8} \right]^{1/0.5}$$

$$T = 39{,}655 \text{ hours}$$

Notice that with a reduction in resources, test duration becomes unreasonably long. The result is that the product may need to be launched prior to demonstrating the target MTBF of 600 hours. If this is not acceptable, it may be possible to increase the sample size to increase the total accumulated time.

Crow–AMSAA NHPP Model

The Crow–AMSAA NHPP[5] model assumes that the failure rate is not constant over test time. To model this nonlinear failure rate, we use Weibull distribution. Similar to the Duane model, this model assumes that we continuously test samples. As failures are found, design flaws are resolved and testing continues with units that are repaired with design improvements. This approach tracks reliability growth within the same phase or build, based on test–find–fix–test and not across phases or different builds. Unlike the Duane model, this method provides a probabilistic estimate of the projected failure rate (or MTBF) in terms of confidence bounds. Additionally, it provides a projection of the expected number of failures (f) at any given time, t.

This model begins by asserting the instantaneous failure rate may be calculated as follows:

$$\lambda_{\text{Instantaneous}} = \hat{\lambda}\hat{\beta}T^{\hat{\beta}-1}$$

where $\hat{\lambda}$ is the cumulative failure Intensity, $\hat{\beta}$ is the maximum likelihood estimate of the failure intensity decline rate (reliability growth index), and T is the test time at any point for projection rate. $\lambda_{\text{Instantaneous}}$ is the instantaneous failure rate achieved at any point during the test at time T.

We begin by computing $\hat{\beta}$, the maximum likelihood estimation (MLE) value of the reliability growth index. This is calculated by finding the value of the parameter $\hat{\beta}$ that fulfills the model given the following equation. This is done by iterations. For time-terminated tests, we have

$$\hat{\beta} = \frac{n}{n \ln(T) - \displaystyle\sum_{i=1}^{n} \ln(T_i)}$$

For failure-terminated tests, $\hat{\beta}$ becomes

$$\hat{\beta} = \frac{n}{(n-1)\ln(T_n) - \sum_{i=1}^{n-1}\ln(T_i)}$$

T is the test termination time (in case of time-terminated tests). n is the total number of failures. T_n is the time of the last failure (in case of failure-terminated tests) and T_i is the individual time to each individual failure; the subscript i refers to each individual failure and the subscript n refers to the total number of failures at test termination.

Depending on the value of $\hat{\beta}$, the reliability of the design is either growing, remaining constant, or declining:

1. For $\hat{\beta} < 1.0$, we have reliability growth. In other words, failure rates drop and MTBF increases.

2. For $\hat{\beta} = 1.0$, reliability remains a constant, i.e., no growth or decline.

3. For $\hat{\beta} > 1.0$, we are observing a reliability decline. In other words, failure rates increase and MTBF decreases.

Once $\hat{\beta}$ is computed, then we calculate the maximum likelihood estimates of failure intensity:

$$\hat{\lambda} = n / T_K^{\hat{\beta}}$$

In this equation T_K is the test termination time, or time to failure at termination.

Our goal of reliability growth modeling is to develop the ability to calculate the instantaneous failure rate at any time, T. Thus, we have

$$\lambda(T) = \hat{\lambda}\hat{\beta}T^{\hat{\beta}-1}$$

$$\text{MTBF}_T = \frac{1}{\lambda(T)}$$

The expected number of failures at time T is calculated as

$$E[N(T)] = \hat{\lambda}T^{\hat{\beta}}$$

Case Study: Crow–AMSAA NHPP

A design team is conducting reliability testing on two engines with the focus on finding issues, correcting them, and eliminating the root cause. As shown in Table 9.2, the time to failure is listed separately for each test sample. A total of 15 failures occurred on both engines, and the test was terminated at 3000 hours of total test time on both engines. In this design, 750 hours of accelerated testing would simulate the design life of this product. The team has an 8-week period of achieving a minimum of 1000 hours MTBF. We need to let management know whether any change of strategy or more resources are needed.

TABLE 9.2

Engine Reliability Growth Testing and Time to Failure Data

Engine 1 Run Time	Engine 2 Run Time	Cumulative Run Time (Hours), T_i	$Ln(T_i)$
4*	1	5 (= 4 + 1)	1.61
12	17*	29	3.37
13	18*	31	3.43
14*	19	33	3.50
30	39*	69	4.23
31*	40	71	4.26
72*	76	148	5.00
85*	82	167	5.12
90	86*	176	5.17
105*	103	208	5.34
120	110*	230	5.44
170*	183	353	5.87
246	290*	536	6.28
412	516*	928	6.83
1012*	1324	2336	7.76
1500	1500	3000	8.01
		$\sum Ln(T_i)$	73.21

* Failure.

We apply the Crow–AMSAA NHPP for time terminated to determine the reliability growth index. The Weibull parameter, $\hat{\beta}$, is

$$\hat{\beta} = \frac{n}{n \ln(T) - \sum_{i=1}^{n} \ln(T_i)}$$

$$\hat{\beta} = \frac{15}{15 \ln 3000 - 73.21}$$

$$\hat{\beta} = 0.3199$$

The failure rate, $\hat{\lambda}$, is

$$\hat{\lambda} = n/T_K^{\hat{\beta}}$$

$$\hat{\lambda} = 15/3000^{0.3199}$$

$$\hat{\lambda} = 1.1582$$

Thus, the instantaneous failure rate at any time T is

$$\lambda(T) = 1.1582 \times 0.3199 \, T^{0.3199-1}$$

TABLE 9.3

Projected Growth of MTBF over Testing Time (Including 3000 hours of Initial Test Time) and Expected Number of Failures

Time, T (Hours)	3500	4000	4500	5000	5500	**6000**	6500
$\lambda(T)$	0.00144	0.00132	0.00121	0.00113	0.00106	0.00100	0.00095
MTBF(T)(Hours)	694	760	824	885	944	**1002**	1058
E[N(T)]	16	16	17	18	18	**19**	19

Finally, the expected number of failures at any time, T may be calculated:

$$E[N(T)] = \hat{\lambda} T^{\hat{\beta}}$$

$$E[N(T)] = \widehat{1.1582} \, T^{0.3199}$$

Table 9.3 shows the expected number of failures and the instantaneous MTBF at any time, T, including the accumulated testing time of 3000 hours.

As shown in Table 9.3, the current design team efforts and engine design configuration under testing are capable of meeting the minimum MTBF of 1002 hours within the additional 3000 hours of testing, and specifically at a total 6000 hours of testing. This corresponds to 4 weeks of continuous testing at 750 hours accelerated every week. From a resourcing point of view, there should be no changes to the team during the 4-week test duration.

Grouped Data Reliability Growth Model

The two reliability growth models discussed so far assume that failures are found and design flaws are removed while testing continues. In reality, this is not common during product development; rather, design fixes for failures found in a design configuration are bundled and applied to the configurations to come.

A somewhat different, and yet related, point is that during system integration, design robustness and subsystem interactions are the focus of reliability testing. Shortcomings and flaws in these two areas constitute the majority of failure modes. During the development process, design configurations (hardware and software) evolve and change. Throughout this period, system-level testing should not focus on discovering component degradation and wear-out. Rather, system testing should focus on repeatedly running test cases to demonstrate functionality of the product.

During early development phases, the absence of wear-out as a factor implies that these repeated runs are statistically independent of each other. In Chapter 6, we mentioned that should the failures for a given system configuration be independent of time,[6] the proper approach to model reliability is the use of binomial distribution. This means that neither the Duane nor Crow–AMSAA are proper models for predicting reliability growth of systems when the test is not running continuously or when design updates are bounded and grouped in future builds and configurations. The proper model for this scenario is the AMSAA reliability growth tracking model–discrete (RGTMD), commonly known as the grouped data reliability growth model. It can be applied to simulate the reliability growth across test phases, each repeating independent trials of system functionality on different design configurations.

Structurally, this model is similar to the Crow–AMSAA approach; however, the maximum likelihood estimate of the Weibull shape factor is found by solving

$$\sum_{i=1}^{N} n_i \left[\frac{t_i^{\hat{\beta}} \ln\left(T_i\right) - t_{i-1}^{\hat{\beta}} \ln\left(T_{i-1}\right)}{T_i^{\hat{\beta}} - T_{i-1}^{\hat{\beta}}} \right] = 0.0$$

Similar to the Crow–AMSAA approach, n_i is the number of failures occurring in the time interval $T_i - T_{i-1}$, where T_i is the accumulated test time at the end of the test stage, i. Depending on the value of $\hat{\beta}$, the reliability of the units under test is either growing, remaining constant, or declining:

1. For $\hat{\beta} < 1.0$, we have reliability growth. In other words, failure rates drop and MTBF increases.
2. For $\hat{\beta} = 1.0$, reliability remains a constant, i.e., no growth or decline.
3. For $\hat{\beta} > 1.0$, we are observing a reliability decline. In other words, failure rates increase and MTBF decreases.

The maximum likelihood estimate of the failure rate, $\hat{\lambda}$, is given by

$$\hat{\lambda} = \frac{\sum_{i=1}^{N} n_i}{T_i^{\hat{\beta}}}$$

And the instantaneous failure rate $\lambda(T)$ at any time, T,

$$\lambda(T) = \hat{\lambda}\hat{\beta}T^{\hat{\beta}-1}$$

$$\mathrm{MTBF}_T = \frac{1}{\lambda(T)}$$

The expected number of failures at time, T, may be calculated as follows:

$$E[N(T)] = \hat{\lambda}T^{\hat{\beta}}$$

The main features and advantages of this model compared to the previous two may be summarized as follows:

1. This model enables calculations of reliability growth across stages or design configurations.
2. Each stage can be aligned with a software build knowing that software is never matured at an early stage of the design and not all functionalities are covered in testing.
3. This approach is applicable for single-use or disposable products.

Case Study: Grouped Data Reliability Growth Model

A medical device company is launching a new product for home use (by the patient). The design team has executed continuous reliability testing across different prototypes and engineering builds before running customer site testing on a limited number of volunteer patients. The design has matured through many configurations. Design changes and improvements were bundled at the end of each design phase and implemented on the design configurations, as shown in Table 9.4.

Table 9.4 below as it shows the number of test hours conducted at each build and the number of failures (issues) found at every stage of the test. The final reliability target for this product is an MTBF of 1000 hours. Management decided to go ahead with the full launch to fulfill its marketing commitment and not lose market share with the latest design configuration. The business leadership wants to know the answers to these questions:

1. Is there a reliability growth and will the target 1000 hours MTBF be achieved?
2. What is the current achieved MTBF in hours?
3. How long should the design team keep the current allocated resources to support the new launch in order to maintain the growth rate to achieve the final target reliability?

As shown in Table 9.4, in these test phases, the number of failures were observed without the exact number of hours to each failure recorded. Also, the data was grouped into phases because the design changes and fixes were bundled and implemented at the end of each phase. We will apply the Crow–AMSAA for Grouped Data method. As shown in Table 9.4, we observed an accumulated total of 53 failures in the test.

The maximum likelihood estimate of the Weibull shape factor, $\hat{\beta}$, is found by solving

$$\sum_{i=1}^{N} n_i \left[\frac{t_i^{\hat{\beta}} \ln(T_i) - t_{i-1}^{\hat{\beta}} \ln(T_{i-1})}{T_i^{\hat{\beta}} - T_{i-1}^{\hat{\beta}}} \right] = 0.0 \tag{9.1}$$

This model is solved by iterations. Table 9.5 lists the calculated values of the elements in Equation 9.1. The solution starts by assuming a value for $\hat{\beta}$, calculating all the elements in Equation 9.1, and then calculating the net value of the left side of the equation until it converges to 0.0. The question we are often asked is what is a good starting value. The truth is that there is really no standard answer. We can assume that $\hat{\beta}$ follows values established by each manufacturer or design team based on experience with previous designs and launched products. In general, if we have confidence that we are staring with a strong reliability growth, we can assume an initial value of $\hat{\beta}$ to be 0.25. Because we believe that reliability may be growing slowly, we can assume a value of 0.75 or higher.

TABLE 9.4

An Example of Device Reliability Testing across Different Builds

System Configuration	Test Phase	Phase Test Hours	Cumulative Test Hours	Test Phase Failures
Engineering Prototype 1	1	985	985	14
Engineering Prototype 2	2	245	1230	2
Manufacturing Build I	3	1417	2647	3
Manufacturing Build II	4	702	3349	5
Field Control Launch	5	9878	13227	26
Field Issues Corrected	6	2556	15783	3

TABLE 9.5

Calculations of Reliability Growth-Related Elements for Home-Use Medical Device Using an Initial Assumed Value of $\hat{\beta}$ of 0.4

System Configuration Test Phase	Phase Test Time, T_i	Cumulative Time, T_i	n_i	$\ln(T_i)$	$t_i^{\hat{\beta}}$	$t_i^{\hat{\beta}}\ln(T_i)$	$T_i^{\hat{\beta}} - T_{i-1}^{\hat{\beta}}$	$t_i^{\hat{\beta}}\ln(T_i)$ $-t_{i-1}^{\hat{\beta}}\ln(T_{i-1})$
Engineering Prototype 1	985	985	14	6.89	15.75	108.6	15.75	108.6
Engineering Prototype 2	245	1230	2	7.11	17.22	122.5	1.46	13.9
Manufacturing Build I	1417	2647	3	7.88	23.39	184.4	6.18	61.9
Manufacturing Build II	702	3349	5	8.12	25.70	208.6	2.31	24.2
Field Control Launch I	9878	13227	26	9.49	44.52	422.5	18.82	213.9
Field Control Launch II	2556	15783	3	9.67	47.78	461.9	3.26	39.4
	T_N	15783						

By applying an assumed value of $\hat{\beta}$ and the values of observed test time and number of failures we can compute Equation 9.1. Table 9.5 shows the first iteration of calculations using an assumed value of $\hat{\beta} = 0.4$, which yielded an output value of Equation 9.1 of 17.47. After using other values of $\hat{\beta}$, the equation converges to zero at a $\hat{\beta}$ value of 0.49 (see Table 9.6):

$$\sum_{i=1}^{N} n_i \left[\frac{t_i^{0.4}\ln(T_i) - T_{i-1}^{0.4}\ln(T_{i-1})}{T_i^{0.4} - T_{i-1}^{0.4}} \right] = 17.47$$

$$\sum_{i=1}^{N} n_i \left[\frac{T_i^{0.49}\ln(T_i) - T_{i-1}^{0.49}\ln(T_{i-1})}{T_i^{0.49} - T_{i-1}^{0.49}} \right] = 0.000025 \sim 0.000$$

Hence, the converged solution is

$$\hat{\beta} = 0.49$$

A $\hat{\beta}$ less than 1.0 means that the failure rate is declining. In other words, reliability is growing. This answers the first question that the business is asking.

To answer the second question, we begin by calculating the maximum likelihood estimate of the failure rate, λ, as given by

$$\hat{\lambda} = \frac{\sum_{i=1}^{N} n_i}{t_i^{\hat{\beta}}}$$

TABLE 9.6

Calculations of Reliability Growth-Related Elements for Home-Use Medical Device

System Configuration Test Phase	Phase Test Time, T_i	Cumulative Time, T_i	n_i	ln (T_i)	$t_i^{\hat{\beta}}$	$t_i^{\hat{\beta}}\ln(T_i)$	$T_i^{\hat{\beta}} - T_{i-1}^{\hat{\beta}}$	$t_i^{\hat{\beta}}\ln(T_i)$ $-t_{i-1}^{\hat{\beta}}\ln(T_{i-1})$
Engineering Prototype 1	985	985	14	6.89	29.23	201.5	29.23	201.5
Engineering Prototype 2	245	1230	2	7.11	32.59	231.8	3.36	30.4
Manufacturing Build I	1417	2647	3	7.88	47.43	373.8	14.84	141.9
Manufacturing Build II	702	3349	5	8.12	53.22	431.9	5.79	58.1
Field Control Launch I	9878	13227	26	9.49	104.27	989.5	51.05	557.6
Field Control Launch II	2556	15783	3	9.67	113.69	1099.0	9.42	109.5
	T_N	15783						

For the total test results, we use T_N for $T_i^{\hat{\beta}}$:

$$\hat{\lambda} = \frac{61}{T_N^{0.49}} = 0.466 \text{ per hour}$$

$$\hat{\lambda} = \frac{61}{15783^{0.49}} = 0.466 \text{ per hour}$$

MTBF is the inverse of the failure rate. So we start by first computing the instantaneous failure rate at time $T = T_N = 15783$ hours:

$$\lambda(T) = \hat{\lambda}\hat{\beta}T^{\hat{\beta}-1}$$

$$\lambda(T) = (0.466)(0.49)T^{0.49-1}$$

$$\lambda(15783) = (0.466)(0.49)\,15783^{0.49-1}$$

$$\lambda(15783) = 0.00165$$

$$\text{MTBF}(15783) = \frac{1}{0.00165} = 606 \text{ hours}$$

We can develop a curve to understand the behavior of the MTBF and project future reliability growth by computing the MTBF value at each test time and then extrapolating the obtained curve into a future date. Figure 9.3 illustrates the growth in the calculated MTBF using the reliability growth model. It indicates that the MTBF goal will be achieved after completing 42,000 hours of testing based on the growth achieved by implementing design improvements across different builds.

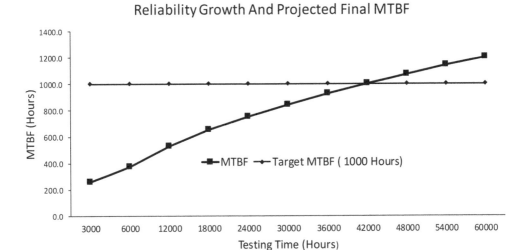

FIGURE 9.3
Reliability (MTBF) growth and projection of goal achievement using reliability growth analysis.

The third question is about the timing of achieving the target MTBF of 1000 hours. Assuming that the same design efforts and resources are focused on fixing issues as in the previous 15,783 hours of testing, we solve the following equation for MTBF = 1000 hours:

$$\lambda(T) = \hat{\lambda}\hat{\beta}T^{\hat{\beta}-1}$$

$$\mathrm{MTBF}\left(\text{at time } T\right) = \frac{1}{\lambda(T)} = \frac{1}{\hat{\lambda}\hat{\beta}T^{\hat{\beta}-1}}$$

$$1000 = \frac{1}{\lambda(T)} = \frac{1}{\hat{\lambda}\hat{\beta}T^{\hat{\beta}-1}}$$

$$1000 = \frac{1}{(0.466)(0.49)T^{0.49-1}}$$

From which, T = 42,000 hours. This result can be shown graphically in Figure 9.3.

Should the business decide to release the product based on this information—by the time a MTBF of 1000 hours is realized—the expected number of failures at 42,000 hours will be

$$E\left[N(T)\right] = \hat{\lambda}T^{\hat{\beta}}$$

$$E\left[N(42,000)\right] = (0.466)\,42,000^{0.49}$$

TABLE 9.7

Calculations of Reliability Growth-Related Elements without Controlled Launch

System Configuration	Test Phase	Phase Test Time, T_i	Cumulative Time, T_i	n_i	$\ln(T_i)$	$t_i^{\hat\beta}$	$t_i^{\hat\beta}\ln(T_i)$	$T_i^{\hat\beta} - T_{i-1}^{\hat\beta}$	$t_i^{\hat\beta}\ln(T_i)$ $-t_{i-1}^{\hat\beta}\ln(T_{i-1})$
Engineering Prototype 1	1	985	985	14	6.89	20.28	139.8	20.28	139.8
Engineering Prototype 2	2	245	1230	1	7.11	22.34	158.9	2.06	19.2
Manufacturing Build I	3	1417	2647	3	7.88	31.22	246.0	8.88	87.1
Manufacturing Build II	4	702	3349	5	8.12	34.60	280.8	3.38	34.8
		T_N	3349			30.01			

Expected number of failures = 85

These are 85 incidents of failures, not 85 unique failure modes. In other words, in addition to the 53 observed failures, an additional 32 failures will be seen.

In this case study, the design team executed reliability testing on six different builds, including the customer site controlled launch. The benefit of a controlled launch is the value derived from real-life usage scenarios. Most failures detected in a controlled launch are indicative of real-life usage under realistic operating conditions. A controlled launch accounts for a wide range of use variability, errors, and abuse that are not accounted for in in-house testing. In Table 9.7, we have removed the controlled launch data to show the difference that it makes.

When we ignore the controlled launch data, the reliability growth index, $\hat\beta$, is calculated as 0.436. This value being less than 1.0 projects a reliability growth; however, since it is less than the previously calculated value of 0.49, it suggests a faster growth.

If we continue our calculations, the estimate of the failure intensity $\hat\lambda$ will be

$$\hat\lambda = \frac{23}{3349^{0.49}} = 0.664 \text{ per hour}$$

Instantaneous failure rate at any time, T, $\lambda(T)$ is given by

$$\lambda(T) = \hat\lambda\hat\beta T^{\hat\beta-1}$$

$$\lambda(T) = (0.664)(0.436)T^{0.436-1}$$

And,

$$\text{MTBF}_T = \frac{1}{\lambda(T)}$$

The answer to the second question about the current achieved MTBF is given by

$$\text{MTBF}(3349) = \frac{1}{(0.466)(0.49)3349^{0.49-1}}$$

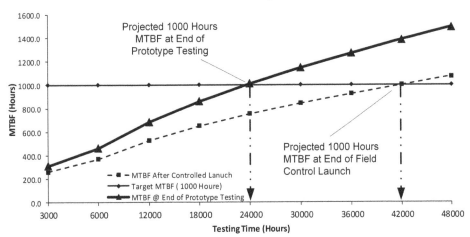

FIGURE 9.4
Difference in projection of the time to achieve reliability target after each testing phase.

$$\text{MTBF}(3349) = \frac{1}{0.003638} = 274.86 \text{ Hours}$$

Notably, this MTBF is substantially lower than the one previously calculated. Figure 9.4 illustrates the projected remaining time of testing to achieve the target MTBF if the testing was terminated at the earlier prototypes and manufacturing built results. As shown in this figure, the results indicate that if the test was terminated after Manufacturing Build II (a total of 3349 cycles and only 5 failures detected in that last phase of testing), the target MTBF of 1000 hours would have been, mistakenly, realized at 24,000 hours. This means that we projected the MTBF after 20,651 total hours of real-life usage in the field. The expected number of failures at the 24,000-hour time frame may be calculated as follows:

$$E\left[N(T)\right] = \hat{\lambda}T^{\hat{\beta}}$$

$$E\left[N(24,000)\right] = (0.664)24,000^{0.436} = 54 \text{ failures}$$

Again, we need to be cognizant of the fact that by ignoring the controlled launch data, our projected number of failures is substantially smaller (i.e., 54 as opposed to 85).

If leadership decided to release the product based on the incomplete set of data, it might have planned for resources to address only 30 more failures and fixes for these failures (54 minus 24 uncovered already) to reach a stable MTBF of 1000 hours. However, the controlled launch demonstrated that the design is not as reliable as was indicated by the four phases of in-house testing as expressed in Table 9.7. This lack of reliability was uncovered by the additional 29 failures in the controlled launch (26 in phase 5 and 3 in phase 6). In addition, the reliability growth model has predicted that there will be 32 more failures in the full launch phase (85 minus 53 uncovered in 6 phase tests).

Only the controlled launch testing conducted by customers and real users detected more failures and issues with the design, which helped the team to change the design and avoid releasing an unreliable product to the field. In a fielded product, design changes and improvements are typically more costly and time consuming. In addition, there is the indirect cost of the damage to the manufacturer's image in the market and financial consequences of low reliability.

Reliability growth tools are very powerful in providing engineering teams and business leaders the data needed to make crucial decisions about product launch, allocation of engineering resources, and avoiding economical losses and potential legal consequences.

Hardware and Software Reliability Growth of an Integrated System

It is difficult to find a product that is purely driven by either mechanical or electrical/electronic components. Embedded software plays a significant and pivotal role in product development activities. As an element within the product development process, software begins its development cycle at about the same time that hardware does. It is then no wonder that when we start our reliability activities, we may discover both hardware and software failures. It is therefore important to entertain the idea of how to manage software reliability and its growth within hardware development phases.

The Rayleigh model (Papoulis 1984) is a well-known software reliability growth prediction tool. Even though this model can predict improvements of software and its reliability during the development phase with some degree of accuracy, it does not have any means to account for customer use. Therefore, it cannot predict the final reliability of the system. In a way, it represents the reliability of the software development process.

In our opinion and experience, system reliability growth is better estimated using a formal system testing. Repairable system reliability growth models such as the basic Crow–AMSAA model represent reliability growth over different engineering builds, which include both hardware and software improvements. The best way to demonstrate this approach is through an example.

Example

Table 9.8 lists different functions required to be performed for a device with embedded software. The first prototype developed for system testing did not have all the functions readily

TABLE 9.8

List of Functions Inclusion and Test Cases Coverage for Reliability Growth Testing of a Device along Software and Hardware Prototype Builds

Functions (Test Cases)	Engineering Prototype I	Engineering Prototype II	Manufacturing Prototype	Manufacturing Build
Function 1	X	X	X	X
Function 2		X	X	X
Function 3	X	X	X	X
Function 4			X	X
Function 5	X	X	X	X
Function 6				X
Function 7	X	X	X	X
Function 8	X	X	X	X
Function 9	X	X	X	X
Function 10	X	X	X	X
Test Coverage Factor	70%	80%	90%	100%

available for testing. As the test progressed, software updates were developed and implemented on subsequent prototypes. With more tests conducted, more failures were identified.

Table 9.9 illustrates the summary of the four phases of testing across builds of our device using different test coverage. Notice that the count of runs is adjusted for the test coverage or test case addressed in each build. Due to a lack of software maturity during early design configurations only a portion of functions are exercised. For this reason, we will be monitoring the number of successful completions of performed cycles as opposed to test duration and times to failure. Each cycle will be interpreted as the run of system mission from beginning to end incorporating only the functions performed by the current design configuration. The cycle duration will differ from one configuration to the next as the design evolves and more functionalities are added. As will be shown, we will adjust the number of cycles at each configuration to the equivalent number of "full cycles." A full cycle is the cycle that contains all functionalities performed on the final design. In this application we will express the reliability of the system in terms of mean cycle between failures (MCBF). If needed, MCBF may be converted to time domain reliability metrics (MTBF) based on the expected mission time in the field usage. In Table 9.9, the functional coverage percentage is calculated by dividing the number of exercised functions at each tested configuration to the final total number of functions at the final configuration.

The same growth analysis conducted previously is applied to both the original and the adjusted number of cycles. Results in terms of MCBF growth and projection are calculated. This estimate does not take into account that early prototypes did not run the full set of test cases that the system would execute in the field. This leads to an overestimation of the number of failures found to the number of or missions in the field under normal usage conditions. To correct for this error, we need to adjust for the test case coverage by reducing the number of cycles at Prototype I by 30% to be ($33 \times 70\% = 23.1$ cycles); Prototype II by 20% to be ($92 \times 80\% = 73.6$ cycles); and Manufacturing Prototype by 10% to be ($115 \times 90\% = 104$ cycles). The adjusted number of cycles is then injected into the reliability growth model.

These numbers of cycles were injected into the reliability model equation to solve for the reliability index similar to the previous example as shown in Table 9.5. The reliability growth index, β, was found to be 0.81 and 0.66 for the original (not adjusted) and the adjusted cycle count data sets, respectively. A higher β is indicative of a lower reliability growth. So adjusting for functionality coverage would suggest higher growth in this case.

As shown in Figure 9.5, if cycle count is not adjusted, the projected timing for achieving the target would be a total of 1550 cycles. This includes the unadjusted 417 cycles, based on a reliability growth index of 0.81. This adjustment may require project management to plan for more resources or test time. As shown in the same figure, if the number of completed cycles is adjusted for the test case or function coverage and based on a growth index of 0.66, the target (MCBF) of 30 will be met after completing a total of 600 cycles. This includes the adjusted 378 cycles on the latest configuration.

System Reliability Demonstration Testing

As our design matures, and as its reliability improves and all design flaws are remediated, the final reliability of the system may need to be demonstrated prior to product launch. The objectives of final reliability demonstration at the system level are twofold. The first is to demonstrate the final reliability of the design prior to launch and identify the remaining system

TABLE 9.9

List of Functions Inclusion and Test Cases Coverage for Reliability Growth Testing of an Electromechanical Device along Software and Hardware Prototypes Builds

Functions (Test Cases)	Function 1	Function 2	Function 3	Function 4	Function 5	Function 6	Function 7	Function 8	Function 9	Function 10	Functional Coverage age%	(Original) Test Cycles	Adjusted Test Cycles	Number of Issues Found
Engineering Prototype I	X		X		X		X	X	X	X	70%	33	23.1	3
Engineering Prototype II	X	X	X		X		X	X	X	X	80%	92	73.6	6
Manufacturing Prototype	X	X	X	X	X		X	X	X	X	90%	115	104	7
Manufacturing Build	X	X	X	X	X	X	X	X	X	X	100%	177	177	6
Total												417	378	22

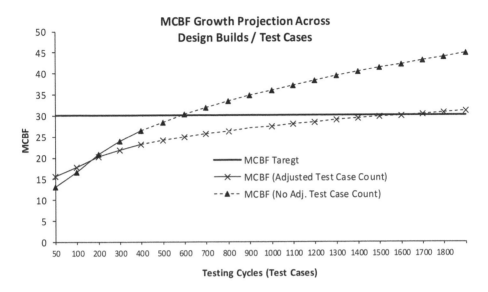

FIGURE 9.5
Repairable device reliability growth across different builds and test case coverage.

failure modes. The second objective is to provide objective evidence that the design will maintain its reliability over its life expectancy. Reliability growth testing (RGT), explained earlier, does not support these two objectives. RGT demonstrates that a design has matured and provides a projection as to whether the target reliability may be achieved. Furthermore, RGT does not provide information on the durability over the design life, neither does it provide information on all failure modes that will be experienced during the life of the product.

A common method to demonstrate reliability, called *sequential reliability testing*, tests for the minimum achievable MTBF. This approach may be used either to launch a product or to acquire a product from a supplier. A second approach is called time-truncated MTBF demonstration (Wald 1945; Epstein and Sobel 1955; MIL-HDBK-781A).

Needless to say, testing a system for reliability over its life expectancy needs to be accelerated to shorten the test duration. Unfortunately, system accelerated life testing is more complex than component testing. The reason is that a system is an assembly of modules and components, each with its own unique failure modes and physics of failure. Therefore, realistically, a single acceleration factor may not exist for a system. One solution may be to identify the critical modules in the system and ensure that they are aged appropriately to reflect their failure modes and rates.

Probability Ratio Sequential Testing

Probability ratio sequential testing (PRST) provides a verification plan for accepting, rejecting, or continuing testing a design in the shortest test time possible. The decision is made based on the relationship between the cumulative number of failures and cumulative testing time. This method does not provide an exact number for the MTBF. Rather, it provides information on whether the design is capable of meeting the MTBF requirements. A pitfall of this technique is that if the design is inherently borderline, the testing could continue indefinitely. For this reason, there may be a need to use test time truncation techniques that could prevent this.

While an in-depth discussion of PRST is beyond the scope of this work, let it suffice to say that it is a statistical technique was originally developed as a tool in quality control and inspection of a given production. In short, first, we develop two MTBF threshold values; one as an acceptance threshold and the second for rejecting the outcome. Then, we start testing our system and periodically calculating the system MTBF based on observed failures and testing time. Finally, we ask whether either of the two thresholds have been reached. On the basis of achieving either of the two thresholds, we terminate the test and accept or reject. If the resultant MTBF has not exceeded either of the two limits, the test continues.

Because PRST is a statistical tool, it involves two factors that we have not yet discussed. The first is called *producer's risk* and is denoted by α. It is the probability that a producer (i.e., manufacturer) rejecting a part that in reality is a good part. In other words, a manufacturer throws a good part away. In the context of reliability testing, producer's risk is the probability of rejecting an MTBF value as not having met requirements when in reality it has. The second is called *consumer's risk* and is typically demoted by β. It is defined as the probability of accepting an MTBF value as having met requirements when in reality it has not. In other words, a manufacturer keeps a good part and sells it to a customer.

Other PRST parameters are as follows:

1. Upper MTBF (denoted as MTBF_0) is the accept threshold with α probability if the true MTBF approaches MTBF_0.

2. Lower MTBF (denoted as MTBF_1) is the reject threshold with β probability if the true MTBF approaches MTBF_1.

3. Discrimination ratio (denoted as $d = \dfrac{\text{MTBF}_0}{\text{MTBF}_1}$).

The decision on whether the test shall continue or terminated is made according to the following formula:

$$\frac{\ln\left(\dfrac{\beta}{1-a}\right)}{\ln(d)} + \frac{\left(\dfrac{1}{\text{MTBF}_1} - \dfrac{1}{\text{MTBF}_0}\right)}{\ln(d)}(T) < r < \frac{\ln\left(\dfrac{(d+1)(1-\beta)}{2da}\right)}{\ln(d)} + \frac{\left(\dfrac{1}{\text{MTBF}_1} - \dfrac{1}{\text{MTBF}_0}\right)}{\ln(d)}(T)$$

where T is the time at which the failures are observed or the decision on the test is made, and r is the number of failures at the point of time T.

Case Study: PRST

A major original equipment manufacturer (OEM) identified an emerging market for a particular type of device. The business leaders capitalized on the opportunity by acquiring an existing device. The acquired product did not have any accurate reliability data and had never been used on a large scale in the identified market. The OEM enjoyed a well-regarded reputation in its main market for a similar device with an MTBF of 24 months. The general wisdom was that a similar reliability may not be needed in this new market and that an MTBF of 8 months would be sufficient. This was partly due to the lower cost of service labor in the emerging market leading to the affordability of more frequent service calls.

We designed a PRST to decide whether to accept the acquired product if the true MTBF is equal or greater than 8 months and possibly even approaching a desirable 24-month period. The rationale for taking this approach was to shorten test time and relax the restriction of setting a single-value MTBF requirement. The opportunity appeared promising and the business leadership was willing to accept some degree of risk with the acquisition. For this reason, we selected $\alpha = 0.05$ as the probability of rejecting a product with true MTBF equal to the upper MTBF of 24 months. Additionally, we chose $\beta = 0.1$ to reflect the probability of accepting a product that has a true MTBF of 8 months. The other parameters in PRST are

$$\text{Upper MTBF} = \text{MTBF}_0 = 24 \text{ months}$$

$$\text{Lower MTBF} = \text{MTBF}_1 = 8 \text{ months}$$

$$\text{Discrimination Ratio} = \frac{\text{MTBF}_0}{\text{MTBF}_1} = \frac{24}{8} = 3.0$$

By applying the accept/reject equation, we have

$$\frac{\ln\left(\dfrac{0.1}{1-0.05}\right)}{\ln(3)} + \frac{\left(\dfrac{1}{8} - \dfrac{1}{24}\right)}{\ln(3)}(T) < r < \frac{\ln\left(\dfrac{(3+1)(1-0.1)}{(2)(3)(0.05)}\right)}{\ln(3)} + \frac{\left(\dfrac{1}{8} - \dfrac{1}{24}\right)}{\ln(3)}(T)$$

Thus

$$-2.049 + 0.07583(T) < r < 2.26186 + 0.07583(T)$$

Table 9.10 and Figure 9.6 show the accept and reject boundaries for the designed PSRT plan. The test results showed that there were 5 failures observed at a cumulative test time of 100 months, which is less than the limit of 5.536 failures to accept at that time frame. Based on the test results, the leadership decided to execute the acquisition since the demonstrated MTBF is approaching 24 months at 90% confidence.

TABLE 9.10

PRST Test Plan with Actual Number of Failures from Test Results

Test Time (Months)	Accept	Reject	Test Results (Cumulative Failures)
0		2.26	0
10		3.02	1
20		3.78	1
30	0.226	4.54	2
40	0.985	5.30	3
50	1.743	6.05	3
60	2.502	6.81	5
70	3.261	7.57	5
80	4.019	8.33	5
90	4.778	9.09	5
100	5.536	9.85	5

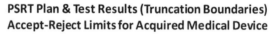

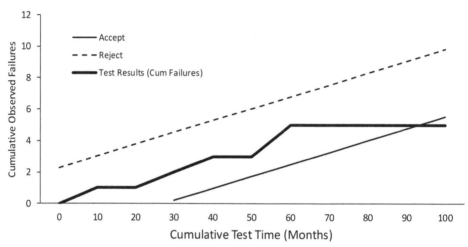

FIGURE 9.6
PRST test plan, accept and reject boundaries, and actual test results and decision.

A χ^2 Approach

We should note that this test could have continued indefinitely if more failures were observe. We need to develop a means of suspending the test should this be the issue. To do this we can employ the χ^2 distribution. The following relationship must hold in order to suspend the test:

$$\frac{\chi^2_{(1-a),2r}}{\chi^2_{\beta,2r}} \geq \frac{\text{MTBF}_1}{\text{MTBF}_0}$$

In this case study, the ratio is equal to 1/3. By looking up values in an chi-square distribution chart as shown in Table 9.11, we find that the ratio 1/3 is at

$$\frac{\chi^2_{(1-0.05),2r}}{\chi^2_{0.10,2r}} \geq \frac{8}{24} = \frac{1}{3}$$

TABLE 9.11

Sample Chi-Square (χ^2) Distribution for Multiple Degree of Freedom, r, $(1-\alpha)$

2r, Degree of Freedom	0.995	0.99	0.975	0.95=1−α	0.9	0.1=β
10	2.156	2.558	3.247	3.94	4.865	15.989
11	2.603	3.053	3.816	4.575	5.578	17.275
12	3.074	3.571	4.404	5.226	6.304	18.549
13	3.565	4.107	5.009	5.892	7.042	19.812
14	4.075	4.66	5.629	6.571	7.79	21.064
15	4.601	5.229	6.262	7.261	8.547	22.307
16	5.142	5.812	6.908	**7.962**	9.312	**23.542**

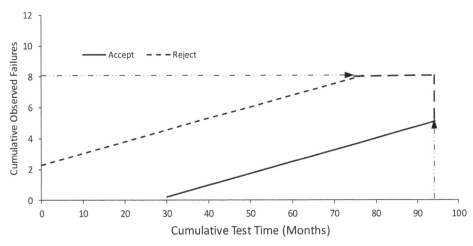

FIGURE 9.7
PRST test plan with truncation boundaries for testing.

If we were to solve this equation for different numbers of failures, we learn that the smallest number of failures that satisfy this equation is eight:

$$\frac{\chi^2_{(1-0.05),16}}{\chi^2_{0.10,16}} = \frac{7.962}{23.542} = 0.338 \geq \frac{1}{3}$$

The implication of eight failures means that once eight failures are observed, the test should be truncated without a need to make an accept or reject decision. This number of failures is then used to calculate the time at which the test is truncated as depicted in Figure 9.7:

$$T = \frac{MTBF_o \chi^2_{(1-a),2r}}{2}$$

$$T = \frac{(24)(7.962)}{2} = 95.54 \text{ Months}$$

As shown in Figure 9.7, the test plan depicted in Figure 9.6 is now modified to be terminated at eight failures or 95.54 hours without reaching an accept/reject decision. Under these circumstances, the question we need to entertain is whether we should accept the product. We determine acceptance or rejection on the basis of the proximity of the last point on the plot to either of the two boundaries. If the vertical position of the last point of the actual test results is closer to the accept line, the decision would be to accept. If the vertical position of the last point of the actual test results is closer to the reject line, the decision would be to reject.

PRST is a powerful tool for shortening a reliability test time; however, it does not provide an exact estimate of the true MTBF. Also, we need to keep in mind that the quality and accuracy of the decision is dependent on the setting of the discrimination ratio between the minimum and maximum MTBFs. Figure 9.8 shows the PRST accept and reject charts

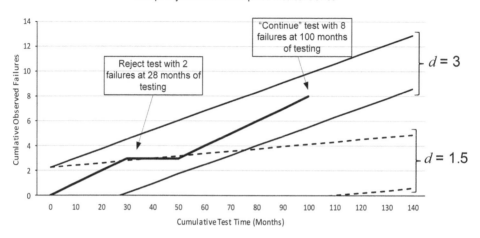

FIGURE 9.8
Different decisions for different discrimination ratios.

for two different test plans. As shown, the decision is to reject for a tighter discrimination ratio of 1.5, while it is to continue the test for the wider discrimination ratio of 3 for the same test results. The discrimination ratio is a business decision, and its selection should account for all business and technical risks.

Time-Truncated MTBF Demonstration Testing

PRST plans provide the lowest cost of testing, but due to the uncertainty of test suspension, they do not have a defined cost. Time and failure-truncated test plans have the advantage of known test cost and engineering resource needs. Based on an exponential distribution reliability model, we can calculate the test time required to demonstrate a target MTBF at a statistical confidence level:

$$T \leq \frac{\text{MTBF} \cdot \chi^2_{(CL, 2 \cdot r + 2)}}{2}$$

where T is the cumulative test time on all samples, MTBF is mean time between failures, CL is the confidence level, r is the number of failures (expected or allowed), χ^2 is the chi-square distribution factor. Even though χ^2 values may be easily retracted from Internet sources or from statistical software, in Table 9.12, for convenience, we have presented the testing time—as a multiple of the target MTBF—for a different number of failures and confidence levels.

Example

We are asked to design a test plan for demonstrating an MTBF of 100 hours for an engineering system that is assumed to have a constant failure rate over time. Reliability requirements indicate that MTBF should be tested at 90% statistical confidence. The design team has implemented design improvements during development and is confident that the

TABLE 9.12

Testing Time Required as Multiple of Target MTBF for Time-Terminated Testing Plan

Degree of Freedom = 2r + 2	Confidence Level					
	0.6	0.7	0.8	0.9	0.95	0.99
0	0.916	1.204	1.609	**2.303**	2.996	4.605
1	2.022	2.439	2.994	3.890	4.744	6.638
2	3.105	3.616	4.279	5.322	6.296	8.406
3	4.175	4.762	5.515	6.681	7.754	10.045
4	5.237	5.890	6.721	7.994	9.154	11.605
5	6.292	7.006	7.906	9.275	10.513	13.108
6	7.343	8.111	9.075	10.532	11.842	14.571
7	8.390	9.209	10.233	11.771	13.148	16.000
8	9.434	10.301	11.380	12.995	14.435	17.403
9	10.476	11.387	12.519	14.206	15.705	18.783
10	11.515	12.470	13.651	15.407	16.962	20.145

system is reliable and robust enough to demonstrate the MTBF without observing any failures in this final test.

Based on Table 9.12, the required testing time, T, is calculated as by

$$T = 100 \times 2.303$$

$$T = 230.3 \text{ Hours}$$

Unlike components and less complicated subsystems where we planned on success run reliability testing, at the system level, we prefer to observe failures during the demonstration testing. A lack of observed failures may lead to significant overestimation of the MTBF value.

To demonstrate the impact of the overestimation, consider the following example: Assume that we are testing to demonstrate a system MTBF of 12 months. We design and execute a "zero" failure test as explained earlier and terminate the test at 230.3 hours as calculated without observing any failures. The MTBF at 90% confidence is calculated by

$$\text{MTBF} \geq \frac{2T}{\chi^2_{(0.9,2)}}$$

$$\text{MTBF} = \frac{2(230.3)}{4.606} = 100 \text{ hours}$$

It is possible that we could have observed the first failure had we run the test for just one more hour, but we terminated the test. If we assume that the first failure does, in fact, happen in the next hour (i.e., test time of 231.3 hours) the new MTBF will be calculated as follows:

$$\text{MTBF} = \frac{2T}{\chi^2_{[0.9, \, 2(1)+2]}}$$

$$MTBF = \frac{2(231.3)}{7.78} = 59.46 \text{ hours}$$

This means that by missing the first failure in the test, the MTBF is overestimated by

$$\% \text{ Error in MTBF} = \frac{(100 - 59.46)}{59.46} = 0.678 \text{ or } 68\%$$

The implication of overestimated MTBF is that the manufacturer's planning for required resources to maintain the product in the field would be underestimated by 68%. Table 9.13 shows the worst-case error of estimating demonstrated MTBF when the test is terminated before an imminent additional failure would be observed in the next moment after test termination.

The error estimates in Table 9.13 are very conservative, as we assumed that the imminent failure is expected to occur in the next moment after test termination. Should the unobserved failure occur further out in testing, the calculated error would be less since the testing time is greater than listed in Table 9.13. For example, if the first failure occurs after 32 hours of test termination at 230.3 hours, then the true MTBF would be calculated as

$$MTBF = \frac{2T}{\chi^2_{(0.9, 2 \times 1 + 2)}}$$

$$MTBF = \frac{2(230.3 + 32)}{7.78} = 67 \text{ hours}$$

This means that by missing the first failure in the test, the MTBF is overestimated by

$$\% \text{ Error in MTBF} = \frac{(100 - 67)}{67} = 0.48 \text{ or } 48\%$$

TABLE 9.13

Error in Overestimating MTBF at Different Numbers of Observed Failures

Test Time (Hours)	Observed Failure	MTBF @ 90% Confidence (Hours)	Error in Overestimating MTBF
230.3	0	100	68%
231.3	1	59	36%
232.3	2	44	25%
233.3	3	35	19%
234.3	4	29	16%
235.3	5	25	13%
236.3	6	22	11%
237.3	7	20	10%
238.3	8	18	9%
239.3	9	17	8%
240.3	10	16	7%

General Considerations in System Reliability and Life Testing

System-level reliability testing is as much about the strategy of testing as it is about what techniques and equations we need to employ. What we present next is a discussion on optimization of the test plan in terms of proper sample size, acceleration factors for different failure modes in the system, and the number of failures found.

Accounting for Manufacturing Defects

System-level testing may not capture all expected field failures, especially those related to customer usage, environment disparity, and variation of the level of customer awareness. This is because in-house testing is often conducted by highly skilled technicians. The final report and analysis should account for those unseen failures and add the expected failure rate to that calculated from test data.

One of these unseen failures is out-of-box failures (OBFs) where, due to a manufacturing issue, a defective product is shipped. Typically, system-level reliability tests are not designed to detect manufacturing defects and lot-to-lot variations. This is partly due to a limited sample size and partly due to an absence of a full-scale manufacturing operation. Yet, we may be able to calculate the probability of detecting manufacturing defects in a given sample size.

For example, let us calculate this probability for a sample size of 10 units that may have a 1% manufacturing defect. We will then repeat the same calculation for a sample size of 100 samples.

To make this calculation, we use the binomial distribution. The probability of detecting only one bad unit in a population that has a 1% manufacturing defect ($p = 0.01$), in a test sample size of 10 or 100 ($n = 10$ or 100), is given by

$$Pr(x = 1) = \binom{n}{1} p^1 (1-p)^{(n-1)}$$

For $n = 10$, the probability of finding one defect in the test is

$$Pr(x = 1) = \frac{10!}{(10-1)!} 0.01^1 (0.99)^9 = 0.0913 \text{ or } 9.13\%$$

For $n = 100$, the probability of finding one defect in the test is

$$Pr(x = 1) = \frac{100!}{(100-1)!} 0.01^1 (0.99)^{99} = 0.3697 \text{ or } 36.97\%$$

Our final report should include this information.

Sample Size and Test Duration

System and subsystem samples are usually very expensive, and are often in short supply and needed for design verification and validation testing and marketing demonstrations. In addition, there is pressure to shorten the test period for demonstrating our system reliability. Sample size and test duration should be optimized so that an adequate number of issues are found to reflect the actual reliability of the system during the short time available for testing. There is often a myth that if we increase the sample size, we can reduce the test time. For example, let us consider a system with an MTBF of 60 hours and determine the chance of revealing its actual reliability if many samples are tested for 2 or 3 hours each.

The probability that any single system, with an MTBF of 60 hours, will have a random failure in 3 hours of testing is given by

$$F(t = 3) = 1 - e^{-3/60} = 0.048 = 4.8\%$$

Thus, when testing ten samples, the probability of finding any failures can be calculated:

$$F(x > 0) = 1 - \sum_{x=0}^{0} \binom{10}{0} 0.048^0 (1 - 0.048)^{(10-0)} = \binom{10}{0} 0.048^0 (1 - 0.048)^{10}$$

$$F(x > 0) = 1 - 0.611 = 0.389 \text{ or } 38.9\%,$$

Based on this equation, we have a 61.1% chance of seeing zero failures. This may be misleading. The flip side is testing fewer samples for a longer period of time. In a demonstration test, this approach may produce misleading results. The reason is that a smaller sample size will not expose the failure modes of all subsystems if we have a particularly good build or give erroneous results if we have a particularly bad build.

A mistake that some reliability owners make is to calculate the probability of failures of ten systems by adding the probability of failures of each single systems. For instance, in the previous example, they would calculate the probability that ten samples would produce failures as

$$F(x > 0) = 10 \times 0.048 = 0.48 \text{ or } 48\%$$

We believe that this is overestimating the probability of obtaining a failure in the test. This is because adding failure rates of the ten samples implies that they are associated with each other in operation. This would be similar to a system that is composed of ten components each with a probability of failure of 4.8%. Physically, this assumption is not valid when testing ten different samples of a system.

System Demonstration Test Plan: "Zero" versus "r" Failures

It is a common practice by manufacturers to design an MTBF demonstration test plan with "zero" failures. This is based on economical or compliance reasons in order to provide objective evidence of the reliability requirements. We believe that any system MTBF demonstration plan should include "r" failures—where r is at least three to four failures—to have a more accurate estimate of the true MTBF for reasons that we mentioned earlier.

We are only too aware that procuring enough test samples for reliability testing is a task to be closely managed. For this reason, when computing the sample size for system reliability demonstration testing we need to keep two main objectives in mind:

1. Optimize the sample size so that test costs do not become prohibitive. Also the sample size should not be so small that the test misses piece-to-piece variations.
2. The test should produce large enough failure counts to build confidence in the calculated MTBF.

Example

A manufacturer wants to demonstrate an MTBF of 12 months at 90% statistical confidence. The test team would stop the test with a pass result if the samples under test accumulate 27.6 months of testing with zero failures. Should they observe a failure, they would continue to 46.7 months if they do not observe a second failure. Should they observe a second failure, they would test to 63.9 months provided a third failure is not seen. Needless to say, the number of failures produced at any test depends on the number of systems under test and its duration.

Let us walk through some sample calculations. Assuming that system time to failure in general follows the exponential distribution, the probability of failure, $F(t)$, of any sample at test time t is

$$F(t) = 1 - e^{-t/\text{MTBF}}$$

For a 12-month MTBF system under test, the probability of failure of different sample sizes for different test setups and the required testing time T is shown in Table 9.14. The calculations are done as follows (note that r is the number of failures):

$$\text{MTBF} = \frac{2T}{\chi^2_{(0.9, 2r+2)}} \tag{9.2}$$

In Table 9.14, the probability of obtaining "r" number of failures is calculated using binomial distribution. For example, the probability of detecting more than two failures in the 2-failure test plan for a test executed on six samples is calculated as follows:

Reliability Requirements: MTBF = 12 Months to be tested @ 90% Statistical Confidence
Samples available for test: 6 samples
Testing time for two failures:

$$T = \frac{\text{MTBF} \; \chi^2_{(0.9, 2r+2)}}{2} = \frac{12 \; \chi^2_{(0.9, 2 \times 2+2)}}{2} = \frac{12 \times 10.532}{2} = 63.2 \text{ months}$$

Then, each sample runs on average for 63.2/6 = 10.6 months, i.e., the probability of failure of the average sample, assuming true MTBF is 12 months, is

$$F(10.6) = 1 - e^{-10.6/12} \approx 59\%$$

Thus the probability of obtaining two or less failures in the test plan is given by

$$Pr(r =< 2) = \sum_{x=0}^{2} \binom{6}{x} p^x (1-p)^{(10-x)}$$

where $p = 59\%$

$$Pr(x \leq 2) = \binom{6}{0} 0.59^0 (1-0.59)^{(10-0)} + \binom{6}{1} 0.59^1 (1-0.59)^{(10-1)} + \binom{6}{0} 0.59^2 (1-0.59)^{(10-2)}$$

$$Pr(x \leq 2) = 0.1933 \text{ or } 19.33\%$$

TABLE 9.14

Probability of Failure per Sample for Different MTBF Demonstration Test Plans for 12-Month MTBF Target @ 90% Statistical Confidence

Zero-Failure Test Plan — Total Test Time $T = 27.6$ Months

Probability of Failure, P_f per Sample at Time, T/n	Test Time/Sample, T/n	Sample Size, n	Probability of Zero Failure
0.90	27.6	1	90.0%
0.68	13.8	2	89.8%
0.54	9.2	3	90.3%
0.44	6.9	4	90.2%
0.37	5.5	5	90.1%
0.32	4.6	6	90.1%
0.28	3.9	7	90.0%
0.25	3.5	8	90.0%
0.23	3.1	9	90.5%
0.21	2.8	10	90.5%

1-Failure Test Plan — Total Test Time $T = 46.7$ Months

Probability of Failure, P_f per Sample at Time, T/n	Test Time/Sample, T/n	Sample Size, n	Probability of $r > 1$ Failure
0.86	23.3	2	74.0%
0.73	15.6	3	82.1%
0.62	11.7	4	84.3%
0.54	9.3	5	85.9%
0.48	7.8	6	87.1%
0.43	6.7	7	87.7%
0.39	5.8	8	88.3%
0.35	5.2	9	87.9%
0.32	4.7	10	87.9%

2-Failure Test Plan — Total Test Time $T = 63.2$ Months

Probability of Failure, P_f per Sample at Time, T/n	Test Time/Sample, T/n	Sample Size, n	Probability of $r > 2$ Failures
0.83	21.3	3	57.2%
0.74	16.0	4	72.1%
0.66	12.8	5	78.0%
0.59	10.6	6	80.7%
0.53	9.1	7	82.0%
0.49	8.0	8	84.2%
0.45	7.1	9	85.0%
0.41	6.4	10	84.8%

4-Failure Test Plan — Total Test Time T 92.4

Probability of Failure, P_f per Sample at Time, T/n	Test Time/Sample, T/n	Sample Size, n	Probability of $r > 4$ Failures
0.74	16.0	5	22.2%
0.67	13.4	6	35.8%
0.61	11.5	7	44.2%
0.57	10.0	8	52.4%
0.52	8.9	9	54.9%
0.49	8.0	10	59.8%

Sample Size Calculations for Service Life System Demonstration Testing

Assuming that system time between failure follows exponential distribution, the only difference here is that we will track the number of failures for an interval of 12 months, i.e., the time, T, in Equation 9.2 will be equal to $n \times 12$, where n is the sample size. Thus, the expected number of failures in 12 months of n system testing that observed r failures will be calculated as

$$ \mathrm{MTBF} = \frac{2T}{\chi^2_{(0.9, 2r+2)}} $$

For target MTBF of 36 months, how many systems should be included in a life test to demonstrate the MTBF consistently over 12 months with 90% statistical confidence, if the error of overestimating MTBF is not to exceed 15%?

As shown in Table 9.13, to minimize the error in MTBF to 16%, we should obtain four failures within 12 months of testing. From the preceding equation, and at a degree of freedom ($d = 2 \times 4 + 2 = 10$) and at $1 - a = 1 - 0.9 = 0.1$ in Table 9.11, χ^2 is

$$ \chi^2_{(0.9,\, 2 \times 4 + 2)} = 15.989 = \frac{2n12}{36} $$

Thus, $n \approx 24$

This means that we have to run 24 system samples for the life test in order to obtain four failures so that the demonstrated MTBF has a 90% confidence over its expected service life.

Repairable Systems Reliability over Service Life

Certain customers and end users may have an interest in knowing the expected service life (ESL) of their product. This is a question typically asked by those who have fleets of products. Their primary interest is knowing how long a manufacturer is willing to support servicing a product line once its production has come to an end. In the field of medical equipment, ESL is considered a safety concern and is mentioned in the third edition of IEC 60601-1 and later versions. In this standard ESL is defined as the "time period specified by the manufacturer during which the medical device is expected to remain safe for use (i.e. maintain basic safety and essential performance). Note: maintenance can be necessary during the expected service life" (IEC 60601 2012). It is implied in this mandate that the MTBF of the medical device is required to be stable and consistent over the ESL to ensure that the rate of occurrence of failure modes with safety-related consequences is maintained at a certain acceptable level.

When the customer is interested in the ESL, the mission of the design team and the reliability engineer is no longer to demonstrate a minimum MTBF before launching the product. Rather, it is to demonstrate that the design will survive a given ESL with a consistent failure rate or MTBF. The manufacturer can manage consistency of the MTBF by using service actions such as reactive or proactive measures.

Since the objective is to demonstrate a consistent MTBF over an expected service life, the mean cumulative function (MCF) approach is a suitable method to monitor stability of reliability over time for a repairable engineering system. The MCF is a plot using the non-homogeneous Poisson process that is valid to represent an increasing, constant, or decreasing failure rate of a system. The trend of the failure rate changes over time is indicated by a shape factor value that could be 1.0 for constant, greater than 1.0 for decreasing, or less than 1.0 for increasing failure rates over time. The model fits the relationship between the cumulative arrival time of failures versus the cumulative tome in testing.

MCF is a nonparametric statistical method that provides a simple method to present event counts, costs, and maintenance downtimes (an indicator of availability), among other reliability elements. This metric often follows a staircase curve with unequal step rises that denote the behavior of the product in the field. At any given age or time t, the corresponding distribution of the metric curves has a mean $M(t)$ that is denoted by MCF. In testing of n systems, the average MCF at any time, t, is the average of the MCF of individual systems calculated as follows:

$$M(t) = \frac{\sum_{i=1}^{N} M_i(t)}{n}$$

where $M_i(t)$ is the number of failures observed on sample i at time t.

Our objective is to demonstrate that the average number of r failures of the design over the expected service life remains below a certain threshold. This would satisfy both customer and business needs. Customers are satisfied because they do not need to deal with the headache of device failures; businesses are satisfied because they do not need to deal with the burden of the cost of service accrued on warranty contracts.

The parameter $M_i(t)$ is calculated using the ratio of the number of samples passed through the test at interval points of time, t, indicated either by failure occurrences on one of the systems or retirement of the system from the test as follows:

$$M_i(t_i) = \frac{1}{n_i} + M_i^*(t_{i-1})$$

where n_i is the number of samples under test at time interval t_i; and $n_i = n_{i-1}$, if t_i is the time at failure. However, $n_i = n_{i-1} - 1$, if t_i is the time retirement of the sample from testing (last time recorded for the sample at this test)

$$M_i^*(t_i) = \frac{1}{n_i}$$

Example

Repairable system life demonstration testing on four samples yielded the time to failures that required repair and service, as shown in Table 9.15. The test was conducted to demonstrate reliability over an expected service life of 16 years. It was terminated at 192 months on each system. One should note that at each stage of testing these four systems have not accumulated equal test time. The reason is that in a real environment a number of different events, such as operational checks or repair of failures, may take time that would not allow "equal" test time.

As can be seen in Table 9.15, the MCF calculations indicate that the expected number of failures at the service life of 192 months is 2.75 failures per unit.

System Accelerated Life Testing

System life testing is usually set and executed at an accelerated fashion to shorten test time and reduce test cost. System accelerated life testing (ALT) differs from that for components

TABLE 9.15

Time to Failure (in Months) of Individual Samples in System Life Testing and Calculations of MCF (Expected Number of Failures at Service Life of 192 Hours)

Test Data				MCF Calculations					
Sys 1	Sys 2	Sys 3	Sys 4	Sys ID	Age at Failure	State	Number of Samples at Test, r_i	$1/r_i$	MCF = M*(t_i)
4	5*	4	4	2	5	F	4	0.25	0.25
21	22	22*	22	3	22	F	4	0.25	0.50
39*	40	42	40	1	39	F	4	0.25	0.75
55	58	57	59*	4	59	F	4	0.25	1.00
67*	76	76.5	75	1	67	F	4	0.25	1.25
83	94	95*	93	3	95	F	4	0.25	1.50
99	112*	113	111	2	112	F	4	0.25	1.75
115	130	131	129*	4	129	F	4	0.25	2.00
131	148	150	147*	4	147	F	4	0.25	2.25
147	166*	168	165	2	166	F	4	0.25	2.50
163	184*	186	183	2	184	F	4	0.25	2.75
192**	192**	192**	192**	1	192	S	0		2.75
				4	192	S	0		2.75

* Failure at this time of this sample.

** Test termination time (sample retirement time).

in many ways, as it involves multiple accelerating techniques and stresses that are usually involved in accelerating the system life. Due to the complexity of engineering systems, different subsystems age at different rates during the same test. Additionally, system-level failure modes are usually *soft* as opposed to *hard* failures, by which we mean there is a degradation or instability of performance or output that occurs at much lower stress levels due to interactions between different nodules than the hard failures of components.

Acceleration factors (AFs) of system life testing may be achieved by increasing the duty cycle by reducing or eliminating idle time or dwell, or by increasing the speed of operations or the number of actuations. Other types of acceleration are increasing stress levels such as temperature, pressure, or operating voltage. We need to keep in mind that electromechanical products consist of different types of subsystems and components[7] with their unique physics of failure. These varying physics of failures present a challenge in calculating acceleration factors for the entire system. We are of the opinion that we need to calculate the acceleration factors for each module and subassembly separately. Table 9.16 shows example AFs for different types of modules in the accelerated life testing of a device (Atua 2019). It is recommended to calculate the individual acceleration factors on each of the impacted subsystems.

The acceleration factors, AF_i's, in Table 9.16 are defined as follows. Chapter 8 provides details of this information.

AF_1 is the ratio of the test duty cycle to the real-life duty cycle for each subsystem.

AF_2 is the acceleration factor associated with Arrhenius equation:

$$AF = \exp\left(-\frac{E_a}{K_b}\left(\frac{1}{T_{\text{acc.}}} - \frac{1}{T_{\text{normal}}}\right)\right)$$

TABLE 9.16

Example of Applicable Accelerant Stresses in a Medical Device System ALT

Subsystem Category	Duty Cycle	Temperature	Power Cycle	Flow Rate	Manual Actuation	Number of Cycle Missions	Cleaning
Electronic and electrical assemblies	AF_1	AF_2	AF_4			AF_7	
Pneumatics and pumping system	AF_1	AF_2		AF_5		AF_7	
Mechanical and moving parts	AF_1				AF_6	AF_7	
Enclosure	AF_1	AF_3			AF_6	AF_7	AF_8
User interface (keypad)	AF_1	AF_2			AF_6	AF_7	AF_8
User interface (screen)	AF_1	AF_2	AF_4			AF_7	AF_8
Heating system	AF_1	AF_2		AF_5		AF_7	

AF_3 represents aging of polymer-based materials; using the applicable acceleration factor formula (ASTM F1980-07 2011), we will have

$$AF_3 = Q_{10}^{(T_2 - T_1)/10}$$

AF_4 is the ratio of test power cycling to the real life power cycling.

AF_5 is the ratio of the test flow rate to the normal flow rate.

AF_6 is the ratio of the number of manual activation in the test to the normal number of manual activation.

AF_7 is the ratio of the number of cycles per day in the test to the number of cycles per day in typical usage.

AF_8 is the ratio of the number of cleaning cycles per day in the test to the number of cleaning cycles per day in typical usage.

One factor to keep in mind is the effectiveness of our system reliability demonstration test. We begin our journey on the reliability assessment and assurance of our new product by developing a baseline model. In other words, we have developed a theoretical understanding of what components would fail under what conditions. If we observe failures at a higher rate than expected and yet have confirmed their acceleration factors to be realistic and in line with their physics of failure, the target reliability may need to be revised. On the contrary, if the test produces unexpected failure modes, the test accelerants and/or the assumed physics of failure analyses need to be reviewed or revised.

We cannot emphasize enough that it is essential to a system-level ALT to accurately identify various acceleration factors for meaningful results. When designed properly, system ALT is a useful technique that eliminates the need for subsystem life demonstration testing by providing accurate estimates of life for each. This is because system ALT demonstrates the life of each subsystem while in actual interaction with other subsystems and components. This does not happen when each subsystem runs in isolation on a test bench.

System ALT can also be used to calibrate our testing method effectiveness by comparing the actual to the expected number of failures for each subsystem based on the associated acceleration factors. Major discrepancies between the test results and the reliability model should be resolved, often by updating the model in favor of test results. System ALT is also used to define the needed or required *preventive maintenance* for each subsystem based on time to failure and trend of failure occurrence over product expected service life.

Notes

1. While a detailed description is beyond the scope of this work, a configuration is a total sum of all the elements that define a system or subsystem. This includes both hardware and software. For more detail on configuration and its management, see Jamnia (2017).
2. This is true in theory but it may be different in practice. It is important, however, to understand the reliability and failure behavior of modules that may significantly impact system reliability.
3. These records of history are important in cases where reliability units were used for other verification activities prior to be used for reliability testing.
4. What we really mean is that cumulative MTBF versus cumulative operating time when plotted on a log–log scale chart has a linear relationship.
5. AMSAA stands for Army Material System Analysis Activity. NHPP stands for nonhomogeneous Poisson process.
6. In other words, a given failure is not dependent on previous failures or the ones that would happen in the future.
7. Examples of various modules are electronics and electrical assemblies, mechanical moving parts, and structural elements, fluidic or pneumatic circuits, plastic enclosures, or housings.

10

Reliability Outputs

Introduction

In this book, we have not discussed various product development review gates explicitly, even though we alluded to them in Chapter 3. These gates are mechanisms by which the design team reviews that tasks are completed in accordance with the goals of various design stages. Two major tasks within the design process are the *design input review* and the *design output review*. The former ensures that product-level requirements along with all design and development activities are completed, whereas the latter confirms that all design activities and associated documents are completed and approved.[1] Documentation and deliverables from reliability activities are no exception and should be prepared for the design output review.

Once reliability analysis and testing of components, subsystems, and systems are completed, the results are analyzed and captured in various reliability outputs. Figure 10.1, which summarizes design for reliability (DfR), provides an overview of these outputs (also known as deliverables). In this chapter, our goal is to provide an overview along with examples of these deliverables. Additionally, we will discuss how to combine the test results along with probabilistic failure rate calculations to develop what we term the *final reliability model*.

The final reliability model is an important document for two distinct reasons. First, based on the information that it contains, we can estimate not only the rate at which fielded products fail, but also which modules (or components) fail more frequently than others. For a repairable product, the service organization uses this data to place initial orders for its inventory. This is an essential input in logistic planning, particularly for large organizations with multiple service locations. The second reason for the importance of the final reliability model is that it provides thresholds for field monitoring of the product failures particularly at the module or component levels. As we have mentioned in Chapters 8 and 9, we can never replicate in full how a product might be used in the field, or how the specific environments of use may influence a product's reliability. These thresholds and monitoring at module levels, particularly if done at regional levels, will enhance our understanding of the product.

It may come as a surprise to some why we are including deliverable such as verification reports as a part of reliability deliverables. The reason is that on the one hand, the reliability and robustness domains encompass the full design. It is not appropriate to say that we have met our reliability goals but have missed some or all of our verification of requirements. On the other hand, any time any one person or function examines the product for any reason, what they have learned or experienced may be used to enhance the final reliability model.

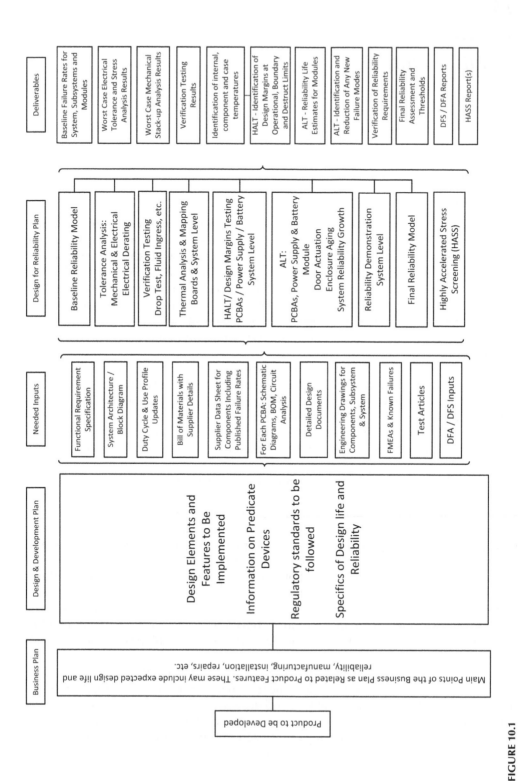

FIGURE 10.1
DfR process with typical outputs deliverables including HASS.

After the product launch and during production, we continue to ensure product reliability by monitoring and detecting manufacturing defects. One of the commonly used tools in production reliability screening is the highly accelerated stress screening test (HASS). We mentioned the HASS in Chapter 8, but in alignment with the product development process and as an output of DfR activities, it will be discussed here.

Reliability Outputs

Baseline Model and Its Application

In Chapter 4, we discussed reliability requirements, along with allocation, and feasibility analyses of these requirements. Although we did not provide detailed instructions, in Chapter 7 we provided examples of developing basic reliability models using databases and generic failure rates that may be available using MIL-HDBK-217 or commercially available software packages such as WindChill or ReliaSoft. Basic reliability data can also be obtained from each manufacturer from legacy data of similar products, or the predicate products, where carryover modules and components are used in the new design. This initial reliability model is used as an input to ask whether reliability goals were feasible. It may also be used to plan reliability testing should we identify areas of the design with little or no known reliability data, or areas of the design that seem to be clearly lower in reliability than other areas. Additionally, this baseline model may be used as a tool in design for serviceability by arranging repairable components in such a way as to enable short repair times.[2] The most important application of the baseline reliability model is in developing a reliability test design by providing data on the expected number of failures.

As explained in Chapter 4, when parts are selected early in the design phase, the baseline reliability model is used to ensure inherent basic reliability is aligned with total system reliability requirements or targets. The reliability model in this phase can also be used to compare and differentiate between different design concepts or different suppliers for the same parts, and provide a feasibility study for the design case, expected cost of service, etc. This prediction-based model provides a powerful tool to assess the expected reliability of the proposed design. Generally, the first iteration of this model is rather coarse and may not have a high degree of accuracy; however, as the product design matures and modules and component designs are refined, the model accuracy increases.

Reliability Model Case Study: Baseline Model

Table 10.1 and Figure 10.2 provide a simple predicted reliability model of a device with six subsystems. In this example, the system target reliability requirement is a failure rate of 4.8×10^{-5} per hour.

In Table 10.1, shows the baseline failure rates of the new design subsystems from legacy products and similar use applications by the manufacturer. The failure rates in Table 10.1 are listed in terms of a mean value and a standard distribution from actual usage data of these subsystems and components. All failure rates in Table 10.1 are provided in failure per hour. The failure rate apportionment of these subsystems is calculated as a percentage of the total mean failure rate of the system and is shown in Figure 10.2. Figure 10.2 represents the baseline model or failure rate budget that we think that the system failure frequency

TABLE 10.1

An Example of a Baseline Failure Rate for a Device with Six Subsystems

Subsystem	Mean Failure Rate	Standard Deviation of Failure Rate	Failure Rate Apportionment
Tray	5.60E–06	2.90E–06	11.67%
Power supply	1.22E–05	6.00E–06	25.42%
Control system	5.00E–07	5.00E–07	1.04%
Mechanical subsystem	2.42E–05	8.80E–06	50.42%
User interface	4.80E–06	2.10E–06	10.00%
Housing	7.00E–07	4.00E–07	1.46%
Total	4.80E–05	1.13E–05	100.00%

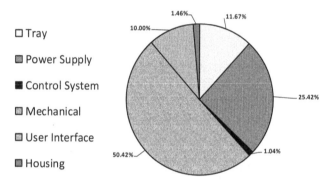

FIGURE 10.2
System baseline reliability model apportionment.

should follow at this point of the product development cycle. Figure 10.2 provides a pictorial representation of the apportionment table. In this example, the mechanical subsystem has the highest reliability budget. The last column in Table 10.1 provides the failure rate *apportionment*, which is the portion of the total reliability budget consumed by each module or subsystem. It is calculated by dividing the failure rate of each subsystem by the total failure rate. For instance, in the case of the disposable tray, the predicted failure rate is 0.0000056 and the total failure rate is 0.000048.

$$\text{Disposable Tray Apportionment} = \frac{0.0000056}{0.0000484} = 0.1167 \text{ or } 11.67\%$$

The standard deviation values in Table 10.1 are useful when failure rate distributions are assumed and the reliability model is updated with test data from the verification phase.

The baseline model as presented in Table 10.1 can be used to design the system reliability and life test based on the expected number of failures of each subsystem. Clearly, the test design should include the appropriate acceleration factor for each subsystem along with the appropriate sample sizes, as discussed in Chapter 9, in order to gain acceptable confidence in the results.

Design-Related Analyses

As explained in Chapter 7, reliability can be calculated and predicted using various techniques and tools such as tolerance and stress analyses. These tools are applied to critical design elements in the early design phase and provide quantitative assessment of the probability of occurrence of the threshold values of critical design parameters. The understanding of the behavior of critical design elements through reliability analysis may be used to make decisions on design changes and/or supplier selection for these components. Inputs of these activities can also be used to supplement the reliability prediction report for custom design components and key components with unique use, applications, or stresses in the new design.

Reliability Testing Output

The first category of reliability testing is qualitative testing (e.g., highly accelerated limit testing [HALT]). The outputs of these tests are the identification of design margins, operating and destruct limits, and their associated design improvements.

The second category of reliability testing is *quantitative* in nature. These are tests such as accelerated life testing, or mean time between failure (MTBF) demonstration testing. Quantitative test reports should provide objective evidence of the life expectancy and/ or reliability of the design. They may also include information that can be used for the determination of the preventive and proactive maintenance of components with a limited life expectancy.

Another major output of quantitative reliability tests is the update to the baseline reliability model to create the final model used to monitor performance in the field after product launch.

Reliability Model Case Study: Test Data

For the baseline model shown in Table 10.1 and Figure 10.2, the reliability tests executed before launch revealed three failures as shown in Table 10.2. The reliability tests executed lasted for a total of 1000 hours. The failures observed are allocated to the subsystem, along with the calculated failure rate in the test as shown in Table 10.2. All listed values are converted to failures per million hours.

Some manufacturers use the test data as the updated reliability model for their new design in lieu of the baseline model developed earlier. This approach could be effective if the design is new and the baseline model is either not available or there is low confidence

TABLE 10.2

An Example of Test Output Failure Rates for a Device with Six Subsystems

Subsystem	Number of Failures	Mean Failure Rate Failure per 10^6 Hour
Tray	1	0.001
Power supply	0	0.0
Control system	1	0.001
Mechanical	0	0.0
User interface	0	0.0
Housing	1	0.001
Total	3	0.003

in its validity. The pitfall of this is that, in most cases, the sample size and test duration is sufficient enough to produce both design-related and age-related failures on all subsystems as well as user-related and experience failures. Therefore, a better approach may be to merge reliability demonstration test data with predictions using data on the legacy products. In other words, to combine the test data with the baseline model. We refer to the combination of these two pieces of information as the final reliability model.

Final Reliability Model

The last reliability deliverable is the *final reliability model*, which is constructed once all reliability tests and evaluations are completed and reports developed. This model combines the reliability test data—typically failure rates—with the baseline reliability information. The premise of this hybrid and final model is that the baseline provides one set of idealized data and the test results provide another set of data. While the true reliability of the product is *not* represented by either of the two models, it is not too far either.

In the field of statistics, Thomas Bayes presented a theorem (Bayes and Price 1763) that explained an approach that today it is referred to as Bayesian statistics. This method seeks to refine a model by combining previously known information and data (albeit incomplete) with current information to predict a future state. The advantage of this method is that the final reliability model may be further improved using the same approach once the product is launched and field data becomes available.

Bayes' Theorem

In the Bayesian method, the failure rate is a random quantity with a probability distribution. The failure rates from the baseline model are used as the prior model, which is updated with additional information from the test data to produce a posterior distribution of the failure rate (Lee 2012). The gamma distribution is used to model the failure rate with a shape factor (α) and a scale parameter (β). Given the nominal failure rate (λ) and standard deviation of the failure rate (σ) from the baseline, the shape and scale parameters for the baseline model are calculated, respectively, as follows:

$$a = \left(\frac{\lambda}{\sigma} \right)^2 \tag{10.1}$$

and

$$\beta = \frac{\sigma^2}{\lambda} \tag{10.2}$$

Reliability test data for the design are combined with the gamma prior model to produce a gamma posterior distribution. The test data for each subsystem will be denoted here by r, number of failures, and T, the combined time on all units under test. We have

$$a' = a + r \tag{10.3}$$

and

$$\beta' = \beta + T \tag{10.4}$$

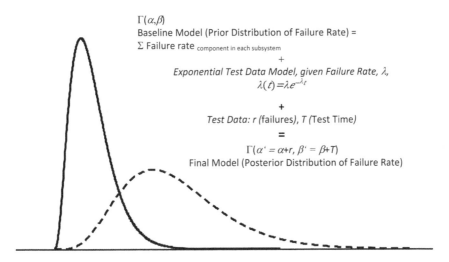

$\Gamma(\alpha,\beta)$
Baseline Model (Prior Distribution of Failure Rate) =
Σ Failure rate _{component in each subsystem}

+

Exponential Test Data Model, given Failure Rate, λ,
$$\lambda(t) = \lambda e^{-\lambda t}$$

+

Test Data: r (failures), T (Test Time)

=

$\Gamma(\alpha' = \alpha + r, \beta' = \beta + T)$
Final Model (Posterior Distribution of Failure Rate)

FIGURE 10.3
Prior gamma distribution process with test data using Bayes' theorem.

The updated mean value of the failure rate of the final reliability model is given by

$$\lambda' = \Gamma^{-1}\left(0.5, \alpha', \frac{1}{\beta'}\right) \tag{10.5}$$

Figure 10.3 illustrates the process of updating a prior gamma distribution with new reliability information to produce the posterior gamma distribution (i.e., the final reliability model).

Reliability Model Case Study: Final Reliability Model

We can now combine the baseline model failure rates (Table 10.1) and test data (Table 10.2). Table 10.3 shows the result of calculations of the gamma distribution parameters using Equations 10.1 and 10.2 for the failure rates presented in the baseline model.

The reliability test was conducted for 1000 hours and revealed only three failures on the tray, control system, and the housing subsystems. Table 10.4 shows the calculated parameters of the posterior gamma distribution using Equations 10.3 and 10.4. Keep in mind we need to use the associate failures of each subsystem along with the total test time. For example, for the tray subsystem, the number of failures is 1 and the test time is 1000 ($r = 1$, $T = 1000$). Therefore, the parameters of the posterior gamma distribution for the *final* failure rate λ' is given by

$$\alpha' = 3.74 + 1 = 4.47$$

and

$$\beta' = 668934 + 1000 = 669934$$

By using the mean and standard deviation of the disposable tray, the failure rate is calculated as

TABLE 10.3

Calculations of the Prior Distribution (Baseline Model) of the Predicted Values of the Failure Rates

Subsystem/System	Baseline Model Prior Gamma		Prior Gamma	
	Failure Rate, Mean	Failure Rate, Standard Deviation	Shape (α)	Scale (β)
Tray	5.6E–06	2.9E–06	3.74	668934
Power supply	1.2E–05	6E–06	4.09	334691
Control system	5E–07	5E–07	0.99	2179329
Mechanical	2.4E–05	8.8E–06	7.63	314695
User interface	4.8E–06	2.1E–06	5.12	1061102
Housing	7E–07	4E–07	3.31	4994563
Total			24.88	
System[a]	4.8E–05	1.1E–05	18.17	378518

[a] If the system is combined into one system and numbers are calculated for one entity, instead of six subsystems.

TABLE 10.4

An Example of Calculations of the Final Reliability Model (Posterior Distribution) of the Failure Rate of Subsystems after Combining Test Data with the Baseline Model

Subsystem/System	Test Results		Posterior Gamma		Final Model	
	Number of Fails	Test Hours	Shape (α')	Scale (β')	Mean Failure Rate	Percentage Failure Rate
Tray	1	1000	4.74	669934	6.57E–06	14.00%
Power supply	0	1000	4.09	335691	1.12E–05	23.87%
Control system	1	1000	1.99	2180329	7.65E–07	1.63%
Mechanical	0	1000	7.63	315695	2.31E–05	49.20%
User interface	0	1000	5.12	1062102	4.51E–06	9.61%
Housing	1	1000	4.31	4995563	7.96E–07	1.69%
Total	3	1000	27.88		4.69E–05	100%
System[a]	3	1000	21.17	379518	5.49E–05	100%

[a] If the system is combined into one system and numbers are calculated for one entity, instead of six subsystems.

$$\lambda' = \Gamma\left(0.5, 4.47, \frac{1}{669934}\right) = 6.57 \times 10^{-6}$$

Table 10.4 and Figure 10.4 show the final model with the identified failure rate budget for each subsystem based on the update of the baseline model and test data. Notice that in the final model the percentage contribution of the failure rate of each subsystem has shifted after incorporating the testing data. The final model is to be used to monitor the reliability performance of each subsystem in the field, and when any subsystem exceeds the assigned threshold, review and action to be taken as identified by the manufacturer.

Table 10.5 shows the calculations of the reliability thresholds to which the actual field reliability of the product will be monitored after launch. If the product reliability requirement is a failure rate of 4.7×10^{-5} per hour (equivalent to an MTBF of 36 months), the failure

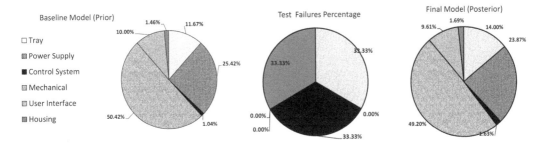

FIGURE 10.4
Final reliability model using prior distribute (baseline model) and test data.

TABLE 10.5

An Example of the Final Reliability Thresholds for a Device with Six Subsystems

Subsystem/System	Failure Rate Percentage	Failure Rate	Threshold MTBF
Tray	14.00%	0.003889	257
Power supply	23.87%	0.006629	151
Control system	1.63%	0.000453	2208
Mechanical	49.20%	0.013668	73
User interface	9.61%	0.002669	375
Housing	1.69%	0.000471	2125
Total	100%	0.027778	36

rate of each subsystem may be calculated by using its appropriate apportionment. For example, the final apportionment for the tray failure rate is 14% of the system. The tray MTBF threshold may be calculated as

$$\text{Allowable system failure rate} = \frac{1}{\text{MTBF}} = \frac{1}{36} = 0.0277$$

$$\text{Tray threshold failure rate} = 14\% \times 0.0277 = 0.0032$$

$$\text{Tray MTBF threshold (per month)} = \frac{1}{0.0032} = 312$$

This final model can be used to monitor reliability in the field upon launch. Upon collection of field data, we can once again update the final model to get more accurate representation of failures and apportionments.

Production Screening Testing

As mentioned earlier, HASS is a production quality screening that is designed to flush out manufacturing defects.[3] It takes place at elevated stresses based on the output of HALT. However, we need to be careful in its design so that it will not consume any more than 5% of the product's life. A poorly designed HASS detects little or no defects and will not help reduce potential field failures.

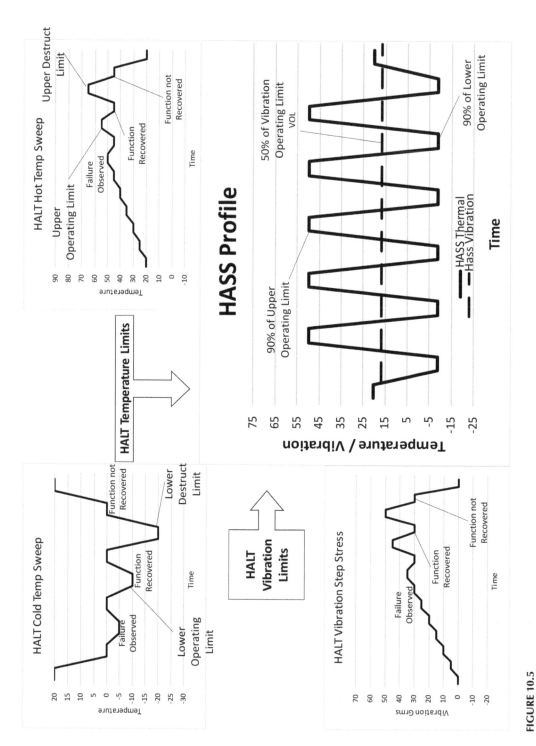

FIGURE 10.5
HASS test setup as an output of HALT results and findings.

Typically HASS runs for a few hours using a combination of thermal cycling and vibration. Figure 10.5 shows a typical HASS profile based on a previously run HALT. As a general rule of thumb, thermal limits are set at 80% of HALT temperatures (both cold and hot) where failures were observed. Vibration levels for HASS are usually set at 50% of the level of the first failure observed during the HALT vibration profile.

HASS may provoke hard and intermittent (soft) failures. Hard failures are indicative of major manufacturing shortcomings and should be investigated and resolved. Intermittent failures may have a variety of root causes and while they may need to be investigated initially, they are not as serious as hard failures. Decisions on treating soft failures are usually based on engineering judgment on the manner on which the product failed. We recommend executing HASS on 100% of the production for new products and production lines. By monitoring the number of defects found over time, we can reduce the number of samples tested to 10% or even 5% of the production.

To design a proper HASS profile, we should run the selected profile 20 or sometimes 30 times on 8 to 10 samples of the product in good working conditions. Then, we need to check their operation to make sure that they still perform within specifications. This is called proof of screen (POS) and is designed to ensure the effectiveness of HASS in detecting issues and defects without excessively removing the remaining life of the product after passing the test. The POS process can be summarized as follows:

- Repeat the HASS profile on 10 "known good" samples 15 to 30 times.
- If failures occur, reduce stress levels.
- If no failures are found, then HASS stress levels are accepted.
- Test "seeded" samples with defects to test HASS effectiveness of detection.

Figure 10.5 depicts the theory behind establishing the HASS profile, however, we need to verify its effectiveness. To do so, the following steps may be used:

- Run four or five cycles of HASS on 100% of the production.
- If failures are not detected in the first cycle, increase the stress levels at HASS.
- Repeat and rerun four to five cycles until all failures are detected in the first cycle.
- If no failures are found, then revisit HASS stresses based on HALT findings.

Impacts and outputs of HASS will be discussed in more detail in Chapter 11 as one of the fielded product reliability monitoring tools.

Preventive Maintenance

Preventive maintenance (PM) of engineering equipment is a strategy to manage the operation and durability of equipment while maintaining safety, cost effectiveness, and reliability over the expected service life of the product. For the old and traditional engineering equipment, preventive maintenance is generally based on activities for maintaining, inspecting, and testing described within an overall maintenance program. Often, traditional PM is based on age or time interval recommendations based on test data or monitoring of field failure trends.

Time- or age-based PM is designed based on cost–reliability trade-offs. The tighter the PM interval, the higher the reliability, but also the higher the cost of service. A commonly used formula to define the optimum PM interval that achieves both acceptable reliability and minimum cost of service is derived from the cost of unscheduled failures and the cost of scheduled service as follows (Glasser 1969):

$$C(t) = \frac{C_{PM}\, e^{-\left(\frac{t}{MTTF_{No\,PM}}\right)^{\beta}} + C_F\left(1 - e^{-\left(\frac{t}{MTTF_{No\,PM}}\right)^{\beta}}\right)}{\int_0^t e^{-\left(\frac{t}{MTTF_{No\,PM}}\right)^{\beta}}} \tag{10.6}$$

where $C(t)$ is the total cost per device per unit time at time (t). C_F is the cost of failure (unexpected/unscheduled replacement). C_{PM} is the cost of scheduled replacement (PM). t is the PM interval time. $MTTF_{No\,PM}$ is the mean time to failure (MTTF) if no PM is implemented at all. β is the Weibull shape factor.

In Equation 10.6, the optimum PM interval (t) is determined by calculating the average total cost per unit time. The total cost at any time, t, is the cost of planned replacement (PM) up to that time plus the cost of unplanned replacement (failure) up to that time divided by the average time to failure with that time t. The elements of this equation are defined as follows.

1. Cost of PM on the percentage of the population that survive up to PM interval t.

   a. $C_{PM}\, e^{-\left(\frac{t}{MTTF_{No\,PM}}\right)^{\beta}} = \text{Cost of PM} \times \text{Reliability at time } t$

2. Cost of repairing the percentage of the population that did not survive at PM interval t.

   a. $C_F\left(1 - e^{-\left(\frac{t}{MTTF_{No\,PM}}\right)^{\beta}}\right) = \text{Cost of failure} \times \text{Probability of failure at time } t$

3. The denominator is the average time to failure of the population at the time t.

   a. Average time to failure: $MTTF = \int_0^t e^{-\left(\frac{t}{MTTF_{No\,PM}}\right)^{\beta}}$

In the preceding equations, the shorter the PM interval, the lower the cost of failure, C_F, and the higher the cost of PM. Figure 10.6 illustrates the typical variation of total cost $C(t)$, as the PM interval increases for a component that has wear-out failure mode. Increasing the PM interval decreases the component reliability $R(t)$.

Case Study

A project manager wants to reduce the sustaining cost of a newly launched product without impairing the new product's quality, reliability, or the customer experience and satisfaction. Reliability testing had revealed that a certain pump in the fluidic circuit wears out prior to reaching the expected service life. The pump is used in three different locations in the circuits. The pump supplier recommended a proactive replacement of the pump every

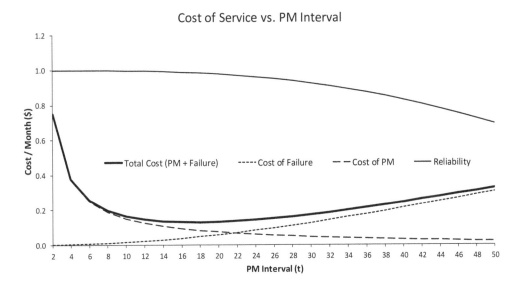

FIGURE 10.6

Reliability–cost trade-off analysis for different PM intervals.

12 months. It appears that the recommended PM interval is too short leading to excessive service costs. An accelerated life testing of the pump indicated that the pump has a time to failure following a Weibull distribution with a shape factor of 2.5 and a scale factor of 96 months. The cost of unscheduled service of this unit is $500 including expedited shipment of the device back and forth to the customer. The proactive replacement cost is $45 per device because scheduled maintenance may take advantage of bulk shipments in various regions. System reliability testing data indicated the device MTBF is 36 months when the pump PM is not implemented. By applying Equation 10.6, we have

$$C(12) = \frac{45\,e^{-\left(\frac{12}{96}\right)^{2.5}} + 500\left(1-e^{-\left(\frac{12}{96}\right)^{2.5}}\right)}{\int_0^{12} e^{-\left(\frac{12}{96}\right)^{2.5}}} = \$3.959 \,/\, \text{Device} \,/\, \text{Month}$$

$$C(32) = \frac{45\,e^{-\left(\frac{32}{96}\right)^{2.5}} + 500\left(1-e^{-\left(\frac{32}{96}\right)^{2.5}}\right)}{\int_0^{12} e^{-\left(\frac{32}{96}\right)^{2.5}}} = \$2.29 \,/\, \text{Device} \,/\, \text{Month}$$

$$C(36) = \frac{45\,e^{-\left(\frac{36}{96}\right)^{2.5}} + 500\left(1-e^{-\left(\frac{36}{96}\right)^{2.5}}\right)}{\int_0^{12} e^{-\left(\frac{36}{96}\right)^{2.5}}} = \$2.293 \,/\, \text{Device} \,/\, \text{Month}$$

Therefore, the optimum PM interval that reduces the total cost of service to a minimum is about 34 months:

$$C(34) = \frac{45\, e^{-\left(\frac{34}{96}\right)^{2.5}} + 500\left(1 - e^{-\left(\frac{34}{96}\right)^{2.5}}\right)}{\int_{0}^{12} e^{-\left(\frac{34}{96}\right)^{2.5}}} = \$2.286\,/\,\text{Device}\,/\,\text{Month}$$

Figure 10.7 shows the variation of the total cost of service with the PM interval for minimum total cost of service to be around 34 months, which reduces the reliability of the pump by 35% from 99.45% to 92.81%. This PM interval also reduces the annual total cost of pump service per device from the current $3.96 to $2.29.

In this case, the average MTTF of the pump after implementing the PM interval of 34 months will be calculated as follows:

$$MTTF_{PM} = \frac{t}{1 - e^{-\left(\frac{t}{MTTF_{No\,PM}}\right)^{\beta}}}$$

$$MTTF_{PM\,@\,34\,months} = \frac{34}{1 - e^{-\left(\frac{34}{96}\right)^{2.5}}}$$

Pump $MTTF$ = 473 Months

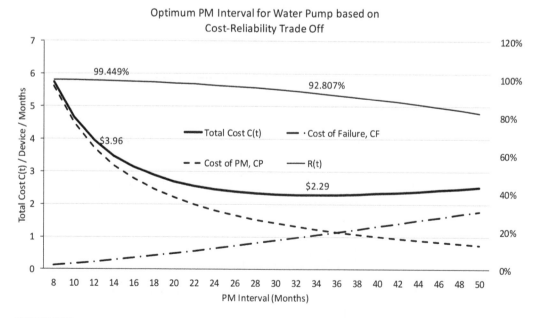

FIGURE 10.7
Optimum PM interval for the water pump based on reliability–cost-of-service trade-off.

The impact of the pump PM on the device MTBF without PM can be expressed as follows:

$$MTBF_{Device} = \frac{1}{Failure\ Rate_{Device}} = \frac{1}{\left(\dfrac{1}{MTTF_{Pump}} + Failure\ Rate_{Other\ System\ Parts}\right)}$$

From which,

$$36 = \frac{1}{\left(\dfrac{1}{96} + Failure\ Rate_{Other\ System\ Parts}\right)}$$

$$Failure\ Rate_{Other\ System\ Parts} = \frac{1}{36} - \frac{1}{96} = 0.017361\ per\ Month$$

The device MTBF after implementing the pump PM is calculated as

$$MTBF_{Device} = \frac{1}{\left(\dfrac{1}{473} + 0.017361\right)} = 51.34\ Months$$

Now, let us assume that the new product will be leased by the customer and may not be accessible for a 100% implementation of the PM at exactly 34 months. This means that PM will be implemented only when the product is returned for service when it is in need of repairs or for other reasons such as the end of the lease. The OEM is interested in assessing the impact of the new PM strategy on the device's overall reliability in the field since the pump is the single most influence on its unreliability.

The implementation of the pump PM will be opportunistic and based on the frequency of return of the device to the service facility. The device MTBF without any PM is averaging 36 months, which is close to the identified PM interval of 34 months. Recall that MTBF is the time when 63.2% of the population would experience failure and have to be returned to the service facility. This means that 63.2% of the pumps would be available for implementing the PM, while the remaining 36.8% of the pumps would be running in the field beyond the PM interval and following the original reliability curve of pump without the PM, which has an MTTF of 96 months. The pump MTTF when the PM is opportunistically implemented can be calculated as follows:

$$MTTF_{Pump\ with\ Opportunistic\ PM} = MTTF_{Pump\ @\ 100\%\ PM}(63.2\%) + MTTF_{Pump\ with\ No\ PM}(36.8\%)$$

$$= 473 \times 63.2\% + 96 \times 36.8\%$$

$$= 334\ Months$$

In turn, the device MTBF after implementing the pump PM, opportunistically, is calculated as

$$MTBF_{Device} = \frac{1}{\left(\dfrac{1}{334} + 0.017361\right)} = 49.12\ Months$$

Thus, implementing the opportunistic PM would increase the pump reliability and the leased device MTBF in the field from 36 to 49.12 months without taking any extra measures in service or logistics that requires shipping of the device in the field.

Predictive Maintenance

The maintenance strategy of newer and more modern equipment relies more readily on monitoring the performance of certain key design parameters or outputs, and in a way tries to *predict* when failures may occur and to prescribe *maintenance* activities. This strategy is often called predictive maintenance (PdM); these are a set of tools to evaluate the conditions of in-service products and to determine when to conduct necessary maintenance and proactive actions (see, for instance, Moubray 1992). This strategy is more cost effective than routine or time-based PM because it avoids replacing good components based on a preset time interval. Mainly, PdM depends on the actual condition of equipment, while PM is based on average or statistics of expected life. PdM is beyond the scope of this work and will not be discussed in more detail.

Notes

1. We are referring to the program in its entirety, but the same concept may also be applied for each integration stage within an Agile environment.
2. See Jamnia (2018) for a more detailed discussion of design for serviceability.
3. HASS may not be used to find either design or software flaws.

11

Sustaining Product Reliability

Introduction

Up to this point, our focus has been on new product development, and utilizing tools and techniques to develop a reliable and robust product. In Chapters 7, 8, and 9, we offered techniques and strategies for examining and ultimately testing the reliability and robustness of our product.

Field reliability data differs from test data in three main ways. First, the population is much larger. Second, the product usage is diverse. Third, the product is operated under a wide range of operating conditions and application scenarios. Test data usually represent small sample sizes that are typically from the same lot, which makes it unlikely to account for manufacturing process variations. Additionally, a small sample size does not allow for defects or failures with low occurrence rates to precipitate during the test. In contrast, field data reflects multiple failure modes caused by the following:

1. Inherent design flaws. The degree with which these types of failures are observed has a direct correlation to the level of rigor in the design and development process. The more robust a design, the lower the number of design flaws that are discovered in the field.

2. Manufacturing variations and quality control issues. These cause trigger fluctuations and spikes in failure rates unseen in the limited reliability testing prior to launch.

3. Service effectiveness and quality control issues for repairable devices. Generally, repeat failures are triggered by misdiagnosis of the actual causes of field failures. Root cause misdiagnosis is often due to pressure from customers to reduce the downtime of the product in the field.

4. User-error-induced failures. These failures are mainly seen in the field due to the fact that field usage is more diverse and stressful than in-house testing. Test engineers or "testers" are usually trained to follow certain protocols in a very limited test-case environment.

5. Random or wear-out failures or duty cycle failures. In the field population, usage and duty cycles vary. This impacts component wear-out and their rates drastically, leading to failure modes and mechanisms not seen in limited in-house testing.

In Chapter 10, we offered a means of developing a hybrid or final model for the reliability of a product. We suggested that the final reliability model is suitable for monitoring and trending the behavior of a newly developed product in the field. The service, quality, and reliability teams monitor the reliability of the fielded product and benchmark it against a set of acceptable failure rates or thresholds offered in the final reliability model. Needless to say, this model should be updated regularly based on feedback from the field data under actual usage over the next several years post launch in order to reflect that actual reliability of the product. The objective of monitoring field reliability is to enable the manufacturer to address and manage any economic concerns due to excessive failures, or the possible legal ramifications of harm to the end user due to an unforeseen failure.

Field reliability monitoring is a cross-functional responsibility of the service, quality, and reliability teams. Service or repair data is tracked and analyzed for identifying failures and is used in conjunction with sales data for reliability metrics calculation. Organizations should monitor the product reliability at both system and subsystem or module levels. A product may meet or exceed its reliability thresholds at the system level; however, certain modules or subsystems may be failing at a higher than expected rate. The ramifications of higher-than-expected failure rates could be excessive cost of service or an increased risk to the end user.

Earlier, we discussed the characteristics of field versus test data. Another major difference between the two sets is that test data is clean and field data is often noisy. What we mean by clean and noisy is this. Under test conditions, the population, test duration, and times of failures are well known. On the basis of this information, we can make rather strong judgments about the outcome of our studies. The field data does not have this clarity for a number of reasons. One is that time to failure is not accurate. There is often a lag between when failures occur, when they are reported, and when the repair takes place. Another reason is that the exact population of devices in the field and when they were deployed in the field are often estimated. For instance, an organization sells 6000 units over a span of 3 months. The sales figures are updated in a database on a quarterly basis. These units are shipped at any given time during these 3 months; however, they may be sitting either on a distributor's inventory or a customer's shelf without being used. Now, imagine that service begins receiving complaints and units to repair during each month. Some units arrive at the beginning, some in the middle of the month, and yet others toward the end of the month. The repair of this last batch may spill over to the next month. Now, remember that simply put, the failure rate is calculated as the ratio of failed units to the entire population. How can we make an "exact calculation"? It is nearly impossible from a practical point of view. Yet, somehow, we need to go on with monitoring our product.

The goal of this chapter is twofold. First, we will review data analysis methods used in situations where the field data is noisy, ambiguous, and incomplete. We will discuss sources of noise in fielded product data and recommended tools and techniques to interpret this data. Once the noise is filtered out of the data and its analysis provides clear evidence of undesirable trends, we need to begin a systematic understanding of root causes of these trends through a proper failure analysis process.

Hence, the second goal of this chapter is to provide a systematic tool and approach for failure analysis. When tracking reliability and field failures, we should strive to have comprehensive data on complaints reported by the customer as well as an in-depth review and analysis of the failed units. A systematic process of investigating reliability issues and failures starts by analyzing customer complaints, continues to root cause analysis, and is followed up until corrective actions are identified and implemented. The last step in this approach is to implement a control plan to sustain product reliability throughout the life of the product.

Field-Related Sources of Noise

As mentioned earlier, product field reliability data may be noisy, ambiguous, and unclear. These data characteristics complicate field monitoring and trending particularly during failure investigations when thresholds are violated.

While we cannot hope to cover all sources and causes of noise and ambiguity here in this section, our goal is to present examples and solutions to situations where field reliability data is incomplete or includes "unknowns." Needless to say, if the field data contains accurate accounts of time to failure, number of failures, and number of and age of surviving units, the data analysis is straightforward.

The approach to analyzing noisy data is similar; however, depending on the causes of the noise and ambiguities, data behavior will be different. Fortunately, many of these "misbehaviors" are known and can be explained. In general, root causes of the noise may be attributed to the following areas:

1. Misalignment of field data with the reliability model
2. Mismatch between field data and fitted distributions
3. Use profile discrepancies
4. Manufacturing process variation
5. Mixed failure modes

Let us explore each area in more detail.

Reliability Model and Field Data Misalignment

A major type of noise in field data is manifested in the misalignment between the system reliability and the reliability model and allocation for subsystems and lower-level components. Fielded product reliability should be tracked in terms of the quarterly or 12-rolling-month average of system mean time between failures (MTBF), in terms of the top ten failures in the field, or both metrics. Although system reliability may be the only requirements to fulfill, tracking the failure rate of the lower-level systems and components help the manufacturer identify opportunities for improving reliability and defining the scope of investigation if system reliability is not met. It is preferred to track replaced component part numbers in the field, especially if these parts are identified as the root cause of the customer complaint or the system failure. Some manufacturers track the Pareto of the top failures. It usually depicts 80% of the total field failure counts generally by either top 10 or top 12 failed components.[1] An example is provided in Figure 11.1.

The problem with tracking system reliability without a reliability model is that we would not know how to prioritize the focus of investigating issues in the field based on the Pareto alone. Often, we do not know whether the failure rate of the top component (in Figure 11.1 it is the motor) is acceptable or within the limitation of the inherent motor reliability. It could be that the motor is within its acceptable reliability, and that the second or third elements on the Pareto chart are the ones violating their acceptable failure rates. Reliability modeling and allocation of failure rates on lower-level subsystems or components is an important factor in guiding us to prioritize our efforts to investigate issues and identify their root causes, leading to improvements in the reliability of fielded products.

There are many situations in the field where more than one part is replaced by service to fix the problem (shotgunning). In other situations, a wrong part is replaced (victim part). In case of intermittent failures, the diagnosis may be that no problem was found. Service effectiveness and field service crew training effectiveness drastically impact reliability metrics in terms of repeat visits or service events due to ineffective repairs, and replacing victim parts due to improper troubleshooting and diagnosis. The reliability and service teams have to filter the field data to align the number of units that experience failures with the number of parts replaced under each subsystem, including victim parts. Failure analysis of

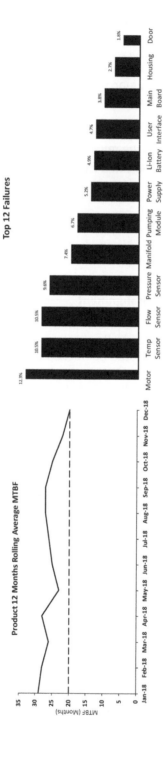

FIGURE 11.1

Fielded product reliability monitoring: MTBF and top 10 failures Pareto.

returned material from the field has to be conducted to define the actual failure rate versus the replacement rate. Another common cause of noise in failed component data is ancillary failures, that is, parts found to require replacement due to degradation or not meeting operating specifications while repairing the complaint-related issue. Maintaining a database of service records and activities helps to identify the root causes of the complaints and to distinguish between primary failure repairs as opposed to ancillary failure repairs.

An alternative means of tracking primary versus ancillary failures is to define two reliability metrics for fielded products, namely, failure rates, and service calls or event rates. The first is used to focus on reliability thresholds and design improvement, while the latter can be used to track and increase customer satisfaction via reducing cost of service, increasing uptime, productivity, improving service effectiveness, and enhancing troubleshooting guides.

Another source of noise in data is the configuration of service or repair kits. Often, the designation of subsystems and modules is not congruent with subassemblies defined or used in either manufacturing or service. Figure 11.2 shows an example of a telecommunication system with a number of SRUs. The printed circuit board assembly (PCBA) *service replaceable units* (SRUs) have functionalities that correspond to different subsystems. For example, the power subsystem resides on parts of three different boards. When a PCBA fails and is replaced, it is not often that we would know with certainty which subsystem present on the card has failed. Service technicians often guess—based on experience—which subsystem to log the failure against causing noise in the reliability data. To eliminate this source of noise, we recommend creation of reliability models based on the breakdown of SRUs rather than on the functionality or design of the system.

Mismatch of Field Data and Selected Reliability Model

Understanding the behavior of product reliability helps identify actions needed to retain or recover the original reliability. A key point in data analysis is to model the data behavior over time by finding a proper mathematical expression. Once this expression is found, it may be used to project future failure rates over the life of the product. Figure 11.3 provides the pictorial view of the steps from collecting field data to generating a model and finally

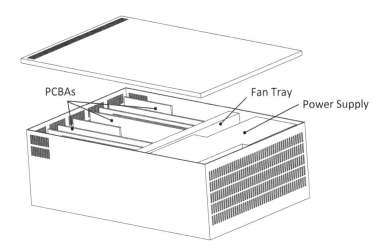

FIGURE 11.2
Electronic PCBA as SRUs in an avionics product.

to predicting future failure rates. This figure depicts fitting a sample data of 120 failures in terms of time to failure to the proper distribution (Weibull). It shows the Weibull plot on a log–log scale with a correlation coefficient of 98.5%. It is crucial to fit the data to a proper distribution model in order to have an accurate diagnosis of the failure trend as well as the projection of the future behavior of the product.

The task of identifying the actual time to failure for fielded products has its complications. Often, there are delays in recording service events or there is ambiguity regarding incident date versus repair date. Another hindrance is selecting the right distribution model to represent the data. These obstacles cause misbehaviors of modeled data. A sign that a model represents data accurately is when the cumulative probability distribution of the data points would align with a straight line when plotted on a log–log scale. The least squares fit of this line yields estimates for the shape and scale parameters of the distribution.

There are many methods of checking goodness of fit of field data to the selected distribution model. One of the methods used when plotting a Weibull distribution is checking the correlation coefficient of the log–log scale chart selected to fit the time to failure on the Weibull chart. Tarum (1999) derived the critical value of the correlation coefficient by applying Monte Carlo

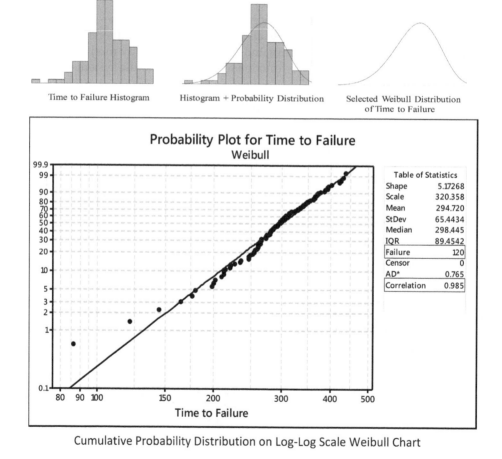

FIGURE 11.3
Fitting time to failure histogram to a continuous distribution model (for field data of 120 failures).

simulation to analyze thousands of data sets for Weibull and lognormal distributions. He recommended that a strong relationship exists between the distribution goodness of fit and the size of the failure set (or sample size). In other words, the correlation coefficient calculated for the fitted distribution should not be less than a critical value for the number of failure points analyzed. Equations 11.1 and 11.2 reflect the outcome of his work, where the correlation coefficient is a function of the number of failures only regardless of the total population size.

Two-parameter Weibull:

$$r^2 = 1 - e^{-(n/0.0399)^{0.177}} \tag{11.1}$$

And for lognormal:

$$r^2 = 1 - e^{-(n/0.0656)^{0.202}} \tag{11.2}$$

where n is the number of failures plotted and r^2 is the *critical correlation coefficient*. The closer the value of r to 1.0, the more accurate is the selected model in representing the data set. Figure 11.4 provides a pictorial representation of Equations 11.1 and 11.2 based on Tarum's analysis.

Example

The Weibull chart in Figure 11.3 shows that the correlation coefficient between a data set of 120 points of time to failure and the selected Weibull distribution to fit these points is 98.5%. Comparing this value to the optimum correlation coefficient from Tarum's analysis in Figure 11.4 for 120 data points, or by applying Equation 11.1, we get

$$r^2 = 1 - e^{-(120/0.0399)^{0.177}}$$

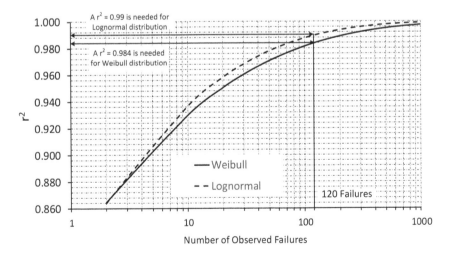

FIGURE 11.4
Critical values of correlation coefficient for reliability data. (Adopted from 104C.D. Tarum, Determination of the critical correlation coefficient to establish a good fit for Weibull and log-normal failure distribution, SAE Paper 1999-01–057, Detroit, March 1999.)

The optimum correlation coefficient should be no less than

$$r^2 = 98.38\%$$

The fitted Weibull distribution in Figure 11.3 has a higher correlation coefficient (98.5%) than the optimum value based on Tarum's analysis (98.39%), indicating that the selected Weibull distribution is a good fit for this data set.

If this data set—for reasons such as a specific failure mechanism or a combination of physical properties—was fit to a lognormal distribution, then based on Figure 11.4 at Tarum's analysis, the correlation coefficient should be no less than 99% for 120 points.

Other Considerations When Fitting a Distribution to Field Data

There are other rules of thumb to keep in mind when selecting a distribution model to fit failure data and calculating reliability metrics.[2]

- When the number of failures is 20 or less, the two-parameter Weibull distribution is to be chosen for the analysis (Wenham 1997).

- When fitting large-size reliability data (greater than 500 failures), use the maximum likelihood estimation (MLE) method. MLE is a method that computes the maximum likelihood estimates for the parameters of a statistical distribution. The mathematics involved in the MLE method is beyond the scope of this book, but discussed in more detail in other works such as Aldrich (1997).

- In general, when fitting less than 500 data points (failures), rank regression is more accurate (Liu 1997).

Use Profile Discrepancies

Compared to the number of samples used for in-house reliability testing, the field population is vastly larger. Operating environments—likely spanning over different geographical locations—and their associated stresses along with varying use profiles are key contributing factors to the rate of occurrence of failure modes. These factors act as accelerants of failure mechanisms. In addition, a large numbers of users of the same product—by necessity—have different use profiles with varying duty cycles. Hence, it is logical to expect that some subpopulations of product may wear out faster than others.

As an example, consider Figure 11.5 depicting a Weibull distribution of a certain failure mechanism of a motor in a fielded product. We note that there are significant changes in the trend of the cumulative failure plot. These changes occur around 100 and 500 weeks. We need to identify the duty cycle distribution between various customers in the field. Some customers have high duty cycles that consume the motor life in a very short period of time. The failure trend for the high usage customers occurs around 100 weeks with an increasing trend. Customers with low duty cycles experience an increasing failure rate trend after 500 weeks. These two different customer populations are depicted in Figure 11.6 at the upper and lower tail of the usage profile distribution.

Another element that impacts use profile along with product age is the fact that products are slowly released into the market. The implication is that at a given month, the service organization may need to repair products that are a few months old to a few years old, each used with varying duty cycles by customers. Different age groups may exhibit

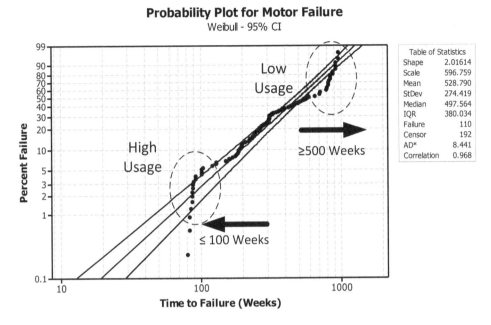

FIGURE 11.5
Typical field data for motor failures with different duty cycles/usage.

different failure rates of the same failure modes. They may also reveal different failure modes. Older products exhibit higher wear-out failures than newer products, which may have different types of manufacturing defects. In analyzing these types of data, we need to exercise extreme caution so as to not develop the wrong conclusions.

We recommend that before deriving any conclusions from a probability-of-failure plot, we should break down the population into usage-based subpopulations and create probability charts for each. For the example shown in Figure 11.5, we segregated the high- and low-usage populations first, and then proceeded in applying the Weibull distribution. Figure 11.7 shows the probability of failure distribution of both populations. The difference in the behavior is clearly obvious; there is a steeper Weibull distribution shape factor for the high-usage population (a shape factor of 21.4 as compared to 2.6).

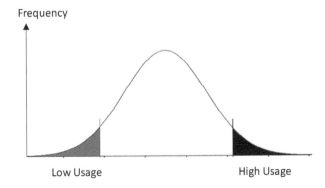

FIGURE 11.6
Typical distribution of different duty cycles/usage for fielded product.

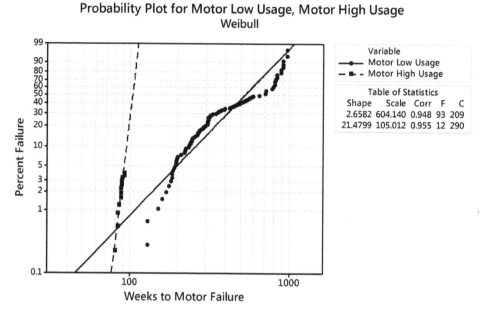

FIGURE 11.7
Probability of failure analysis for motor failures with different duty cycles/usage.

We can look at this particular example from another angle. Motor life may be expressed in terms of time or it may be evaluated in terms of revolutions and cycles. If fact, in similar situations where the life may be expressed in terms of different independent variables such as time, cycles, or revolutions, we can convert the measured field failure time into the other independent variable (such as cycles). We can then establish the probability of failure and derive any conclusion about the data and its behavior.

Manufacturing Quality Variations

As we mentioned, fielded products do not have a uniform age. What this means is that products were produced at different times, possibly at different manufacturing locations and personnel. Therefore, they are subject to manufacturing process quality variations. Some may even have low quality leading to irregularities in rate of occurrence of certain failure modes.

As an example, Figure 11.8 shows a distribution of time to failures of a fielded motor. We notice a bimodal behavior in this graph as indicated by an upward bent around the 100 days mark. This behavior may be due to many factors. First, we may have a mix of two different failure mechanisms. We discussed this with the manufacturer who claimed that all motors in the field have the same failure mode and mechanism. The second cause of a bimodal behavior may be a bad lot or batch in manufacturing. Upon examining the manufacturing dates of failed units, we learned that the failed units within the first 100 days were from the same manufacturing batch. A collaborative investigation with the motor suppliers revealed that bearing alignments in this lot were out of specifications. This manufacturing defect caused motor bearings to wear out faster leading to an excessive wear-out trend.

We separated the data into two groups: one for failures that occurred within 100 days (bad batch), and those that occurred beyond. The results are presented in Figure 11.9.

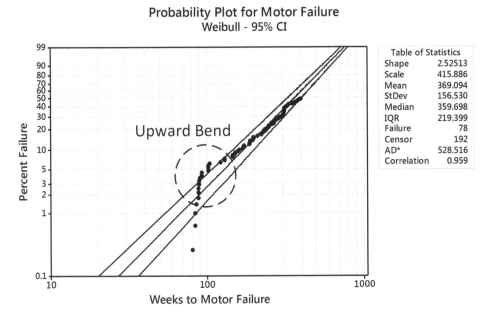

FIGURE 11.8
Typical field data for motor bearing failures with bad batch.

We observed that both populations are experiencing a wear-out failure mode, but at different rates and trends. Figure 11.9 points out that the bad batch failure is occurring at a more severe trend, indicated by a steeper slope and a corresponding shape factor (β) of 13. In contrast, the "good batches" have a smaller slope and a shape factor (β) of 3.83, indicative of the actual reliability of this component.

We recommended checking the serial numbers and the lot numbers of parts replaced in the field when conducting the investigation. This will provide insight into whether severe lot-to-lot variations may exist.

Figure 11.9 reveals another factor worth noting. The failure modes trends curve upward. This may indicate presence of a failure-free period. It is important to realize this because it will influence our choice of model. For instance, a two-parameter Weibull will not reflect a failure-free period accurately; however, a three-parameter Weibull distribution does (Abernethy 2006). We emphasize that we can use our engineering knowledge of a given failure mechanism to ask whether failure-free periods should be expected and use an appropriate model. Clearly, selection of the correct distribution model does influence our conclusions.

As an example, Figure 11.10 shows the probability distribution of the motor using the three-parameter Weibull distribution.[3] This distribution shows that there is a period of 107.135 days for the "good lots" and 78.99 days for the "bad lot" when no failures are present. Now, let's examine the shape factors for the two distributions. For the bad lots, the shape factor is 1.11. This is very close to being a random occurrence ($\beta = 1$), though occurring only after 79 days. We note that the data is better fitted to the three-parameter Weibull distribution with a correlation coefficient of 96.9% instead of 85.6%, as was previously indicated (Figure 11.9). Similarly, the good lots show a better fit with a correlation coefficient of 99.6% (compared to 97.6%) and a lower shape factor (1.68 compared to 3.83). More details and examples on this technique are discussed in Abernethy (2006). This information is

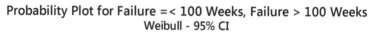

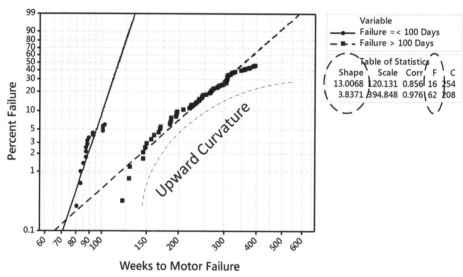

FIGURE 11.9
Probability of failure plots after separating the bad batch data set.

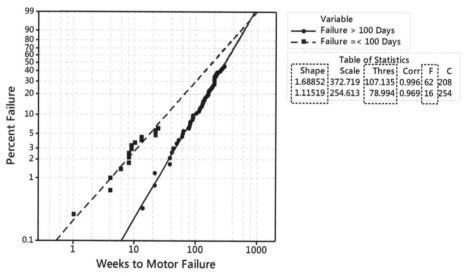

FIGURE 11.10
Probability of failure of the compressor using three-parameter Weibull.

very useful when identifying a proactive replacement (preventive maintenance) in the field, as explained in Chapter 10.

Mixed-Failure Mechanisms

In the previous example, we discussed how to discover and analyze the distribution of a bad manufacturing lot. In that example, the same failure mechanism was common between the two groups of data. However, it is possible to receive failure data containing two or more failure mechanisms. This is a common pitfall in reliability analysis of fielded data when mixed failure mechanisms are not separated. This mistake, if not discovered, may lead to wrong conclusions. We need to keep in mind that Weibull plots (and indeed, any fitted distribution) should be used to represent a single-failure mechanism, not failure modes.[4] Figure 11.11 gives an example of mixed failure mechanisms of a ball bearing installed on a compressor. Mixed-failure mechanisms are usually manifested by a downward curve on the Weibull plot. Another common manifestation of two-failure mechanisms is the presence of a shallow trend of point clusters followed by a steeper trend on the point plot on the Weibull chart. For example, the failure analysis of returned ball bearings revealed that the inner wall of the outer ring of some of the ball bearings was chipped due to the presence of material nonhomogeneities due to manufacturing defects. Figure 11.12 depicts this failure mechanism.

To understand this behavior better, we separated the mixed data and created groups of distributions corresponding to different failure mechanisms. As shown in Figure 11.13, failures occurring up to 200 or 300 hours are due to quality issues that cause these components to fail early in life with a Weibull shape factor of 0.336, while failures occurring after 500 hours represent worn out components with a Weibull shape factor of 4.67.

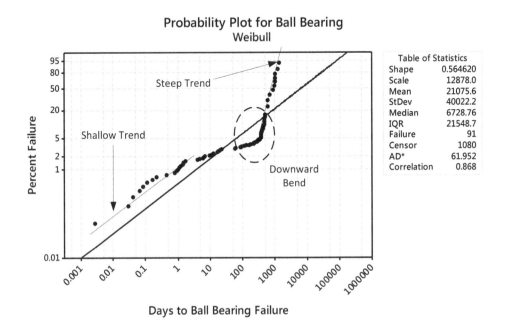

FIGURE 11.11
Typical field data for mixed failure mechanisms of a compressor.

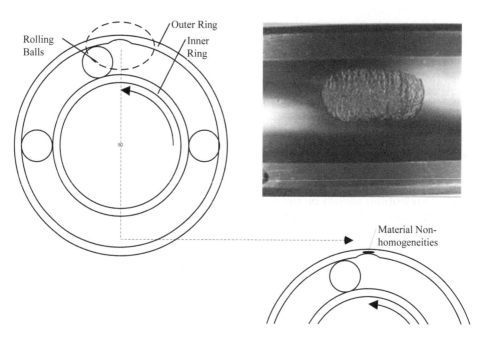

FIGURE 11.12
Outer wall damage due to material nonhomogeneities in manufacturing.

After plotting the early life failure due to manufacturing defects separately, as shown in Figure 11.13, it is to be noted that the early life failure trend is curved upward, which indicates that there might be a failure-free period (threshold) before this failure occurs. Figure 11.14 shows the probability distribution of the compressor early life failure using a lognormal distribution, which indicates a better fit to the distribution with a correlation coefficient of 97.9% instead of 96.4%, as was previously indicated in Figure 11.13. More details and examples on this technique are discussed in Abernethy (2006).

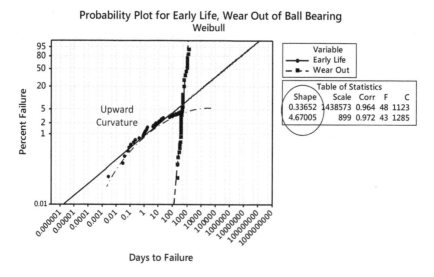

FIGURE 11.13
Separate analysis for field data for mixed-failure mechanisms.

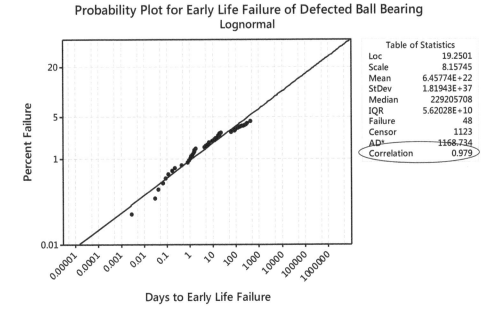

FIGURE 11.14
Early life failure probability distribution analysis using lognormal distribution.

Field Failure Investigation Process

Once products are released to the field, reliability data is monitored, collected, and analyzed. The reliability model is used to define whether products meet their reliability requirements. It is used to identify gaps in the reliability of subsystems and top failures by comparing to reliability thresholds allocated to each subsystem and component.

Reliability, quality, and service teams should develop a postmarket reliability data analysis strategy and work instructions for collecting, analyzing, and publishing reliability data. Work instructions will identify the process of categorizing service data, failures, and collecting reliability data. The postmarket risk management team will be involved in this process to monitor the field data from the risk management perspective. The team will identify when certain failures are violating acceptable risk limits. As we discussed, field data include many sources of noise that, if not properly identified and managed, may lead to erroneous conclusions and/or actions. The postmarket reliability analysis strategy should address crucial issues to ensure the accuracy of reliability monitoring process and metrics calculations. It should also set the rules for the definition and the count of different categories of service events such as proactive service/maintenance, unscheduled events, and customer-performed maintenance/repairs. Additional information includes a count of parts for each service event, and primary and secondary parts replaced.

We need to keep in mind that the main objective of field reliability monitoring is to sustain or to improve product reliability by understanding the root causes of failures. On the basis of this knowledge, we can propose various improvements. The main elements of the field failure investigation process are discussed next.

It is ironic that successful product verification, validation, and ultimately launch yields little to no actionable information on which to base design improvements. Failures, on the other hand, contribute a wealth of data on what to improve or what to design against in subsequent efforts. The nature and underlying causes of failures once identified and corrected lead to product improvements. Various organizations develop their own unique methods of tracking failure metrics (and their rates), which are often monitored through quality reviews and corrective action and preventive action (CAPA) events.

In general, failure reporting is a subset of reliability data review that includes not only reports of the manner in which a product fails but also an understanding of the successful duration of the fleet's operation. Analysis of failure becomes more meaningful when it is done under the umbrella of a reliability analysis where the entire failed units and operational population are considered. Analysis of reliability data has three main purposes:

- To verify that the device is meeting its performance requirements or goals
- To discover deficiencies in the device to provide the basis for corrective action
- To establish failure histories for comparison and for use in future product development activities

Any organization should develop its own guidelines for conducting failure investigations in the field. This investigation should provide essential information on the following:

1. What failed
2. How it failed
3. Why it failed
4. How future failures can be eliminated

A commonly used method for investigating and correcting reliability issues for fielded products is the "define, measure, analyze, improve, and control" approach known as the *DMAIC* methodology (Webber and Wallace 2015). The DMAIC process systematically drives to root cause analysis of failures and improvements, and is composed of five steps, as depicted in Figure 11.15 (Jamnia 2017).

The main five steps of the DMAIC process, along with the detailed tasks and steps to be executed along the process are summarized as follows.

Define

In this step, the problem statement is clearly and concisely defined. This is a crucial step to the success of DMIAC activities. In fact, the team should dwell in this stage as long as possible and come back to it in order to ensure that the problem statement is complete. It should focus on what has failed (or to be improved), who has observed the issue, and its extent. The define stage corresponds to blocks 1 (failure observation) and 2 (failure documentation) in Figure 11.15. Completion of these two blocks enables the investigator to develop a strategy to collect data and/or approach to the manner in which inquiries are to be conducted.

Failure Observation and Determination

The event (e.g., reliability metric trigger) or the decision to investigate should be documented with the following information:

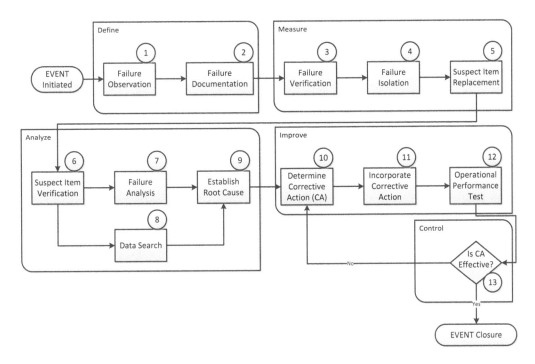

FIGURE 11.15
Reliability-based investigation process flowchart.

1. Date of field data review, when it was determined that an investigation was required
2. Product line and all applicable product codes applicable to the reliability data under investigation
3. The subsystem associated with the trigger item
4. Reliability metric value (i.e., MTBF, failure rate) during the triggering month
5. Reliability metric threshold(s) and the value that was exceeded

This information may be summarized in a problem statement. For example:

> On 17-Dec-2017, during a review of the product performance data, it was observed that the Learning Station Electronics Subsystem mean time between failure (MTBF) value for Dec 2017 was 380 months. Furthermore, this subsystem was exhibiting a negative trend for five consecutive months. Figure 11.16 shows the electronic subsystem MTBF decline below the reliability threshold for both the monthly and the three-month moving average MTBF.

Failure Isolation and Scope Determination

Considering that the primary goal of this type of investigation is to improve product performance and because a subsystem may consist of several components or parts, it is important to determine which components will be included in the scope of the investigation. Therefore, the failure isolation can prioritize the top failed component(s) of the subsystem rather than investigating all subsystem components/parts. This decision should be driven by the reliability metrics, with the goal of making a design change that will improve

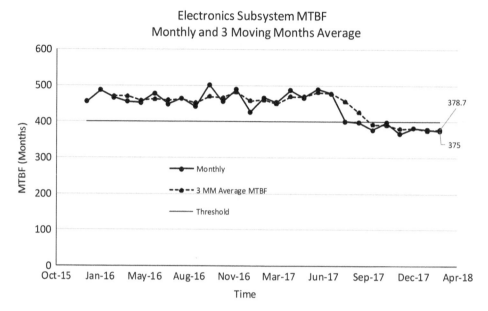

FIGURE 11.16
Electronic subsystem MTBF and threshold triggers based on field data.

reliability metrics. The scope determination must be documented with the rationale as to why certain components/parts were deemed out of scope. Often the use of a Pareto chart of parts failed or replaced helps the decision-making process.

After the scope is determined, the problem statement should be updated to include the components that will be investigated:

> On 17-Dec-2017, during a review of the product performance data, it was observed that the electronic subsystem mean time between failure (MTBF) value for Sep 2017 was 385 months. Furthermore, this subsystem was exhibiting a negative trend for the next five consecutive months. The top most failed component in this subsystem is the power cord, which causes 40% of electronic failures.

Measure

As is determined by its name, in this stage relevant data is collected. This step corresponds to blocks 3 (failure verification), 4 (failure isolation), and 5 (suspect item replacement) of Figure 11.15. If additional facts are identified that affect the problem statement, the define step is revisited and the problem statement is updated. The more comprehensive the problem statement, the richer the collected data.

Failure Verification

Because product performance metrics (such as reliability or failures) are calculated from service and field data, it is important to verify failures for an initiated event. The first step of this process is to obtain failed samples from the service center(s). If possible, we should collect parts from the time periods when the negative trends were observed. Failed samples received from service center(s) should be examined to ensure that they have indeed

A) Good Cord   B) Damaged Cord

FIGURE 11.17
Good and returned sample power cord from the field.

failed, and that the product performance metrics accurately reflect the fielded product. As an example, Figure 11.17 shows a sample of the returned power cords with disconnected strain relief mentioned in the Learning Station example.

Finally, it is important to have a full review of all the changes to the product since the original field launch, such as:

- Product design, manufacturing process, and service process changes
- Design history and changes of the product, especially in the area where the trigger was created
- Complaint and quality records logged to determine the top failure modes associated with the trigger issue
- Service process history and changes to service capacity to point any impact or contribution the service process can make to the reliability measurement in the field
- Supplier history and changes should identify any changes in the manufacturing process, facility and logistics, functional testing process, third-party supplier, etc.
- Pareto of field failures and failure rates, both monthly and 3- or 12-month moving average to help define variation over time and method of data analysis in the analyze phase

Analyze

Once sufficient levels of data are collected, we begin the analysis phase of our investigation. As seen in Figure 11.15, blocks 6 (suspect item verification), 7 (failure analysis), 8 (data search), and 9 (establish root cause) correspond to the analyze stage. The focus should be placed on identifying factors that have a significant impact on functions that are critical to quality. By studying these factors and their impact, a root cause to failure is generally identified.

Failure Analysis and Root Cause Determination

Once failure trends have been confirmed from a reliability point of view, a failure analysis of the defective items should be performed to establish the responsible failure mechanism. One pitfall is our own engineering team insisting that we need to solve problems ourselves. Often, we have the tendency to rely on what we or our own labs may do.

While this is an admirable character, we need to be realistic to ask who may be best suited to help us analyze our data or investigate the root causes of failures. This resource may be found either in-house or at an external partner or supplier.

At times, performing a search of existing data to uncover similar failure occurrences in this or related items may provide a historical perspective of the observed failure mode/failure mechanism. We need to ask whether these failure modes have been evaluated and/or addressed before.

We recommended that the design, manufacturing, or service changes—either in process or tools—that may have caused an increase in failure rates of the component and/or subsystem to be analyzed. Examples of these are

1. Service process changes
2. Manufacturing process changes (of both the subsystem and the component)
3. Supplier changes
4. Design changes to the system and subsystem
5. System and subsystem reliability trending in past 12–24 months
6. Geographic location of products in the field and duty cycle by users
7. Component age

Root Causes

Through a disciplined approach of defining, measuring, and analyzing collected data, potential root causes may be identified. Often failures are due to a product or process change, premature wear-out, heavy usage and duty cycles, or narrow design margins.

Example

The root cause for the power cord failures is the dislocation of the strain relief for the housing. This could be due to insufficient mating between the strain relief, wear, misuse by the customer due to tugging or yanking the power cord, or a combination of all of these.

During brainstorming sessions, the team creates a list of potential root causes that will be investigated to determine if they contributed to the issues under investigation. Table 11.1 lists example potential main categories of failures.

To identify the actual failure mechanism from a list of potential root causes, a number of tools exist. Among them are the fishbone diagram, the 5 whys, and the contradiction matrix. A full description of these tools is beyond the scope of this work.

Table 11.2 shows a contradiction matrix comparing information obtained from the investigation activities to the potential causes, to eliminate or support potential causes based on facts.

TABLE 11.1

Example Potential Categories of Failure Root Causes

Category	Potential Root Cause
Material	Change of supplier
Method	Shipping mishandling, housing dimension quality inspection process changed
Measurement	Time to failure of each failure, data entry (open or close date)
Design	Housing opening diameter, strain relief insert length inside housing
People	Customer misuse (angle of tugging, yanking, etc.)

TABLE 11.2

Example Contradiction Matrix for the Power Cord Failure

Category	Potential Root Cause	Housing dimension tolerance control affects failure rate	Shallower strain relief disconnects easily from housing opening	Entry data upon close of complaint not when opened	Housing dimension quality inspection process changed	Shipping package is not robust	Supplier change did not increase failure rate	Duty cycle difference between users	Physical damage (yanking, tugging) increases failure rate
		Facts							
Material	Change of supplier						No		
Method	Shipping mishandling					No			
Method	Manufacturing assembly procedure				Yes				
Measurement	Time to failure of each failure							No	
Measurement	Data entry (open or close date)			No					
Design	Housing opening diameter	Yes							
Design	Strain relief insert length inside housing		Yes						
People	Customer misuse (angle of tugging, yanking, etc.)								Yes

Improve

On the basis of root causes identified and what has been learned, changes to the design or to the manufacturing or service procedures may be needed. We should be cautious and not jump to conclusions and implement proposed changes without making sure that these changes are in fact the right changes to be made.

We need to start by developing prototypes or pilot programs. Then we can evaluate them for the expected outcomes. Should the need be, designs of experiments (DoEs) may be conducted to evaluate and document the impact of proposed changes. In Figure 11.15, blocks 10 (determine corrective action), 11 (incorporate corrective action), and 12 (operational performance test) belong to improve step of the DMAIC.

Corrective Actions and Verification

The first step of reliability improvement is to determine the necessary action(s) to prevent future failure recurrence (design change, manufacturing process control, supplier change, troubleshooting guide, service training, etc.). The decision regarding the appropriate corrective action should be made by a cross-functional design team.

Once a corrective action has been identified, and prior to its implementation on fielded products, its effectiveness at eliminating the issues at hand should be verified. It is important to authenticate that the financial impact of the proposed corrective actions will justify the resources and time dedicated to implementing them. Another consideration should be to verify that when proposed actions are implemented, no new issues are introduced. The following steps can be used to verify the recommended actions:

1. Incorporate the recommended corrective action into the original test system/ equipment. *If the action is to make a design change, put the new design into a unit.*

2. Retest the system/equipment with the proposed corrective action modification incorporated. *For a design change, expose the product to the same field conditions that resulted in the issue under investigation.* Also, design the test using the appropriate sample size and test duration to ensure the elimination of the failure mode and the reliability requirements, as explained in Chapters 8 and 9.

3. Determine if proposed corrective action is effective at solving the problem that impaired the performance of the system or subsystem. If needed, perform accelerated life testing to ensure that the new design is more reliable than the existing design. If the issue recurs, it can be concluded that the design change is not effective or that the right root cause has not been correctly identified. However, if the issue does not occur, the design change can be considered a success and ready for implementation on a larger scale.

There are many tools in the improve phase to define the weight and contribution of possible multiple contributing root causes to the issue investigation. One of the most commonly used methods is the DoE. The DoE is used to investigate the most critical root cause and also to optimize the design

Example

After investigating the power cord failure shown in Figure 11.17, the design team concluded that the damage to the power cord could be either due to physical abuse by the customer or due to design robustness of the cord. The team identified three main factors affecting the failure of the power cord:

1. Tugging angle, α
2. Housing diameter, d
3. Strain relief insertion, I

A factorial analysis of the contributing root causes to the failure of the power cord was conducted using DOE (concept explained in detail in Chapter 7), which indicated that the design team has to redesign the insertion of the strain relief into the housing and create a better understanding of the control of the housing opening dimensions. The physical abuse by the customer was not the major factor contributing to the failure of the power cord based on the DOE analysis as shown by the relatively small coefficient of the togging angle, α, shown in the response equation (as explained in Chapter 7):

$$Y_{Stress} = 0.737 - 0.218\,I - 0.389\,d + 0.0793\,a$$

Control

The DMAIC process identifies the root cause(s) of the failure. However, if the problem has happened once, it is likely to happen again. To close the loop, control mechanisms must be identified and put in place to prevent a relapse and a return to the state under which failure took place. Step 13 in Figure 11.15 asks this question and ensures that the identified problems are properly addressed.

Effectiveness and Control

Once the changes are applied to the fielded products, they need to be monitored to ensure that they do not decline over time. In order for the improvement to be monitored, it has to be standardized and measurable. It is important to document a standard process or instruction of the implementation of the improvement or changes, i.e., ensure it is a structured process.

Statistical process control (SPC) can be an effective tool to measure and monitor improvement implementation effectiveness in the field.

Keys to Successful Investigations

There are several "keys" that make the failure reporting and investigation process effective. These include the following:

1. The discipline of the report writing itself must be maintained so that an accurate description of failure occurrence and proper identification of the failed items are ensured.
2. The proper assignment of priority and the decision for failure analysis must be made with the aid of design engineers and systems engineers.
3. The status of all failure analyses must be known. It is of prime importance that failure analyses be expedited as priority demands and that corrective action be implemented as soon as possible.

4. The root cause of every failure must be understood. Without this understanding, no logically derived corrective actions can follow.

5. There must be a means of tabulating failure information for determining failure trends and the mean times between failures (MTBFs) of system elements. There should also be a means for management visibility into the status of failure report dispositions and corrective actions.

6. The system must provide for high-level technical management concurrence in the results of failure analysis, the soundness of corrective action, and the completion of formal actions in the correction and recurrence prevention loop.

Maintaining accurate and up-to-date records through the implementation of the data reporting, analysis, and corrective action system provides a dynamic, expanding experience base. This experience base, consisting of test failures and corrective actions, is not only useful in tracking current programs but can also be applied to the development of subsequent hardware development programs. Furthermore, the experience data can be used to

1. Assess and track reliability

2. Perform comparative analysis and assessments

3. Determine the effectiveness of quality and reliability activities

4. Identify critical components and problem areas

5. Compute historical part-failure rates for new design reliability prediction (in lieu of using generic failure rates found in MIL-HDBK-217, for example)

Corrective Action and Preventive Action (CAPA) Core Team

In all systematic failure investigations, a core team is established to oversee the effective functioning of the investigation and proper execution of corrective action (see Figure 11.18). It also provides increased management visibility and control of the investigation. The team typically consists of a group of cross-functional representatives with sufficient level of responsibility to ensure that failure causes are identified with enough detail to generate and implement effective corrective actions that are intended to prevent failure recurrence and to simplify or reduce the maintenance tasks. The main responsibilities of a core team consists of setting priorities, establishing schedules, assigning specific responsibility, and authorizing adequate funding to ensure the implementation of any necessary changes when dealing with complex and difficult problems.

In a performance-based investigation, the core team is especially important and helps the investigator(s) to improve reliability and maintainability of hardware and associated software by the timely and disciplined utilization of failure and maintenance data. A successful core team will include (but is not limited to) higher authority representatives from the following teams: reliability engineering, systems engineering, and quality. It should also contain a product expert for the product under investigation.

The core team should be involved in all major steps of the investigation, with the purpose of providing guidance and expertise to enable the investigator to make concrete decisions based on the available data. The core team is to be consulted during the stages outlined in dashes in Figure 11.18 and as listed next:

- Determine scope
- Root cause of trigger

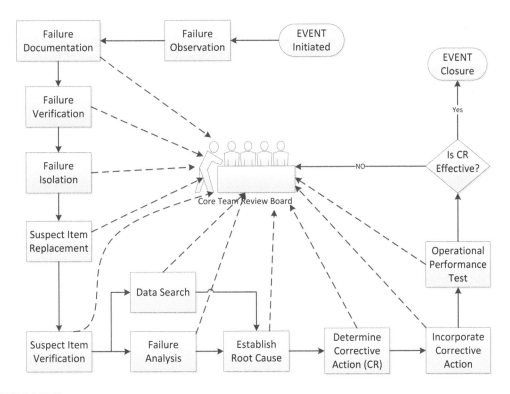

FIGURE 11.18
Failure reporting and the role of the core team.

- Root cause of top failure modes
- Determination of recommended actions
- Verification of actions
- Effectiveness of actions

When consulting the core team members at the various investigation steps, it is important to provide them with adequate yet concise information. Be sure to present only the pertinent data, in an unbiased manner, so that they can make decisions and give advice for how to move forward with the investigation. Tools like a four-blocker can be helpful for communicating the issues that require insight and collaborative decisions from the team.

Warranty and Service Plan Data Review

The need for a reliability distribution that fits the data is crucial for warranty analysis and proactive service strategy. A warranty is a guarantee in a form of contract, in which the manufacturer defines a period of time and service tasks to be conducted on the system to fix issues due to manufacturing defects, improper assembly or construction, etc. Also, a warranty could be conducted by replacing defected parts, or adjusting or recalibrating parts that drifted out of the original setting. In engineering systems and commercial products, there is no such thing as a lifetime warranty, i.e., it is expected that components and parts have limited design life, will be subject to wear-out, and eventually will require replacements. A buyer expecting a car to last for 25 years will experience multiple failures of moving parts and engine parts before the 25 years—expected service life—matures.

It is the responsibility of the reliability engineer to define the design life of critical components that will fail before the expected service life quoted for the system matures. The manufacturer will define the warranty period for a product based on the failure-free period of the design before any wear-out failure trends start to occur. The warranty trade-off analysis is between the maximum warranty period (to make the product appealing to the buyers) and the minimum warranty period (to reduce the cost of service). As discussed in Chapter 8, initial preventive maintenance (PM) can be identified for limited life parts based on testing data. However, after product release, failure data is obtained per real-life usage by a large number of customers in different geographical locations, and a wider range of operating conditions and stresses. Based on field data and failure rate, PM strategy and terms can be reviewed and enhanced for optimization between reliability and cost of service.

The most important tool in the warranty data analysis is fitting the field failure data (time to failure) to the proper distribution that can represent the failure rate change over time and projecting the probability of failure in the future.

Notes

1. Some manufacturers refer to the top 12 failures as the *Dirty Dozen*, a reference to an old movie.
2. Some of the terms used in this list may not be familiar. Unfortunately, a detailed explanation of these terms is beyond the scope of this work.
3. A full description of this term is beyond the scope of this work.
4. If we were to use the parameter diagram to describe failure modes versus failure mechanisms, we would say that failure modes are associated with the error states of a desired function. Failure mechanisms are associated with the physical condition that leads to the failure mode. It is possible that several failure mechanisms lead to the same failure mode. Using a cause-and-effect analogy, we can say that a failure mode is the effect associated with a failure mechanism.

12

A Primer on Product Risk Management

Introduction

In a product development process, there comes a time when the development team should ask some fundamental questions about the design under development. These are

1. What can potentially go wrong with the product?
2. How serious will it be when it happens?
3. How will it impact its users?
4. How often will it happen?

The result of the probability of a failure multiplied with the severity of the outcome of this failure is often referred to as *risk*.[1] Once the risk of failure has been identified, the design and development team should ask whether the benefits that the intended product provides outweigh its potential risk(s), or change the design such that either the failure mode has been eliminated or its likelihood of occurrence has been significantly reduced.

To be clear, this definition of risk on the basis of failure and the likelihood of its outcome is based on a product development mindset. Some may define risk on the basis of unintended consequences of a given action or event and the associated outcome(s). For example, an unintended consequence of walking on an icy sidewalk is slipping, falling, and possibly getting hurt. As we will explain later in this chapter, for risk to exist, *hazards* and *hazardous situations* should be present. In this scenario, an icy sidewalk is a hazard; walking on an icy sidewalk is deemed a hazardous situation. In this context, walking is the intended action; its associated failure mode is falling. The possibility of getting hurt—a consequence of falling—is defined as *harm*. Now, it is easy to argue that not everyone who walks on ice falls; and, not everyone who falls gets hurt; and finally, not everyone who gets hurt is hurt seriously. So, as it is hopefully clear, there has to be a failure (at times defined as an unintended consequence to an event) along with harm (or associated outcome) for risk to exist.

A Risk Management Tool

The focus of this chapter is to provide an overview of two tools needed to understand product risks and means of mitigating them. The first tool is called failure modes and effects analysis (FMEA), which is a method to understand what may go wrong with a product. FMEAs are quite versatile and may be applied to a variety of situations such as product design and how the design may fail. They may apply to manufacturing or service processes. They may also be applied to how a product is used. Principally, FMEAs may even be applied to an analysis of types of hazards that may be inherent in the design of a product. For instance, a scalpel is designed to cut biological tissue. Used properly, its use is beneficial and healing; however, its misuse may lead to serious injury or even death.

The second tool discussed here is called fault tree analysis. This tool is often described as a top-down root cause analysis in an effort to understand the chain of events that may lead to an undesired outcome. It considers a combination of factors or lower-level faults using Boolean logic. In contrast, FMEAs are considered to be single fault, i.e., a failure may be explained with only one cause.

Failure Modes and Effects Analysis

Once both system to subsystem architecture and a physical concept of the new product have been developed, before rushing into finishing the detailed design, it is prudent to ask what may go wrong in manufacturing, use, or even during service. Failure to ascertain design pitfalls often leads to low production yields, excessive field returns, long service-turnaround times, dissatisfied customers, or potential harm to the end users.

A tool to provide a path to product robustness, as well as a systematic approach to risk evaluation, is called concept failure modes and effects analysis (CFMEA). This technique enables the design team to uncover potential shortcomings in the conceptual design by identifying and evaluating

1. Design functions and requirements
2. Foreseeable sequence of events both in design and production
3. Failure modes, effects, and potential hazards
4. Potential controls to minimize the impact of the end effects

Should a CFMEA be conducted at an early stage, it may then become the basis of a more formal and rigorous design failure modes and effects analysis (DFMEA) and process failure modes and effects analysis (PFMEA). The purpose of a CFMEA is to understand what may go wrong and attempt to remove failure modes or reduce their probability of occurrence through making changes to the design. Considering that in this chapter our focus is on exploring product risk and its management, we will focus on developing a DFMEA.

It should be noted that both DFMEA and PFMEA are living (version-controlled) documents that are initiated before design requirements have been fully established. They are certainly started prior to the completion of the design and updated throughout the life cycle of the product including design changes, where appropriate.

The first step in creating the DFMEA charts is to ask about the main and essential requirements and/or functions of the product. The next step is to ask how these functions will fail. Note that functions fail in one or a few of the following ways (through rarely more than three ways):

1. No function. For example, consider that the function of a vacuum cleaner is to produce suction to remove dirt. No function means that you turn a vacuum cleaner on and no dirt is removed.

2. Excessive function. By way of the vacuum cleaner example,[2] the suction is so strong that you cannot move the cleaner around as it is stuck to the surface being cleaned.

3. Weak function. There is hardly any suction present.

4. Intermittent function. Vacuum cleaner works fine, but unexpectedly it loses power but shortly after its power is restored.

5. Decaying function. Vacuum cleaner starts working fine but over time (either short or long term) loses its suction, and it does not recover.

To illustrate the development of a DFMEA, consider a specialized computer system called the Learning Station (see Chapter 6). Three of the functions (or requirements) associated with the Learning Station are shown in Table 12.1. Notice that included with this table, failure modes associated with each function are also indicated.

Table 12.2 contains not only the function and the failure modes shown in Table 12.1, but also the effects associated with the Learning Station. This table also provides a column to capture the severity of the effects along with a second column to provide potential causes of failure.

In a typical FMEA table, the module to be studied is mentioned in the first (or second) column from the left. By suggesting that the module is the system (specified as the Learning Station in Table 12.2), the design team will be concerned with system-level failures. The next column contains the specific function under observation. In this example, three functions are mentioned, namely, protect from the environment, enable developing social skill, and run educational software. Associated with each failure mode, one or two effects are suggested. Finally, a potential cause of failure is also provided. At this point, the "Severity" column is left blank, though it can just as easily be filled out.

We like to make an observation at this moment: we are using the term "failure modes and effects," however, at the early stages of design, we have more interest in potential causes. Why? Because once we can identify the root causes of failures, we can attempt to make changes to the design to either remove the failure mode(s) or attempt to reduce their frequency, or mitigate them in such a way as to reduce their impact. As we will explain later, "effects" impact product risk, whereas "causes" influence product design configuration and reliability.

TABLE 12.1

Learning Station Functions and Failure Modes

General Function	Specific Function	Failure Mode
Protect from environment	Stop liquids from entering internal compartments	Liquid leaks inside
	Resist electrostatic discharge from reaching sensitive internal components	Electrostatic discharge reaches sensitive components
	Stop users from reaching (or touching) internal components	Users reach (or touch) internal components
Develop social skills	Devise a turntable for turn-taking between users	Turntable fails to turn over time
	Devise system operation by tokens to enforce social skills development	Tokens fail to work
		Token operation fails to work over time
Run educational software	Execute commercially available software	Software does not run
		Software begins to run but stops

TABLE 12.2

Learning Station Initial Design Failure Modes and Effects

Item	Module	Function	Potential Failure Mode	Potential Effects of Failure	Severity	Potential Causes of Failure
1		Protect from environment	Liquid leaks inside	Unit stops working temporarily		Lack of gasket and/or drain paths in the design
				Unit stops working permanently		
2			Electrostatic discharge reaches sensitive components	Unit stops working temporarily		Lack of ESD/EMI barriers in the design
				Unit stops working permanently		
3	Learning Station		Users reach (or touch) internal components	Electric Shock		Excessively large openings in the enclosure
				Bodily injury		
4		Enable developing social skills	Turntable fails to turn over time	Unit does not rotate on its axis		Turntable deflects under unit weight
						Turntable bearings have corroded
5			Tokens fail to work	Unit stops functioning		Wrong tokens were used
6			Token operation fails to work over time	Unit stops functioning		Component life is too short
7		Run educational software	Software does not run	Software does not accept commands		Incompatible software with operating system
			Software begins to run but stops	Software does not accept commands		System is overheated

In the context of the Learning Station design, the first item in this initial DFMEA reminds the design team of the possible need for either a gasket or a drain path in the housing (enclosure) to guide liquids away from the inner cavity. The second item ensures that both electromagnetic emission interference (EMI) and electrostatic discharge (ESD) have been properly considered and mitigated. In fact, a clever design may be to integrate both functions into one component.

Another point worth noting is the potential conflict between mitigating overheating (item 7) by having large openings in the unit's housing and mitigating users reaching/touching internal components by having no openings (item 3) at all. Openings are needed to remove heat generated by the electronics, but at the same time, any opening provides an opportunity for users to reach the electronics inside and hurt themselves. Clearly, this conflict should be resolved satisfactorily.

Finally, this initial FMEA provides other design insights to be considered. First, before specifying the turntable willy-nilly, features such as its weight-bearing capacity as well as its resistance to corrosion should be considered. Another insight is that products do not last forever (item 6). Product and component life expectancy and their impact on the service organization should not be overlooked. One last point brought to light by this FMEA is that certain failures may not be avoided altogether and require mitigation. For instance, an identified cause in item 7 is software incompatibility with the operating system. It is next to impossible to design software for all computer operating systems. For this reason, a sensible solution and mitigation is through labeling. Typically, the point of purchase packaging of application software provides information on the appropriate class of operating systems the intended software may run on. To clear any possible misunderstandings, we emphasize that labeling is not just a sticker that is applied to a product. In general, any printed material either attached to the product or included within the packaging or even the packaging material itself is considered as labeling.

In Table 12.3, two additional columns are introduced. The first is the rate of occurrence of the failure mode and the second is the possible means of preventing the failure to take place. In the Learning Station example, the design team conducts a finite element analysis (as specified under the prevention column) to examine whether the gasket warps excessively. Should the analysis indicate excessive warping, design changes should be put in place. Other preventive measures may be done by conducting design reviews and soliciting opinions of subject matter experts, or even conducting reliability analysis or testing.

Hazard, Hazardous Situation, and Harm

When the question of what can potentially go wrong is asked, the real but unexpressed concern is the kind of harm the end user is exposed to.[3] But, on the one hand, just because something can go wrong, it does not always go wrong; and on the other hand, if it goes wrong, it does not necessarily cause harm. By way of example, consider a car with a flat tire. Depending on whether the flat tire is the front or back tire, or if it happens when the car is parked in a driveway or going down the highway at 65 mph, the outcomes could be very different. Another example is a slippery walkway after a rain shower. It is obvious that not every walkway is slippery when it rains, nor does everyone who walks on a slippery walkway slip and fall. So, how should we look at what can go wrong and the potential of harm?

Let us consider this: *harm* (or *mishap* as used in MIL-STD-882E) is defined as "an event or series of events resulting in unintentional death, injury, occupational illness, damage to or loss of equipment or property, or damage to the environment." However, before harm comes to anyone or anything, there has to be a potential source of harm defined as a *hazard* (MIL-STD-882E). Following the same logic, for a harm or mishap to take place, either people or property have to be exposed to one or more hazards. This exposure is called a *hazardous situation*. For instance, lightening is a hazard; being in a storm with lightening is a hazardous situation; harm is getting hit by lightning. Finally, *severity* is the indication of the consequences of harm if it were to befall.

Now that the four terms associated with risk have been defined, we like to note that there are two other factors to be considered in determining risk. The first is the frequency of *occurrence*. This metric provides a measure of the likelihood of occurrence of the specific hazardous situation (or the failure effect). The second factor called *detection* is a measure of how easily the hazardous situation (or failure effect) can be detected. In other words, if a

TABLE 12.3

A Populated Initial Design Failure Modes and Effects Matrix for the Learning Station

Item	Module	Function	Potential Failure Mode	Potential Effects of Failure	Severity	Potential Causes of Failure	Occurrence	Prevention
1	Learning Station	Protect from environment	Liquid leaks inside	Unit stops working temporarily / Unit stops working permanently		Excessive warping of gasket		Finite element analysis
2			Electrostatic discharge reaches sensitive components	Unit stops working temporarily / Unit stops working permanently		Inadequate ESD/EMI barriers in the design		Design review
3			Users reach (or touch) internal components	Electric shock / Bodily injury		Excessively large openings in the enclosure		Human factors study
4		Develop social skills	Turntable fails to turn over time	Unit does not rotate on its axis		Turntable deflects under unit weight		Load analysis
						Turntable bearings have corroded		Material compatibility
5			Tokens fail to work	Unit stops functioning		Wrong tokens were used		Labeling
6			Token operation fails to work over time	Unit stops functioning		Component life is too short		Reliability analysis
7		Run educational software	Software does not run	Software does not accept commands		Incompatible software with operating system		Labeling
			Software begins to run but stops	Software does not accept commands		System is overheated		Thermal analysis

shark is present in the area of the beach with people swimming, how easily can it be seen and pointed out?

Risk may also be measured as the product of severity, occurrence, and detection. On the DFMEA, there is typically a column known as RPN (risk priority number), which is a product of these three metrics. Clearly, the higher the RPN, the more catastrophic the risk may be.[4] The question remains of how to assign a numerical value to the individual constituents of the RPN.

Some sources, particularly in automotive, choose a 1 to 10 scale (Engineering Materials and Standards 1992). For severity, a numerical value of 1 indicates the lowest level of harm typically associated with end user discomfort, or minimal damage to property and/or the environment. A numerical value of 10 indicates death or extremely serious injury, or extreme damage to property and/or the environment.

For occurrence, a numerical value of 1 indicates the lowest level of occurrence typically identified as *almost impossible*. How should we quantify "almost impossible"? Stamatis (2003) suggests using a numerical value of 1 when reliability has to be 98% (or better). In contrast, Engineering Materials and Standards (1992) suggests a probability of occurrence equal to 1 in 1,500,000 opportunities, for a rating of 1. The same source suggests using a rating of 10 where occurrence is almost certain (better than 1 in 3). Again, Stamatis (2003) recommends using the reliability measure; and for a rating of 10, reliability should be less than 1%.

For detection, a rating of 1 means that the defect is easily discernable and that there are proven methods for detection. A rating of 10 refers to defects that are almost always undetectable with any known methods.

On the basis of these numerical values of severity, occurrence, and detection, the RPN may be computed as low as 1 and as high as 1000.

Table 12.4 is an example of how these ratings may be put to use for the air detection sensor of an infusion pump.[5] A clinical risk when infusing medication to a patient is air in the intravenous (IV) line. Often, the dissolved air in the fluid is released due to a variety of reasons and appears as bubbles in the IV line. Should this air be injected into the vein, it has the potential to cause harm to the patient. In Table 12.4, there are two columns titled "Detection." One is a numerical rating and the second is a description of the methodology (Anleitner 2011). In addition there is a "Classification" column that we will explain shortly.

This initial assessment of risk provides an early warning to the design team on mechanisms to prevent the causes of failure and to detect potential effects of the failure (Anleitner 2011). In the air-in-line sensor example, the highest RPN belongs to line number 2 (Sensor does not detect bubbles over time). This only means that the design team, as a part of the detailed design, will focus on conducting reliability analysis and life testing to improve the design in an effort to reduce the occurrence rating from 7 to a lower number. As will be shown later, if this reduction is not possible, then a preventive maintenance of this component should be developed to alleviate the risk.

The "Classification" column is used to mark which design characteristics may have *critical* or *significant* impact on the end user. In this instance, item number 2 may be classified as critical and item 4 (Sensor intermittently detects bubbles) may be considered as significant. The decision where to draw the line between these classifications is generally made by the design team with some input from management and/or the legal team.

Risk Assessment Code Table

Other sources, in particular MIL-STD-882E, choose different scales for severity and occurrence as shown in Tables 12.5 and 12.6. As may be observed, this approach does not easily

TABLE 12.4

An Initial Design Failure Modes and Effects Matrix with RPN

Item	Module	Requirement(s)	Potential Failure Mode	Potential Effects of Failure	Severity	Classification	Potential Causes of Failure	Occurrence	Prevention	Detection	Detection	RPN
1	Infusion Pump	Detect air bubbles larger than 10 ml	Sensor does not detect bubbles	No early warning/ death or serious injury	10		Combined component electronics tolerances may be excessive	3	Tolerance analysis	Verification test	1	30
2			Sensor does not detect bubbles over time	No early warning/ death or serious injury	10		Components have shorter than expected design life	7	Reliability analysis	Life test	2	140
3			Sensor detects false bubbles	False early warning/ delay of therapy	7		Improper transducer has been specified	2	Design review		3	42
							Electronics components have been improperly specified	2	Electronic simulation	Verification test	3	42
							Detection software has not been properly verified	4	Design review		1	28
4			Sensor intermittently detects bubbles	No early warning/ death or serious injury	9		Ultrasonic transducers are not mechanically aligned	3	Tolerance analysis		2	54
							Required gap between transducers varies excessively	3	Tolerance analysis	Verification test	2	54
							Combined component electronics tolerances may be excessive.	3	Tolerance analysis		2	54
5			Sensor only detects bubbles much larger than 10 ml	Unreliable early warning/serious injury	8		Improper transducer has been specified	2	Design review		2	32
							Combined component electronics tolerances may be excessive	3	Electronic simulation	Verification test	2	48
							Detection software has not been verified properly	4	Design review		1	32

TABLE 12.5

A Table of Severity Levels (Adapted from MIL-STD-882E)

Description	Level	Criteria for Harm
Catastrophic	1	Death, extreme bodily injury, permanent environmental damage or extreme financial loss
Critical	2	Substantial bodily injury, significant environmental damage, or extensive financial loss
Marginal	3	Bodily injury, moderate environmental damage, or significant financial loss
Negligible	4	Minor bodily injury, minimal environmental damage, or some financial loss

TABLE 12.6

A Table of Probability of Occurrence Levels (Adapted from MIL-STD-882E)

Description	Level	Specific Individual Item
Frequent	A	Likely to occur regularly
Probable	B	Will occur several times
Occasional	C	Likely to occur sometime
Remote	D	Unlikely, but possible to occur
Improbable	E	So unlikely, it can be assumed occurrence may not be experienced
Eliminated	F	Incapable of occurrence; this level is used when potential hazards are identified and later eliminated

TABLE 12.7

A Risk Assessment Code Table (RACT) (Adapted from MIL-STD-882E)

Severity / Probability	Catastrophic	Critical	Marginal	Negligible
Frequent	High	High	Serious	Medium
Probable	High	High	Serious	Medium
Occasional	High	Serious	Medium	Low
Remote	Serious	Medium	Medium	Low
Improbable	Medium	Medium	Medium	Low
Eliminated	Eliminated			

integrate with the DFMEA matrix. Nonetheless, the risk associated with each system or subsystem may be evaluated in the risk assessment code table (RACT) as shown in Table 12.7. Some (especially in the medical field) prefer this approach to risk evaluation because of the clear classification of risk types. Depending on the industry, you may need to adapt to a commonly used approached.

Briefly, Table 12.7 indicates that if the severity of a particular event is *catastrophic* and its probability is *remote*, then the associated risk is *serious*. Whereas if the severity is marginal and its probability is occasional, then the risk is considered as medium.

The DFMEA matrix may be updated with these ratings, and in place of the RPN, RACT may be used. The recommendation is that any *high* or *serious* risk should be reduced, and any *medium* risk should be investigated for a possibility of reduction. This is done for the air-in-line sensor example as shown in Table 12.8. It should be noted that in this approach,

TABLE 12.8

An Initial Design Failure Modes and Effects Matrix with Risk Assessment Code (RAC)

Item	Module	Requirement(s)	Potential Failure Mode	Potential Effects of Failure	Severity	Classification	Potential Causes of Failure	Occurrence	RAC
1	Infusion Pump	Detect air bubbles larger than 10 ml	Sensor does not detect bubbles	No early warning/death or serious injury	Catastrophic		Combined component electronics tolerances may be excessive	Remote	Serious
2			Sensor does not detect bubbles over time	No early warning/death or serious injury	Catastrophic		Components have exceeded their life span	Probable	High
3			Sensor detects false bubbles	False early warning/delay of therapy	Critical		Improper transducer has been specified	Remote	Medium
							Electronics components have been improperly specified	Remote	Medium
							Detection software has not been verified properly	Remote	Medium
4			Sensor intermittently detects bubbles	No early warning/death or serious injury	Catastrophic		Ultrasonic transducers are not mechanically aligned	Remote	Serious
							Required gap between transducers varies excessively	Remote	Serious
							Combined component electronics tolerances may be excessive	Remote	Serious
5			Sensor only detects bubbles much larger than 10 ml	Unreliable early warning/serious injury	Critical		Improper transducer has been specified	Remote	Medium
							Combined component electronics tolerances may be excessive	Remote	Medium
							Detection software has not been properly verified	Occasional	Serious

prevention and detection do not play an upfront role. For the same reason, a direct comparison of the outcomes of the RACT with RPN may not be appropriate. Once risks are mitigated, the table can be updated to reflect the reduced risk levels.

Controlling Critical or Safety-Related Items

We have made references to critical and/or safety (or significant) characteristics. The xFMEA matrix has a column dedicated to this classification. It should therefore be no surprise that all items associated with either critical or safety characteristics are monitored to ensure that they are in control. There are two aspects or levels of this control.

First, these items are controlled by having proper requirements in place. As requirements are verified initially when a product is launched into a market, features related to safety or criticality are not overlooked. More important, as a part of postlaunch, there are often cost-cutting measures as well as other activities that may alter the initial design and its configuration. By having proper requirements in place, future configuration changes will not have an adverse effect on safety.

Second, variations are part and parcel of every production and service environment. On the one hand, the design team needs to be aware of how variations impact these safety/critical items. On the other hand, plans need to be developed and implemented to ensure that safety/critical items characteristics are maintained within acceptable bounds.

The process of writing a control plan begins with developing an understanding of process variations. This may come about through a short capability study to ensure that the proposed process is realistic. This requires that a knowledge of measurement systems exists and that it is appropriate. By measurement systems, we mean that the techniques and approaches used to measure the process outcome are appropriate to the process.[6] Often, if a measurement system analysis (MSA) is not available, a design of experiments may be conducted (called gage repeatability and reproducibility, or gage R&R) to measure and qualify the levels of variation. A more detailed examination of gage R&R is beyond the scope of this work.

Often a control plan incorporates a statistical sampling method through which, at regular intervals, components are removed and measured, and their results are logged and measured against previous measurements. Once a comparison with previous data is made, production or service personnel should be trained to know what actions are required should negative trends be observed. For more detail on developing a control plan, see Jamnia (2018).

Other Risk Management Tools

FMEA is a widely used tool at every major development stage of a product, be it at the system and subsystem level or at detailed design. This tool may be applied to software, and even to how a product may be used or misused. FMEAs enable the design team to find weak points in the design or the processes of manufacturing or servicing a product. Then, it will help prioritize high-risk areas. In this chapter, we will provide an example of a DFMEA and will allude to a PFMEA. While DFMEAs start with functions or requirements and their associated failures, PFMEAs start with a process flow. This flow may be the manufacturing or service process. However, in principle, they apply to any process flow. For instance, suppose that a software program is needed to operate a machine. The embedded software may interact with its user for input, and based on the given input, the electromechanical product may achieve certain tasks. Similar to other process FMEAs, it is possible to develop a process map of software steps and interactions with the user and identify where in the process failures may be introduced. This special form of FMEA is called use FMEA. These days, FMEAs are applied to cybersecurity to identify weak points of software as it may come under attack from hackers.

FMEAs are considered to be single-fault failure analysis tools. In other words, their focus is on the effect or the impact of a single failure. However, it is possible for several failures to be combined to produce a larger impact on end users. An example is the combination of a wet road, speeding, and an inexperienced driver may lead to a higher probability of an accident than any one of these factors alone. The proper tool for this type of analysis is called fault tree analysis.

Fault Tree Analysis

As we just mentioned, *fault tree analysis* (FTA) may be used to identify a combination of events or failures that lead to other faults. However, this requires an understanding of, and the development of fault scenarios. To do this, we need to keep in mind that failures take place within a system or a system of systems. For instance, in the previous example of a car having a flat tire and its possible consequences, the main players were the environment, the driver, and the car's speed. The road, road condition, and even weather are all elements of the environment. The system needed to develop a proper FTA that consists of these three elements alone. The product, i.e., the car, while instrumental in the outcome, may not be the cause, per se.[7] Should elements of the car (such as maintenance) become relevant to the fault analysis, then we would be dealing with a system of systems.

In typical FMEAs, single-failure points of products or processes are considered and analyzed. The only FMEA that considers the role of human–machine interface is use FMEA. However, machine failures are not taken into account. In contrast, in FTA, the entire system, including people, machine (or material), and environment along with any other factors may be included. This is an acknowledgment of the fact that components, assemblies, subsystems, and systems are all interrelated with the use environment and end user, and a failure of one may have a ripple effect on others.

Having said this, FTAs are used to conduct functional analyses of highly complex systems. FTAs first develop scenarios that start by breaking down chains of failures. This approach enables the analyst to evaluate the system reliability and assess safety requirements and specifications in lieu of human, software, and hardware interfaces. As a part of a new product development process, FTAs will lead to the identification of potential design defects and safety hazards. In the case of failures of a fielded product, FTAs will lead to root causes and possible corrective actions.

One should note that we are using FTAs as opposed to FTA. Typically, in a new product development process, fault tree analyses are conducted for the top critical or safety events of the product. This implies that unlike FMEAs, only a subset of failures are studies—albeit, in a much higher degree of detail.

The foundation of a fault tree analysis is based on a scenario of a set of events and their logical relationship with one another. Before diving into the steps needed to develop a fault tree, it may make sense to walk through a simple FTA. At the beginning of this section, we spoke of an example in which a combination of a wet road, speeding, and an inexperienced driver may lead to a higher probability of an accident than any one of these factors alone. Let us walk through this example.

Case Study

Prior to starting an FTA, it is prudent to develop a block diagram of the system that is being analyzed. The more complete this diagram, the better the interactions between various components can be understood. However, at some level, the assumption must be made that lower levels will not provide additional details. Figure 12.1 depicts the block diagram for the car accident. This figure indicates that the levels of the engine, as well as the brake system, will be scoped out of this analysis.

Figure 12.2 depicts the top undesired event of a car accident. It further shows that this accident might have been caused by "driving speed not being suitable for the road conditions" or the driver "failed to stop or slow down." The OR *gate* shows a logical relationship between events, namely, the driving speed or the inability to slow down. Either of the two lower events could lead to the undesired car accident.

If we were to examine the driving speed, two lower-level events may be considered, as shown in Figure 12.3. One may be "excessive rainfall" and "driver failed to adjust to the road conditions." Similar to the OR gate, the AND gate shows a logical relationship

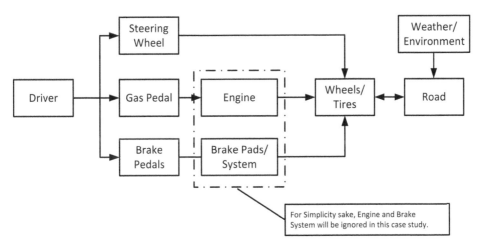

FIGURE 12.1
A block diagram for a fault tree analysis example.

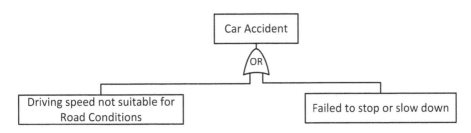

FIGURE 12.2
The top events in a fault tree analysis example.

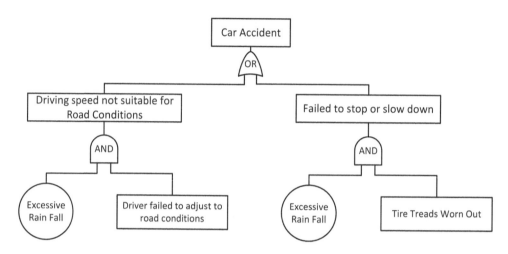

FIGURE 12.3
The top two event layers in a fault tree analysis example.

between various events. In contrast to OR, the AND gate requires that all the presented events need to be present for the top event to happen. In this example, the presence of both rain and the failure to adjust to road conditions leads to speeds being too high or not suitable. We see that on the right-hand branch of the fault tree, excessive rainfall along with worn tire treads leads to a failure to stop or slow down.

One may notice that at the lower levels, two of the events are shown within a circle as opposed to a box. The significance of this change is that it is not fruitful to break the excessive rainfall event further. This type of event is called *basic* and is depicted by a circle.

Finally, Figure 12.4 shows another layer of the FTA. On the left branch, we learn that the failure of the driver to adjust to the road conditions is caused by an *inhibit* gate based on the driver being in a hurry. This gate (in the shape of a hexagon) is an AND logical relationship between two events when one of them is outside of the control of the chain of events. In this example, the driver's lack of experience is an external factor; however, the fact that he is in a hurry leads to a lack of proper judgment of the road conditions. If the driver was not in a hurry, his judgment would not have been hindered.

The last layer of the right branch in Figure 12.4 depicts two diamonds. One points to an inadequate maintenance schedule and the other calls out a missed service event. Either of these two events may lead to a failure to detect the excessive wear of the tire treads. The presence of a diamond indicates that there may be underlying causes for these failure but they were not explored further.

The next logical step in this process is to assign probabilities of occurrence to each event at the lowest level, and then, by using Boolean algebra and the logic diagram, calculate the probability of occurrence of the next level up. We will examine this approach in Chapter 13. In lieu of not having the required probabilities, it is possible to assign qualitative values such as low, medium, and high to estimate the likelihood of the top undesired event.

Fault Tree Diagram Symbols

As shown in Figure 12.5, there are three basic types of fault tree diagram notations: events, transfers, and logic gates. There are five event types, one transfer type, and six gate types.

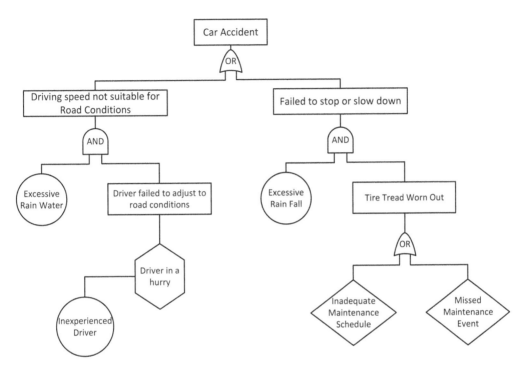

FIGURE 12.4

The fault tree analysis of car accident example. Note that the details of the engine and break system have been ignored in this case study.

The undesired top level event is usually depicted as a rectangle. The primary or basic failure event is usually denoted with a circle. A house (or external house) event is usually depicted with a symbol that looks like a house. It is an event that is normal and guaranteed, or expected to occur. An undeveloped event usually denotes something that needs no further breakdown or investigation, or an event for which no further analysis is possible because of a lack of information. A conditional event is a restriction on a logic gate in

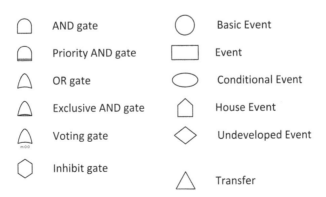

FIGURE 12.5

Symbols in a fault tree analysis.

the diagram. A transfer symbol (a triangle) is used to transfer information from one area (or event) to another. Gate symbols can be the following:

1. OR gate—An event occurs as long as at least one of the input events takes place.
2. AND gate—An event occurs only if all input conditions are met.
3. Exclusive OR gate—An event occurs only if one of the input conditions is met, not if all conditions are met.
4. Priority AND gate—This is probably the most restrictive scenario when an event occurs only after a specific sequence of conditions.
5. Inhibit gate—An event will only occur if all input events take place as well as whatever is described in a conditional event.
6. Voting gate—In a voting gate, the output event occurs if m or more of the input events occur.

These gate symbols describe the Boolean relationship between outcomes. This will be further described in Chapter 13.

FTA Process

In a way, a proper FTA should start with a solid understanding of the system within which a failure has happened or may take place. The easiest means of ensuring that major aspects of the system (or subsystem/assembly, if more appropriate) have been considered is to develop a visual block diagram. This diagram will help communicate to others what is being analyzed, and in a sense, what assumptions have been made. This step also helps fill any gaps that may exist in our knowledge and understanding and its functions.

The next step depends to a large extent on whether FTA is being conducted as part of new product development (NPD) or a response to a field failure and its consequences. Regardless, we need to identify the adverse effect or the undesired outcome. The only difference between what we may do as a part of NPD is the number of outcomes: we try to account for as many as possible, whereas, with the fielded products, we are more interested in the one that had occurred. Having said this, for each top-level undesired outcome, a separate fault tree may be constructed.

Through a brainstorming session, various root causes may be ascertained. It is important to only consider immediate causes that would stem from a layer lower. Other causes that are farther down will eventually be identified at their proper layer. The block diagram helps maintain focus in this area. This activity may be repeated at lower levels until a layer is reached that the events considered are basic and further decomposition is not needed, or information is not available for further expansion.

We recommend that the fault tree be constructed as failures and causes are identified; however, it may be prudent to assign the logical gates when the tree structure is somewhat defined. Once the logical gates are assigned, they may be reviewed by the team of experts to ensure that any errors are identified and removed. The next step is to assign probability of occurrence to each event—either qualitatively or quantitatively—and evaluate the probability of occurrence of the top undesired outcome. The FTA helps identify major contributing factors; however, it is up to the product team to develop appropriate solutions to move forward. In brief, the steps of an FTA may be enumerated as follows:

1. Develop a block diagram of the system under evaluation and collect detailed information for each block.

2. Define the top-level undesirable outcome(s). For each top-level outcome, a separate tree must be constructed.

3. Capture the possible causes for undesirable events layer by layer down to basic events or to when no additional information about an event is available.

4. Once the tree is fairly well developed, assign the logical gates and relationships between various events.

5. Provide the probability of occurrence for each event either qualitatively or quantitatively.

6. Evaluate the probability of occurrence of the top undesired event, again either qualitatively or quantitatively. Use of Boolean operations for quantitative calculations will be discussed in Chapter 13.

7. Brainstorm on possible means to develop possible redesign solutions or corrective actions, as appropriate.

Notes

1. More specifically, risk may be defined as the combination of the probability of occurrence of harm and the severity of that harm. It is hard to imagine that harm comes without a failure. ISO 31000 defines risk as the effect of uncertainty on objectives.

2. We can use any example here including the Learning Station that we have used previously, but for simplicity sake, we believe we can all relate to the functions of a vacuum cleaner easier than other more complicated engineering systems.

3. Regulated industries such as avionics and medical devices are very focused on risk. Recent events such as toys that are harmful to children or hoverboards that catch fire are reminders that even consumer product developers should be mindful of the impact of their products on the public.

4. It should be noted that the RPN is used to measure the relative risk of one failure mode compared to others. It is not an absolute measure.

5. For brevity, the final FMEA is presented here. The reader may develop an understanding of functions, failure modes, and causes by studying Table 12.4.

6. Here is an example to clarify our point. A common grade-school ruler is not a good tool to measure the precision of a machined surgical tool. This example may be sound silly, but often we do not stop and think about whether our measurement approach fits what is being measured. Another flaw in measurement is that what may be measured may not be relevant to the required function of the component. For instance, measuring velocity when mass flow rate may be needed.

7. One may play the devil's advocate role and say that a "smart" car with artificial intelligence may indeed prevent accidents of this nature. This is true, but this conclusion is an outcome of an FTA, nevertheless.

13

Relating Product Reliability to Risk

Introduction

In Chapter 12, we entertained the idea that product failures are considered significant when there is either impact to the well-being or property of the end user.[1] We mentioned that when the question of what can potentially go wrong is asked, the real but unexpressed concern is the kind of harm the end user may be exposed. Alternatively, one may raise the question of the impact of a particular failure to the overall functionality of a system, and try to ascertain or quantify the risk of a given failure to complete a given system-level task. For instance, consider the functionality of an automobile. The functionality of this system is to transport a payload from one point to another under a number of environmental and use conditions. In this regard, certain failures will have little or no impact, whereas other failures may have significant or catastrophic impact. For example, failure of the entertainment system has a different *criticality* than, say, failure of the timing belt. In the same light, should the power steering fail, one may still be able to drive, albeit the vehicle would be hard to maneuver. A flat tire, a somewhat simple failure, will significantly hinder the primary function of the automobile.

Reliability, Risk, and Safety

As we have mentioned, the field of reliability focuses on failures, their modes, and mechanisms along with their rates. A component failure, regardless of its mode of failure or mechanism, is considered a hazard. Risk analysis is concerned with how a hazard may lead to a hazardous situation and ultimately the associated harm. Safety analysis focuses on identifying, evaluating, and controlling hazards (de Vasconcelos et al. 2009). Recall that harm is the product of the probability of the occurrence of an adverse event with the severity of that event. One can then imagine that if a product or system is reliable, i.e., it has a very low failure rate, then it may be deemed as safe.

Leveson (2011) warns us not to confuse safety with reliability. She suggests that "a system can be reliable but unsafe. It can also be safe but unreliable" (Leveson 2011, p. 7). We believe it is important to consider her point to some extent. A system may be reliable, but this does not mean that failures will never take place. *Reliable* means that failures are few and far between. However, the manner in which a system fails is important. For instance, lithium batteries are very reliable, but one of their failure modes is that they catch fire under certain conditions. Hence, the use of these types of batteries in electronic devices necessitates the use of design features that monitor the state of the battery temperature and charge conditions to ensure that the battery is operating safely. In other words, the conditions in which the battery would ignite are avoided. Similarly, for safety reasons, shipping lithium batteries is highly regulated (see, for instance, http://www.iata.org/whatwedo/c argo/dgr/Documents/lithium-battery-shipping-guidelines.pdf).

If we consider "reliable but unsafe" as one side of the coin, the other side may be "safe but unreliable." The latter means that should the system fail, it will remain safe and it should not cause bodily injuries, damage to the environment, or financial loss.[2] It is possible that such systems fail quite frequently for a variety of reasons, but none of the failures have harm levels designated above low.

Most design engineers attempt to develop fail-safe products. Standards such as IEC 60601 series or ISO 13485 are regulations to ensure safety and quality of products. Nevertheless, it is possible that in certain circumstances failures lead to harm. At times, failures may lead to a loss of mission, i.e., failure of the intended function. In other scenarios, failure may not cause harm directly; however, the associated loss of function may lead to harm. For instance, consider an LCD display. This display may be integrated into a gaming console. Should the display fail, the only impact is that the gaming console is inoperable. There is no harm associated with this failure. Now, consider the same LCD display is used as a part of the interface subsystem to operate heavy earth-moving machinery. Should the display fail during the operation of this machinery, it is possible that the control of the system is lost leading to possible loss of life or other catastrophic events.

The second pertinent point made by Leveson (2011) is an overreliance on the reliability of the components of a system without considering the interactions between them. The point that she makes is that a *system* is often larger than the combination of its hardware pieces. Consider that many of our current electromechanical devices are software driven. The intricacies of the interactions of the software with the users (i.e., human–machine interface and interaction) may not be realized by the system designers. If the software workflow is not properly designed and vetted, it is quite possible that the end user may not clearly follow a "safe" path. In other words, the user may provide a series of inputs to the machine in such a way that each step is processed correctly; however, due to oversight, errors, and software bugs, the outcome may be contrary to the user's goals. In more complicated systems where software interacts and receives input from a variety of people, truly understanding possible failure modes from the use of products becomes extremely important. The news is fraught with reports of people using a product inappropriately or misusing a product. This analysis falls in the domain of use failure modes and effects analysis (FMEA) and an in-depth review of this aspect of safety is beyond the scope of this work.

Relating Reliability to Risk

In the course of this book, we have often referred to risk as the product of the probability of occurrence of hazardous situation times the probability of harm. In the language of mathematics, probability of occurrence is expressed as P_1 and probability of harm is expressed as P_2. Before discussing how reliability relates to hazardous situations, it would be helpful to understand the relationship between hazards, hazardous situations, harm, and risk.

Risk Analysis

Consider Figure 13.1. In this figure, the hazard is an undesired event that takes place. To analyze risk, we first need to start by evaluating the hazard as well as what the exposure to this hazard may be. This exposure may be evaluated relative to people, property, or even finances. The reason is that just because an undesired event has taken place, it does not automatically impact people, property, or finances. For instance, for someone who is not in the stock market, a sharp downturn in the market leaves no impact. Another example may be presence of swift ocean currents in a particular area of an ocean. Should no boats be present, no concerns will be raised. Hence, the next question that we need to explore is what hazardous situations may exist. Hazardous situations refer to the conditions when people or property are exposed to given hazards. Exploration of hazardous situations enables us to calculate their probability, often referred to as P_1. Let us now consider the example of the ocean currents again in an effort to explain how the probability of a hazardous situation may be calculated. We will then return to the risk analysis as depicted in Figure 13.1.

Suppose that swift currents happen randomly 10 days a month; meanwhile, in any given week there may be four boats passing through that area. These currents constitute a hazard, and the presence of boats and small ships in these areas become a hazardous situation. From the analysis of the hazardous situation, it is possible to calculate its probability of occurring:

$$P_1 = \frac{4 \text{ boats}}{7 \text{ days (or 1 week)}} \times \frac{10 \text{ days}}{30 \text{ days (or 1 month)}} = 0.19 \text{ per day,}$$

where P_1 is the probability of a hazardous situation.

This calculation shows that, for this example, for any given day, the probability of a boat experiencing swift currents is 19%. This is the probability of a hazardous situation occurring. However, we should note that just because on any given day a boat may experience rough waters, it does not mean that harm has befallen anyone or anything. We need to look at harm next.

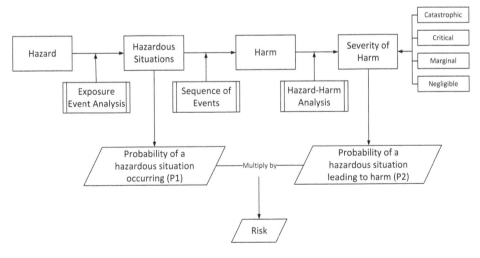

FIGURE 13.1
A basic flow diagram for risk analysis.

TABLE 13.1

Severity of Harm Example

Boat Size	Severity of Harm (P_2)			
	Catastrophic	Critical	Marginal	Negligible
Small	10%	20%	33%	37%
Medium	5%	10%	15%	70%
Large	0%	3%	12%	85%

The next level of analysis as shown in Figure 13.1 is to evaluate the *sequence of events*. In this regard, we may consider factors such as the size of the boat and the speed of the ocean currents along with environmental conditions. As an analysis of the sequence of events, we need to ask what combination of events must be aligned so that harm will befall. In this example, a sequence of events may be to have the combination of an inexperienced captain and first mate with a small boat and 10-foot-high waves. We need to keep in mind that different sequences may lead to different levels of harm.

The next step is to examine the harm. If the conclusion is that a boat may be ripped apart and its passengers drown, this harm classifies as *catastrophic*. If the damage is only a few frightened passengers, harm may be considered as *negligible*. This classification is called hazard–harm analysis and leads to a determination of the severity of harm (see Figure 13.1). The harm severity analysis relies to a large extent on past data and subject-matter-expert opinion. The outcome is typically an estimate of the probability of severity of harm should a subject experience a hazardous situation. This probability is often depicted as P_2. The probability of risk is the product of P_1 and P_2:

$$\text{Risk} = P_1 \times P_2$$

To clarify, let us again consider the ocean current example. In the hazard–harm analysis, experts suggest that only three types of boats travel in the region: large, medium, and small. Additionally, they determine the severity of harm for each type of boat (Table 13.1). For instance, 10% of small boats may experience some type of catastrophe when they enter these waters, but 37% of the harm may be negligible. In contrast, only 5% of medium-size boats experience catastrophe and 70% negligible severity of harm. Large boats may have 3% critical damage but 0% catastrophic. Between 37% (small) and 85% (large) boats pass through the rough waters with nothing significant to report.

To evaluate risk, the P_1 values (i.e., probability of a hazardous situation occurring) should be multiplied by P_2 values (i.e., the severity of harm). Earlier, we calculated P_1 to be 0.19 or 19%. For this example, the values are summarized in Table 13.2.

TABLE 13.2

Risk Calculation Example

Boat Size	Risk per Day ($P_1 \times P_2$)			
	High	Serious	Medium	Low
Small	19% × 10% = 1.9%	3.8%	6.3%	7.0%
Medium	1.0%	1.9%	2.9%	13.3%
Large	0.0%	0.6%	2.3%	19% × 85% = 16.2%

Reliability Concerns Leading to Hazards

Now that we have developed a feel for conducting a risk analysis, let's explore how reliability relates to risk. For this purpose consider Figure 13.2. From this figure, we can ascertain that product reliability concerns or failures contribute to the hazards, and their failure rates contribute to the probability of a hazardous situation occurring (P_1). The challenge here is to correctly identify the hazards associated with the failure modes of various components. For instance, a fuse may fail short or it may fail open.[3] Should a fuse, which is designed to prevent an overcurrent, fail short, the corresponding hazard may be overheating of the unit leading to smoke and fire. In contrast, should the same fuse fail open, the corresponding hazard may stem from the inability to operate the product. In many circumstances, this may be a benign effect; however, should the product be the navigation system of an airplane or a medical device, the outcome and the hazard may not be as benign.

Modarres et al. (2017) suggested a list of general hazards that contain such items as chemical, thermal, mechanical, electrical, and biological hazards. These general hazards have to be tailored for each product and product family. For example, in the case of medical devices such as infusion pumps or dialysis machines, a hazard may be air being injected into a patient. Other hazards may be inadequate volume or excessive levels of fluid being delivered to a patient. In the nuclear industry, a hazard may be radiation exposure; and in avionics, loss of cabin pressure may be considered as a hazard. As we mentioned, there are no one-size-fits-all set of hazards.

A tool for linking failure modes and their rates to specific hazards is called failure modes and effects criticality analysis (FMECA). This tool may be considered as an extension of a DFMEA. The following sections focus on FMECA.

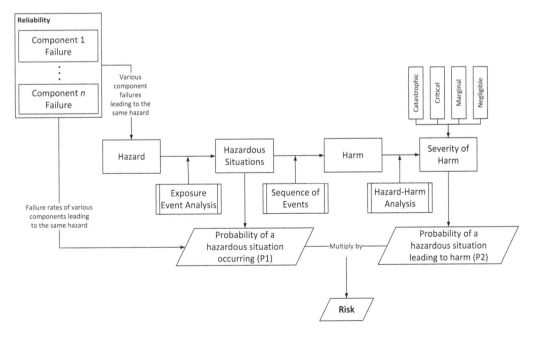

FIGURE 13.2
A flow diagram relating reliability analysis to risk.

FMECA and Criticality Index

Clearly, the goal of each design engineer is to mitigate the high risks and to develop a fail-safe product. The tools used for this purpose are the DFMEA and fault tree analysis (FTA). By properly understanding the effect of each failure, it may be possible to develop mitigations that would lead to a fail-safe product.

Another approach is to evaluate the impact (or criticality) of the failure within the intended delivery of the system function. For instance, consider a car with a flat tire. If the car is parked on a driveway, the impact on the intended system function is insignificant, i.e., the car is not going anywhere. Now assume that the car is traveling at 70 miles an hour when the failure (flat tire) occurs. This failure may have a more critical consequence. Additionally, the mode of failure becomes significant as well. In this scenario, the failure mode is which of the four tires may develop the defect and blow out. The significance of the failure mode and its impact on the consequences is as follows. The failure of one of the front tires may lead to loss of control, whereas the failure of the back tires may not have a significant impact on controlling the vehicle.

Another example of failures and failure modes is the fuse example that we cited a short while ago. As mentioned, a fuse may fail short or it may fail open; yet, each failure mode may have a different criticality depending on the outcome. One outcome may be smoke and fire, whereas another outcome may be that the system shuts down.

This type of analysis where the criticality of component failure on the overall functionality of the system or its outcome is considered is called *failure modes and effects criticality analysis*. This process was first outlined in MIL-STD-1629A in 1980. Later, in 1993, the Reliability Analysis Center (RAC), a Department of Defense Information Analysis Center, adopted this approach as a reliability evaluation/design technique. However, in 2006, IEC 60812, second edition, applied this methodology to risk analysis.

As a reliability tool, FMECA provides data for developing fault tree analyses and reliability block diagrams. It also provides a means of identifying root failure causes. Finally, by ranking each failure based on severity and impact on the system's mission success or failure, it highlights the top single-point failures requiring corrective action.

As a safety and risk analysis tool, FMECA provides mission-based hazards and hazardous situations. Furthermore, by considering the phases of the mission and failure modes, the probability of specific hazardous situations may be calculated. Ultimately, by integrating the probability of the hazardous situations with the probability of harm, the risk may be evaluated.

Case Study

To better illustrate this point, consider the following example. A device is designed to detect air in an intravenous tube that delivers vital medicine to a patient. The mission of this system is to detect air bubbles larger than 10 ml. The device is equipped with an alarm when bubbles larger than 10 ml are detected. A previous DFMEA had identified the four failure modes and their potential effects presented in Table 13.3. The failure rate of this device has been evaluated and is determined to be $\lambda = 3 \times 10^{-6}$ per hour.

The potential hazards of this device have been identified as follows. For the "false early warning" failure mode, the hazard is "delay of therapy." Why? Because, as the device falsely alarms, it also stops delivery of the medicaments, requiring a nurse to restart the delivery of the therapeutic drugs. Hence, the therapy is delayed. As the patient is in need of therapy, the delay is considered to be a hazardous situation.

For the "no early warning" or "unreliable early warning" failure modes, the hazard is "air bubble enters vein." The justification for this potential hazard is that a nurse may

TABLE 13.3

Failure Modes and Effects of an Ultrasonic Sensor

Item	Requirement(s)	Potential Failure Mode	Potential Effects of Failure
1	Detect air bubbles larger than 10 ml	Sensor detects false bubbles	False early warning
2		Sensor does not detect bubbles	No early warning
3		Sensor intermittently detects bubbles	No early warning
4		Sensor only detects bubbles much larger than 30 ml	Unreliable early warning

not realize that there is air in the intravenous tube. Air in a human vein can be deadly; however, the body absorbs certain small volumes of air in the vein. The volume of air absorption varies from person to person. We would like to determine the overall risk of such a device after one year of operation. Table 13.4 provides an outline of the calculation of probability of a hazardous situation occurring during use time (P_1).

For this example, as reflected in Table 13.4, the contribution of "sensor detects false bubbles" to the total count of failures is 64%. This and other failure mode contributions may be calculated using theoretical modeling when the product is in early design stages; from a combination of testing and modeling in late stages of design and development; or from field data, should the product have been launched. In fact, this table should be updated as the product matures. Similarly, the contribution of "sensor does not detect bubbles" is only 9%, and for the other two failure modes, 11% and 16%, respectively. Note that the sum of the contributions of the failure modes[4] is 100%.

Probability of impact to mission (β) is somewhat more subjective, as it requires a judgment call by subject matter experts. For instance, in the case of detecting false bubbles, will the clinicians get tired of a device that alarms for no reason and send it for repair, or will they put up with it? In this scenario, the subject matter experts believe that there is a 50% chance that the device will be sent for repair. In the case of "sensor does not detect bubbles," there is no way that bubbles can be detected, so they enter the vein unless someone is present and sees the bubble moving in the intravenous tube. So, in this scenario, there is 100% impact to the mission of the device, which is to sense air bubbles. The mission impact of the rest of the failure modes have been determined the same.

The next column in Table 13.4 is use time (depicted as t) during which the probability of hazardous situation occurring is calculated.

Once all the input data, i.e., λ, α, β, and t, is available, the criticality index[5] (P_d) may be calculated:

$$P_d = \lambda \alpha \beta t$$

The last column of Table 13.4 shows the P_1 values for each hazardous situation. In this case study, there is only one contributor to delay of therapy, its $P_1 = 8.41 \times 10^{-3}$. In general, the contributions of all failure modes have to be taken into account. In other words,

$$P_1^i = \sum_{j=1}^{n} P_d^j$$

In this equation, i refers to the hazardous situation and j refers to the contribution of the criticality (P_d) of the failure mode j to the hazardous situation i.

TABLE 13.4

Calculation of Probability of Hazardous Situation Occurring during Use Time (P_1) for an Ultrasonic Sensor

Item	Requirement(s)	Potential Failure Mode	Potential Effects of Failure	Failure Rate of the Component per Hour (λ)	Contribution to Each Mode (α)	Probability of Impact to Mission (β)	Use Time (t, hr.)	$P_d = \lambda \alpha \beta t$	Potential Hazard	Probability of Hazardous Situation Occurring during Use Time (P_1)
1	Detect air bubbles larger than 10 ml	Sensor detects false bubbles	False early warning	3.00E−06	0.64	0.5	8760	8.41E−03	Delay of therapy	8.41E−03
2		Sensor does not detect bubbles	No early warning		0.09	1.0	8760	2.37E−03	Air bubble enters vein	4.78E−03
3		Sensor intermittently detects bubbles	No early warning		0.11	0.4	8760	1.16E−03	Air bubble enters vein	
4		Sensor only detects bubbles much larger 30 ml	Unreliable early warning		0.16	0.3	8760	1.26E−03	Air bubble enters vein	

TABLE 13.5

Risk Calculations $(P_1 \times P_2)$ for an Air Bubble Sensing Device

		Catastrophic	Critical	Marginal	Negligible
	P_1 P_2	1.00E–06	1.50E–05	5.00E–03	9.95E–01
Delay of Therapy	8.41E–03	8.41E–09	1.26E–07	4.20E–05	8.37E–03
		Catastrophic	Critical	Marginal	Negligible
	P_1 P_2	1.00E–02	1.50E–01	5.00E–01	3.40E–01
Air Bubble Enters Body	4.78E–03	4.78E–05	7.17E–04	2.39E–03	1.63E–03

There are three contributors to the "air bubble enters vein" hazard, therefore,

$$P_1 = 2.37 \times 10^{-3} + 1.16 \times 10^{-3} + 1.26 \times 10^{-3}$$

$$P_1 = 4.78 \times 10^{-3}$$

We should be mindful that these are the probabilities of a hazardous situation occurring and not of the harm itself. To calculate the probability of harm, P_2 values are needed. Determination of these values is subjective and requires the input of subject matter experts. The P_2 values and the calculated risk values for this case study are given in Table 13.5. Note that the P_2 values are different for the two hazardous situations.

Table 13.5 shows us that even though the probability of "delay of therapy" occurring is higher than "air bubble enters body," its catastrophic (as well as critical and marginal) risk is substantially lower (8.41×10^{-9} as opposed to 4.78×10^{-5}). Another implication of the calculated risk is our acceptance of what may happen in the course of one year that was the period for risk calculation. For the risk of "air bubble enters vein," we are effectively suggesting that for every 100,000 patients who are treated using this device in a year, nearly 5 people are subject to catastrophic harm. In a realistic situation, the design team has to entertain whether it can reduce this harm by making design changes or providing a justification why the benefits of the device outweigh its risks.

Fault Tree Analysis

Up to this point in this chapter, we relied primarily on FMEAs to either calculate risk or relate reliability to risk. As we discussed in Chapter 12, another effective tool is the fault tree analysis. In this segment, we will review how to calculate risk if the tool being used is FTA. Considering that FTA is founded on the Boolean logic of AND as well as OR gates, first, we need to review the mathematics of these two Boolean operations.

Boolean Operation

In the case of an AND gate, the probabilities of each event need to be simply multiplied in order to obtain the resulting fault. Mathematically, this is expressed as follows:

$$P_1^{\text{Fault}} = \prod_{i=1}^{n} P_1^{\text{event } i}$$

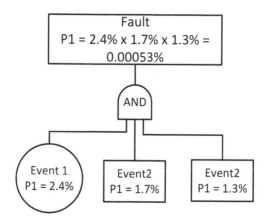

FIGURE 13.3
An example of AND operation calculation.

One may note that in FTA, we are calculating P_1, i.e., the probability of occurrence of the hazardous situation and not the risk. Once P_1 values are calculated, risk may be calculated as before.

As an example, consider the simple FTA diagram as shown in Figure 13.3. It consists of three events that are related to the top fault by means of a simple AND gate. The probabilities of these events are 2.4%, 1.7%, and 1.3%, respectively. To calculate the outcome of an AND gate, we need to multiply these probabilities. The result is 0.00053% or 5.3 incidents in 1,000,000 opportunities.

In the case of an OR gate, the outcome is calculated by multiplying the survivability of each event (i.e., 1 minus probability of occurrence or P_1) and then subtract the value from 1. Mathematically, this is expressed as follows:

$$P_1^{\text{Fault}} = 1 - \prod_{i=1}^{n} (1 - P_1^{\text{event } i})$$

As an example, consider the simple FTA as shown in Figure 13.4. This figure depicts three events that are related to the fault through the OR gate. Similar to Figure 13.3, the probabilities of these events are 2.4%, 1.7%, and 1.3%, respectively. To calculate the outcome of the OR gate, we need to first calculate the quantities of $(1 - 0.024)$, $(1 - 0.017)$, and $(1 - 0.013)$. Then, we need to multiply these quantities and finally subtract the results from 1. The result is 5.3% or 5.3 incidents in 100 opportunities.

FTA P_1 Calculations

It is not hard to imagine that a fault tree analysis is really a combination of braches that are either AND or OR gates. That being the case, we need to start from the lowest levels of the FTA and work our way up to resolving the lower events into higher ones.

To illustrate this approach, recall the FTA conducted in Chapter 12 and depicted in Figure 12.4. For convenience, this figure is repeated here in Figure 13.5. Additionally, we have provided the probability of occurrence at the lowest layer.

In here, the probabilities of an inadequate maintenance schedule and missed maintenance events have been provided as 3% and 5%, respectively. As a result, the probability of the tire treads being worn out has been calculated as 7.85%. To calculate the probability

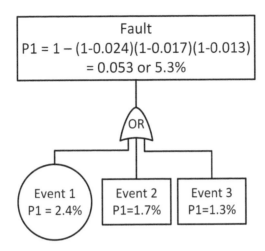

FIGURE 13.4
An example of OR operation calculation.

that the driver failed to slow down or stop, we need to have the probability of excessive rainfall. This value along with the probability of other branches is provided in Figure 13.5. Additionally, this figure shows the calculated probabilities for the entire fault tree. For the sake of simplicity, an explanation of these additional calculations is not provided and readers are encouraged to verify these calculations on their own.

Based on the FTA conducted here and demonstrated in Figure 13.6, the probability of a car accident under these circumstances is 0.0116% or 116 incidents in 1,000,000 opportunities. Note that this is only the P_1 value. To calculate risk, we need to have the probability of harm.

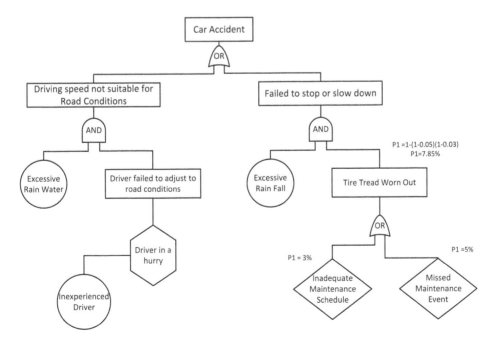

FIGURE 13.5
An example of P_1 calculation within a fault tree.

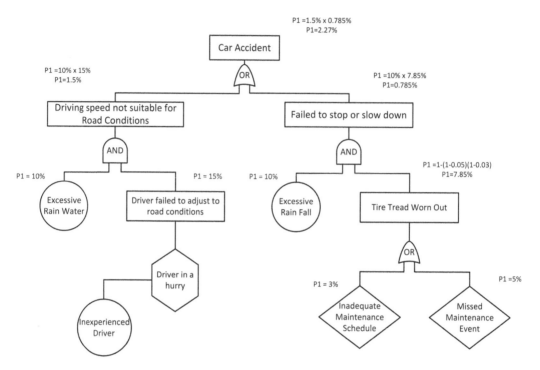

FIGURE 13.6
An example of P_1 calculation for an entire fault tree.

TABLE 13.6

Risk Calculations $(P_1 \times P_2)$ for Car Accident

		Catastrophic	Critical	Marginal	Negligible
P_1 \ P_2		1.40E−05	2.30E−04	6.50E−03	9.93E−01
Car Accident	1.16E−04	1.62E−09	2.67E−08	7.54E−07	1.15E−04

Note: These numbers are for illustration only and have no clinical basis.

The probabilities of harm and the risk calculations are provided in Table 13.6. These calculations and their explanations are similar to what has been already provided and for simplicity are not repeated here.

Notes

1. One may argue that excessive product failures impact an organization's financial or reputation negatively. This argument is correct but beyond the scope of this discussion.
2. An example of this type of system is the air brake in a truck. The brake system is designed such that they are normally closed and press against the wheel rotors to prevent wheel rotation. When the truck driver initiates the motion, air pressure forces the brake pads to open and allow wheel rotation. At any time, should the pneumatic system fail and air pressure be lost, the brakes are activated and the truck stops.

3. Failing short means that the electrical resistance between two points drops to zero. Failing open is the opposite of failing short, meaning that the electrical resistance between two points is infinite.
4. Contribution of failure modes is often designated as α.
5. In both MIL-STD-1629A and IEC-60812, the criticality number is designated as C_m.

Appendix

Reliability Standards and Guidelines

The following is a list of various standards and guidelines relative to reliability.

Standards on General Design Practices

Reliability in Design Task	Standard or Tool
Derating for electronic components	SAE International, "Derating of Electronic Components," http://standards.sae.org/geiastd0008/
Tolerance analysis	MIL-STD-785, Military Standard: Reliability Program for Systems and Equipment Development and Production
Worst-case analysis	Electrical Design Worst-Case Circuit Analysis: Guidelines and Draft Standard (REV A), http://aerospace.wpengine.netdna-cdn.com/wp-content/uploads/2015/04/TOR-2013-00297-Electrical-Design-Worst-Case-Circuit-Analysis-Guidelines-and-Draft-Standard-REV-A.pdf
Sneak circuit analysis	BSR/AIAA S-102.2.5-201X, American National Standard, Performance-Based Sneak Circuit Analysis (SCA) Requirements
Drop test	ASTM D2463 – 10b (ASTM D2463 – 10b Standard Test Method for Drop Impact Resistance)IEC60601-2-24 1st Ed: Medical electrical equipment – Part 2-24: Particular requirements for the basic safety and essential performance of infusion pumps and controllers
Radiated and conducted emission and immunity	IEC 60601-2-1, 4th ed., Medical electrical equipment

Reliability Standards and Guidelines

Category	Title
Reliability (general)	MIL-HDBK-263A, Electrostatic Discharge Control Handbook for Protection of Electrical and Electronic Parts, Assemblies and Equipment (Excluding Electrically Initiated Explosive Devices)
Reliability (general)	MIL-HDBK-338, Electronic Reliability Design Handbook
Reliability (general)	MIL-HDBK-189, Reliability Growth Management
Reliability (general)	Technical Report No. TR-2011-24, Design for Reliability Handbook, August 2011

Category	Title
Definitions	MIL-STD-721C, Definitions of Terms for Reliability and Maintainability, 1981
Definitions	MIL-STD-2074, Failure Classification for Reliability Testing, 1978
Prediction	MIL-HDBK-217F, Reliability Prediction of Electronic Equipment
Modeling	MIL-STD-756b, Reliability Modeling and Prediction, 1981
FMEA	MIL-STD-1629, Procedure for Performing Failure Mode Effect and Criticality Analysis, 1984
Requirements	MIL-STD-785B, Reliability Program for Systems and Equipment, Development and Production
Components and part selection	MIL-HDBK-965, Acquisition Practices for Parts Management, 1993
Components and part selection	MIL-STD-883, Test Method Standards Microcircuits, 2006
Maintainability	MIL-STD-471, Maintainability Verification/Demonstration/Evaluation, 1973
Maintainability	MIL-HDBK-472, Maintainability Prediction, 1963
Testing	MIL-HDBK-781, Reliability Test Methods, Plans and Environments for Engineering Development, Qualification and Production
Testing	MIL-STD-781D, Reliability Design Qualification and Production Acceptance Tests: Exponential/Distribution
Testing	JESD22-A104D – Temperature Cycling, JEDEC, Ed. D, 2006
Testing	JESD201 – Environmental Acceptance Requirements for Tin Whisker Susceptibility of Tin and Tin Alloy Surface Finishes, JEDEC, March 2006
Testing	JESD22A121 – Measuring Whisker Growth on Tin and Tin Alloy Surface Finishes, JEDEC, May 2005
Testing	European Standard EN 50419, European Committee for Electrotechnical Standardization
Testing	J.T. Duane, "Learning curve approach to reliability monitoring," *IEEE Transactions on Aerospace*, Vol. 2, No. 2, April 1964
Sampling	MIL-HDBK-H108, Sampling Procedures and Tables for Life and Reliability Testing (Based on Exponential Distribution)
FRACAS	MIL-STD-2155, Failure Reporting, Analysis and Corrective Action System (FRACAS)
Environmental stress screening	MIL-HDBK-344, Environmental Stress Screening of Electronic Equipment
Environmental stress screening	MIL-STD-2164, Environment Stress Screening Process for Electronic Equipment
Guidelines	MIL-HDBK-454b, General Guidelines for Electronic Equipment, 2007
Testability	MIL-HDBK-2165, Testability Handbook for Systems and Equipment, 1995
Reliability-centered maintenance	MIL-STD-3034, Reliability-Centered Maintenance (RCM) Process, 2011
Safety	MIL-STD-882E, Safety System, 2012

References

Abernethy, R.B. *The New Weibull Handbook: Reliability and Statistical Analysis for Predicting Life, Safety, Supportability, Risk, Cost and Warranty Claims*, 5th edition. North Palm Beach, FL, Robert B. Abernethy, 2006.

Abernethy, R.B., and Fulton, W. New methods for Weibull & Lognormal Analysis, *ASME Winter Annual*, 92-WA/DE-14, 1992.

Aldrich, J. R. A. Fisher and the making of maximum likelihood 1912–1922, *Statistical Science*, Vol. 12, No. 3, pp. 162–176, 1997.

Alippi, C. *Intelligence for Embedded Systems*, Springer International Publishing, Cham, Switzerland, 2014.

Anleitner, M.A., *The Power of Deduction, Failure Modes and Effects Analysis for Design*, ASQ Quality Press, Milwaukee, WI, 2011.

ASTM F1980-07. Standard Guide for Accelerated Aging of Sterile Barrier Systems for Medical Devices, ASTM International, West Conshohocken, PA, 2011.

Asrar, N., Vancauwenberghe, O., and Prangere, S. Tin whiskers formation on electronic products: A case study, *Journal of Failure Analysis and Prevention*, Vol. 7, pp. 179–182, 2007.

Atua, K. *System Life and Reliability Testing Design Optimization*, ARDC, Indianapolis, 2019.

Atua, K.I., Ayyub, B.M., and Assakaff, I. Statistical characteristics of strength and load random variables of ships, presented at ASCE Specialty Conference on Probabilistic Mechanics and Structural Reliability, Worcester, MA, August 7–9, 1996.

Ayyub, B., and McCuen, R. *Probability, Statistics, and Reliability for Engineers*, CRC Press, New York, 1997.

Azarkhail, M., and Modarres, M. The evolution and history of reliability engineering: Rise of mechanistic reliability modeling, *International Journal of Performability Engineering*, Vol. 8, No. 1, pp. 35–47, 2012.

Barlow, R.E., and Proschan, F. *Statistical Theory of Reliability and Life Testing*, Holt, Rinehart and Winston, New York, 1975.

Bayes, Mr., and Price, Mr. An essay towards solving a problem in the doctrine of chances, by the Late Rev. Mr. Bayes, F. R. S. communicated by Mr. Price, in a letter to John Canton, A. M. F. R. S, *Philosophical Transactions of the Royal Society of London*, Vol. 53, pp. 370–418, 1763, retrieved on March 19, 2019, from www.stat.ucla.edu/history/essay.pdf.

Bazovsky, I. *Reliability Theory and Practice*, Dover, New York, 2004.

Bellcore TR-332. Reliability prediction procedure for electronic equipment, Issue 5, Piscataway, NJ, December 1995.

Bethea, R.M., and Rhinehart, R.R. *Applied Engineering Statistics*, Marcel Dekker, New York, 1991.

Cooper, R.G. New products—What separates the winners from the losers and what drives success, in K.B. Kahn (Ed.), and G. Catellion and A. Griffin (Assoc. Eds.), *The PDMA Handbook of New Product Development*, 2nd edition, pp. 3–28, John Wiley & Sons, Hoboken, NJ, 2005.

Cooper, R.G. *Winning at New Products, Accelerating the Process from Idea to Launch*, 3rd edition, Basic Books, New York, 2001.

Cooper, R.G., Edgett, S.J., and Kleinschmiddt, E.J. *Portfolio Management for New Products*, 2nd edition, Basic Books, New York, 2001.

Coppola, A. Reliability engineering of electronic equipment: A historical perspective, *IEEE Transactions on Reliability*, Vol. R-33, No. 1, pp. 29–35, 1984.

Condra, L.W. *Reliability Improvement with Design of Experiments*, 2nd edition, Marcel Dekker, New York, 2001.

Cox, D.R. Some simple approximate tests for poisson variates, *Biometrika*, Vol. 40, pp. 354–360, 1953.

Creveling, C.M. *Tolerance Design: A Handbook for Developing Optimal Specifications*, Addison-Wesley, Reading, MA, 1997.

De Feo, J.A. Quality planning: Designing innovative products and services, in J.M. Juran, and J.A. De Feo (Eds.), *Juran's Quality Handbook: The Complete Guide to Performance Excellence*, 6th edition, pp. 83–136, McGraw Hill, New York, 2010.

de Vasconcelos, V., da Silva, E.M.P., da Costa, A.C.L., and dos Reis, S.C. Safety, reliability, risk management and human factors: An integrated engineering approach applied to nuclear facilities, presented at 2009 International Nuclear Atlantic Conference (INAC 2009), Rio de Janeiro, Brazil, September 27 to October 2, 2009.

Del Vecchio, R.J. *Understanding Design of Experiments*, Hanser/Gardner Publishers, New York, 1997.

Dodson, B., and Schwab, H. *Accelerated Testing: A Practitioner's Guide to Accelerated and Reliability Testing*, SAE International, Warrendale, PA, 2006.

Duane, J.T. Learning curve approach to reliability monitoring, *IEEE Transactions on Aerospace*, Vol. 2, pp. 563–566, 1964.

Dudley, B. *Electronica Derating for Optimum Performance*, Reliability Information Analysis Center, Utica, NY, 1999.

Edson, L. Applying physics of failure and reliability statistics to understand the needed design margins—Example, Larry Edson Consulting Inc. Force Technology Workshop Notes, Denmark, August 30, 2018.

Epstein, B., and Sobel, M. Sequential life tests in the exponential case, *Annals of Mathematical Statistics*, Vol. 26, No. 1, pp. 82–93, 1955.

Forsberg, K., and Mooz, H. System engineering for faster, cheaper, better, Center for Systems Management, Inc., reprinted by SF Bay Area Chapter of INCOSE, 1998.

Glasser, G. Planned replacement: Some theory and its applications, *Journal of Quality Technology*, Vol. 1, No. 2, pp. 110–119, 1969.

Griffin, A. Obtaining customer needs for product development, in K.B. Kahn (Ed.), and G. Catellion and A. Griffin (Assoc. Eds.), *The PDMA Handbook of New Product Development*, 2nd edition, pp. 211–227, John Wiley & Sons, Hoboken, NJ, 2005.

Hatley, D., Hruschka, P., and Pirbhai, I. *Process for System Architecture and Requirements Engineering*, Dorset House Publishing, New York, 2000.

Hicks, C.R. *Fundamental Concepts in the Design of Experiments*, 2nd edition, Holt, Rinehart and Winston, New York, 1973.

Hnatek, E.R. *Practical Reliability of Electronic Equipment and Products*, Marcel Dekker, New York, 2003.

Hobbs, G. Reflections on HALT and HASS, *Engineering Evaluation*, December 2005, retrieved on July 12, 2019, from, http://evaluationengineering.com/archive/articles/1205/1205reflections.asp.

Hooks, I.F., and Farry, K.A. *Customer Centered Products: Creating Successful Products through Smart Requirements Management*, AMACOM, New York, 2001.

IEC 60601-1, 3rd edition, Medical Electrical Equipment—Part 1: General Requirements for Basic Safety and Essential Performance, Geneva, Switzerland, 2012.

IEC 60812. Analysis Techniques for System Reliability—Procedure for Failure Mode and Effects Analysis (FMEA), National Standards Authority of Ireland, Dublin, Ireland, 2006.

IEC 61164. Reliability Growth—Statistical Test and Estimation Methods, 2nd edition, International Electrotechnical Commission, March 2004.

IEEE Std. 1624. IEEE Standard for Organizational Reliability Capability, IEEE Reliability Society, New York, 2008.

ISO/IEC 15288. Systems and Software Engineering; System Life Cycle Processes, International Organization for Standardization, 2015.

Jamnia, A. *Design of Electromechanical Products: A Systems Approach*, CRC Press, Boca Raton, FL, 2017.

Jamnia, A. *Practical Guide to the Packaging of Electronics: Thermal and Mechanical Design and Analysis*, 3rd edition, CRC Press, Boca Raton, FL, 2016.

Jamnia, A. *Introduction to Product Design and Development for Engineers*, CRC Press, Boca Raton, FL, 2018.

Jeon, M.S., Song, J.K., Lee, E.J., Kim, Y.N., Shin, H.G., and Lee, H.S. Failure analysis of thermally shocked NiCr films on Mn-Ni-Co spinel oxide substrates, *Materials Research Society Symposium Proceedings*, Vol. 875, retrieved on November 28, 2007, from www.mrs.org/s_mrs/bin.asp?CID=2740&DID=149393&DOC=FILE.PDF.

Johnson, N.L., and Kotz, S. *Distributions in Statistics: Continuous Univariate Distributions*, Wiley, New York, 1972.

Kececioglu, D. *Reliability and Life Testing Handbook*, Vol. 2, revised edition, DEStech Publications, Lancaster, PA, 2002.

Klinger, D., Nakada, Y., and Menendez, M. *AT&T Reliability Manual*, Van Nostrand Reinhold Company, New York, 1990.

Lawless, J.F. *Statistical Models and Methods for Lifetime Data*, John Wiley & Sons, New York, 1982.

Lee, P.M. *Bayesian Statistics*, John Wiley & Sons, New York, 2012.

Leemis, L. M. *Reliability: Probabilistic Models and Statistical Methods*, Prentice-Hall, Englewood Cliffs, NJ, 1995.

Leveson, N. G. *Engineering a Safer World: Systems Thinking Applied to Safety*, MIT Press, Cambridge, MA, 2011. Online version available at https://app.knovel.com/hotlink/toc/id:kpESWSTAS6/engineering-safer-world/engineering-safer-world.

Lipson, C., and Sheth, N.J. *Statistical Design and Analysis of Engineering Experiments*, McGraw Hill, New York, 1973.

Liu, C. A comparison between Weibull and lognormal models used to analyze reliability data, Doctoral thesis, University of Nottingham, UK, August 1997.

Martin, J.N. *System Engineering Guidebook: A Process for Developing Systems and Products*, CRC Press, Boca Raton, FL 1997.

Mathews, P.G. *Design of Experiments with MINITAB*, ASQ Quality Press, Milwaukee, WI, 2005.

McLeish, J.G. Transitioning to physics of failure reliability assessments for electronics, DfR Solutions, retrieved on May 15, 2018, from www.dfrsolutions.com/hubfs/Resources/sherlock/Transitioning-to-PoF-Reliability-Assessments-for-Electronics.pdf?t=1514473946162.

McLinn, J. A short history of reliability, retrieved on September 18, 2018, from https://kscddms.ksc.nasa.gov/Reliability/Documents/History_of_Reliability.pdf.

MIL-HDBK-189. Reliability Growth Management, US Department of Defense, February 13, 1981.

MIL-HDBK-217. Reliability Prediction for Electronic Systems, December 2, 1991. Available from the National Technical Information Service, Springfield, Virginia.

MIL-HDBK-217F. Reliability Prediction of Electronic Equipment, US Department of Defense, 1991.

MIL-HDBK-217F. Reliability Prediction of Electronic Equipment, US Department of Defense, 1991. Notice 1 (1992) and Notice 2 (1995).

MIL-HDBK-338B. Electronic Reliability Design Handbook Equipment, US Department of Defense, 1988.

MIL-HDBK-338B. Electronic Reliability Design Handbook, US Department of Defense, October 1, 1998.

MIL-HDBK-781A. Reliability Test Methods, Plans, and Environments for Engineering Development, Qualification, and Production. US Department of Defense, April 1, 1996.

MIL-HDBK-189. Reliability Growth Management, US Department of Defense, February 19, 1981.

Modarres, M., Kaminsky, M.P., and Krivtsov, V. *Reliability Engineering and Risk Analysis: A Practical Guide*, 3rd edition, CRC Press, Boca Raton, FL, 2017.

Montgomery, D.C. *Introduction to Statistical Quality Control*, 7th edition. John Wiley & Sons, Hoboken, NJ, 2013.

Moubray, J. *Reliability-Centered Maintenance*, 4th edition, Industrial Press Inc., New York, 1992.

Morris, G. DFSS for thermal management, part IX: DFSS for thermal management: regression – part 1, retrieved on July 7, 2004, from https://www.coolingzone.com/Guest/News/NL_JUN_2004/Garron/Garron_June_04.html.

Morris, S., Dudley, B., Caroli, J., MacDiarmid, P., Nicholls , D., Coppola , A., Criscimagna , N., and Farrell , J. *Reliability Toolkit: Commercial Practices Edition*, Rome Laboratory/ Reliability Analysis Center, Rome, NY, 1993.

Myers, R.H., and Montgomery, D.C. *Response Surface Methodology: Process and Product Optimization Using Designed Experiments*, 2nd edition, John Wiley & Sons, New York, 2001.

NASA Systems Engineering Handbook, NASA/SP-2007-6105 Rev 1, 2007, retrieved on September 22, 2016, from http://ntrs.nasa.gov/archive/nasa/casi.ntrs.nasa.gov/20080008301.pdf.

Nelson, W. *Applied Life Data Analysis*, John Wiley & Sons, New York, 1982.

Nelson, W.B. *Accelerated Testing, Statistical Models, Test Plans, and Data Analysis*, John Wiley & Sons, UK, 2004.

Nelson, W.B. *Recurrent Events Data Analysis for Product Repairs, Disease Recurrences, and Other Applications*, ASA-SIAM Series on Statistics and Applied Probability, Alexandria, VA, 2002.

NIST/SEMATECH. Cumulative distribution function of the standard normal distribution, in *NIST/ SEMATECH e-Handbook of Statistical Methods*, retrieved on May 5, 2012, from www.itl.nist.gov/ div898/handbook/eda/section3/eda3671.htm.

NIST/SEMATECH. *e-Handbook of Statistical Methods, Notes on Bayesian Theorem and Its Applications*, retrieved on Decembr 8, 2016, from www.itl.nist.gov/div898/handbook/.

NPRD-91. Nonelectronic Parts Reliability Data. USAF Rome Air Development Center. The National Technical Information Service. Springfield, Virginia, 1991.

O'Connor, P.D.T., and Kleyner, A. *Practical Reliability Engineering*, 5th edition, John Wiley & Sons, Chichester, UK, 2012.

Ogrodnik, P.J. *Medical Device Design, Innovation from Concept to Market*, Elsevier, Oxford, UK, 2013.

Pahl, G., Beitz, W., Feldhusen, J., and Grote, K.H. *Engineering Design: A Systematic Approach*, Springer, Berlin-Heidelberg, Germany, 2007.

Pancake, M.H. Human factor engineering considerations in new product development, in K.B. Kahn (Ed.)., and G. Catellion and A. Griffin (Assoc. Eds.), *The PDMA Handbook of New Product Development*, 2nd edition, pp. 406–416, John Wiley & Sons, Hoboken, NJ, 2005.

Papoulis, A. *Probability, Random Variables, and Stochastic Process*, McGraw Hill, New York, 1984.

Peck, D.S. Comprehensive model for humidity testing correlation, presented at 24th Annual Proceedings, Reliability Physics 1986, Anaheim, CA, April 1–3, 1986.

Pugh, S. *Total Design: Integrated Methods for Successful Product Engineering*, Addison-Wesley, Wokingham, UK, 1991.

ReVelle, J.B. *Quality Essentials: A Reference Guide from A to Z*, ASQ Quality Press, Milwaukee, WI, 2004.

Ross, R.G., and Wen, L.-C. Solder creep-fatigue interactions with flexible leaded surface mount components, in *Thermal Stress and Strain in Microelectronics Packaging*, J.H. Lau (Ed.), Van Nostrand Reinhold, New York, 1993.

Sadlon, R.J. *Mechanical Applications in Reliability Engineering*, Reliability Analysis Center, Utica, NY, 1993.

Saleh, J.H., and Marais, K. Highlights from the early (and pre-) history of reliability engineering, *Reliability Engineering and System Safety*, Vol. 91, pp. 249–256, 2006.

Sloan, J. *Design and Packaging of Electronic Equipment*, Van Nostrand Reinhold, New York, 1985.

Stamatis, D.H. *Failure Mode Effect Analysis: FMEA from Theory to Execution*, 2nd edition, ASQ Quality Press, Milwaukee, WI, 2003.

Steinberg, D. *Cooling Techniques for Electronic Equipment*, 2nd edition, John Wiley & Sons, New York, 1991.

Steinberg, D. *Vibration Analysis for Electronic Equipment*, 2nd edition, John Wiley & Sons, New York, 1988.

Stockhoff, B.A. Research and development: More innovation, scarce resources, in J.M. Juran and J.A. De Feo (Eds.), *Juran's Quality Handbook, The Complete Guide to Performance Excellence*, 6th edition, pp. 891–950, McGraw Hill, New York, 2010.

Taguchi, G., and Clausing, D. Robust quality, *Harvard Business Review*, Vol. 68, No. 1, pp. 65–75, January/February 1990.

Tarum, C.D. Determination of the critical correlation coefficient to establish a good fit for Weibull and log-normal failure distribution, SAE Paper 1999-01–057, Detroit, March 1999.

Tasooji, A., Ghaffarian, R, and Rinaldi, A. Design parameters influencing reliability of CCGA assembly: A sensitivity analysis, retrieved on November 23, 2007, from http://trs-new.jpl.nasa.gov/ dspace/bitstream/2014/39705/1/06-0593.pdf.

Thaduri, A. Physics-of-failure based performance modeling of critical electronic components, Doctoral thesis, Luleå University of Technology, Luleå, Sweden, 2013.

Viswanadham, P., and Singh, P. *Failure Modes and Mechanisms in Electronic Packages*, Chapman & Hall, New York, 1988.

Ullman, D.G. *The Mechanical Design Process*, 4th edition, International Edition, McGraw-Hill, New York, 2010.

Wald, A. Sequential tests of statistical hypotheses, *Annual Mathematica Statistics*, Vol. 16, pp. 117–186, 1945.

Webber, L., and Wallace, M. *Quality Control for Dummies*, Wiley Publishing, Hoboken, NJ, 2015.

Weiss, S.I. *Product and System Development: A Value Approach*, John Wiley & Sons, Hoboken, NJ, 2013.

Wenham, M. Comparison of Weibull PARAMETER Calculation Methods, *GKN International Report No. 5647*, 1997.

White M., and Bernstein, J.B. *Microelectronics Reliability: Physics-of-Failure Based Modelling and Lifetime Evaluation*, JPL publication 08-5, Jet Propulsion Laboratory, California Institute of Technology, Pasadena, CA, 2008.

Williard, N., Baek, D., Park, J.W., Choi, B.O., Osterman, M., and Pecht, M. A life model for supercapacitors, *IEEE Transactions on Device and Materials Reliability*, Vol. 15, No. 4, pp. 519–528, 2015.

Xu, C., Fan, C., Vysotskaya, A., Abys, J.A., Zhang, Y., Hopkins, L., and Stevie, F. Understanding whisker phenomenon, part II: Competitive mechanisms, retrieved on November 23, 2007, from http://www.hkpc.org/hkiemat/mastec03_notes/11.pdf.

Zhang, Y., Xu, C., Fan, C., Vysotskaya A., and Abys, J. A. Understanding whisker phenomenon, part I: Growth rates, retrieved on November 23, 2007, from http://www.hkpc.org/hkiemat/mastec03_notes/10.pdf.

Index

Accelerated life testing (ALT), 3, 4, 22, 48, 83, 185, 189, 260–261, 277
  known stress–life relationships, 190
    Arrhenius model, 190–191
    Arrhenius–Peck model, 193–194
    Black's equation for electrical stress effects, 192–193
    example of how to calculate activation energy, 191–192
    Eyring model, 192
    inverse power model, 193
    Q_{10} equation for polymer aging, 193
  unknown stress–life relationships, 194–198
Acceleration factor (AFs), 163, 189, 191, 193, 198, 205, 211, 213, 214, 216, 261–262
Acceptance sampling test, 199
AFs, see Acceleration factor
Agile product development environment, 175
AGREE, see Agree Group on Reliability of Electronic Equipment
Agree Group on Reliability of Electronic Equipment (AGREE), 2, 3
ALT, see Accelerated life testing
AND gate, 319–320, 322, 334
Apparent frequency, 158
Applied loads, 7, 126, 148, 150
Apportionment, 23
Apportionment table, developing, 65–66
ARINC (Aeronautical Radio Inc.), 2
Arrhenius equation, 211, 213, 261
Arrhenius–Peck equation, 211, 216
Assembly design document, 37, 38–39
AT&T Weibull equation, 168
Availability, 54
Average availability, 54

Baseline reliability model, 83–84, 267, 268, 269
Bathtub curve, 53, 71, 112–114
Bayes formula, 201
Bayesian statistics, 270
Bayes–Lipson equation, 201
Bayes' theorem, 270–271
Bell curve, 90
Bill of materials (BOM), 47, 78, 165
Binomial distribution, 104–105
Blended reliability model, 84
BOM, see Bill of materials

Boolean operation, 333–334
Business plan, 76
Business plan document, 56

CAD modeling, see Computer-aided design modeling
CAPA, see Corrective action and preventive action
CFMEA, see Concept failure modes and effects analysis
Chemical failures, 149–150
Chi-square distribution, 97–98, 218
CL, see Confidence level
Coffin–Manson equation, 160
Cold temperature step stress test, 185
Computer-aided design (CAD) modeling, 41, 73
Concept failure modes and effects analysis (CFMEA), 308
Confidence level (CL), 199
Consumer's risk, 248
Continuous distributions and reliability analysis, 90
  exponential distribution model, 92–94
  gamma distribution, 94–95
    application to cumulative damage, 95
    chi-square distribution, 97–98
    failure-truncated test, 98
    pressure sensor first failure after partial physical damages (case study), 95–97
    time-truncated test, 98
  lognormal distribution, 99–100
  normal distribution model, 91–92
  selecting the right distribution for continuous variables, 102
    exponential distribution, 103
    lognormal distribution, 103–104
    Weibull distribution, 104
  Weibull distribution, 100–102
Continuous variables, 102
  exponential distribution, 103
  lognormal distribution, 103–104
  Weibull distribution, 104
Control factors, 225
Corrective action and preventive action (CAPA), 82, 296, 304–305
Corrosion, 150, 151
CoS, see Cost of service

Cost and return of reliability, 5–7
Cost of service (CoS), 56, 66
Creep–fatigue interactions, life expectancy for, 159–162
Critical correlation coefficient, 287
Crow–AAMSA NHPP model, 229, 233–236, 244
Cumulative damage
    application to, 95
    reliability testing, 210
        case study, 212–213
        humidity aging, 216–217
        random vibration, 217–218
        temperature aging, 213–215
        thermal fatigue, 215–216
        use and environmental profiles, damage caused by, 211–212
Cumulative-damage testing, 58
Cumulative distribution function (cdf), 88, 100

Databases, failure prediction using, 163
    electrical stress effects, 163–164
    environmental factors, 164
    system failure rate, calculating, 164–168
    temperature effects, 163
DDP, *see* Design and development plan
Defect rate, 55
Degree of freedom, 97
Derating, defined, 132
Design and development plan (DDP), 76–78
Design capability, 176
Design failure modes and effects analysis (DFMEA), 14, 18, 24, 47–48, 152, 154, 178, 308, 309, 310, 315, 317, 330
Design for assembly (DfA), 74, 82–83
Design for manufacturability (DfM), 74
Design for reliability (DfR), 2–4, 5, 8, 17, 27, 51, 265, 266
    deliverables, 84–85
    design life, 20
    failure rate, 20
    life expectancy, 20
    plan, 83
        baseline reliability model, 83
        design documents, 83
        final reliability model, 84
        highly accelerated stress screening (HASS), 84
        reliability growth, evaluation, and demonstration testing, 83–84
    product risk, reliability and its association with, 24–25
    product use profile, 18
    reliability allocation, 23

reliability data analysis, 22–23
reliability modeling, 20–21
reliability requirements, planning, and execution in the design process, 18
reliability testing, 21–22
Design for serviceability (DfS), 82–83
Design history file (DHF), 34–35
Design life, 20
Design margins, 126, 176
Design of experiment (DoE), 141, 179, 302
    case study, 142–143
    DoE test design, 143–144
    test runs and output signal strength, 144–147
Design process and V-model, 27
    design documents, 35
        assembly design document, 38–39
        subsystem architecture document, 38
        subsystem design document, 38
    product development models, 27
        engineering activities, 31–32
        life cycle management, V-model for, 29–31
    product requirements development, 39
        defining the product to be developed, 39–40
        measuring what customers need, 40–41
        product requirements document (PRD), 42–43
        quality function deployment (QFD), 43–47
        translation from needs to requirements, 42
    reliability planning and execution in the V-model, 47–49
    V-model in a nutshell, 47
    voice of the customer (VoC) and voice of stakeholders (VoS), 34–35
Destruct limit, 181
Detailed design documents, 81
Device failure rate prediction, 162
    failure prediction using databases, 163
        electrical stress effects, 163–164
        environmental factors, 164
        system failure rate, calculating, 164–168
        temperature effects, 163
    failure rates based on physics of failure, 168
        case study, 169–172
Device master record document, 73
DfA, *see* Design for assembly
DfM, *see* Design for manufacturability
DFMEA, *see* Design failure modes and effects analysis
DfR, *see* Design for reliability
DfS, *see* Design for serviceability

DHF, *see* Design history file
Discrete distributions and their applications to reliability, 104
   binomial distribution, 104–105
   expected service calls (case study), 111
   geometric distribution, 106
   hypergeometric distribution, 106–108
   new product launch (case study), 109–111
   Poisson distribution, 108–109
DMADV (define, measure, analyze, design, and verify), 39
DMAIC (define, measure, analyze, improve, and control) process, 85, 296–303
DoE, *see* Design of experiment
Duane model, 229–233

ECM, *see* Engine control module
Electromagnetic compatibility (EMC), 46
Electromagnetic emission interference (EMI), 46, 154, 310
Electromechanical interface document, 31
Electromechanical requirements, 31
Electronic equipment, 2
Electrostatic discharge (ESD), 46, 154, 310
Embedded software, 244
EMC, *see* Electromagnetic compatibility
EMI, *see* Electromagnetic emission interference
Engine control module (ECM), 210
Environmental stress screening (ESS), 185; *see also* Production reliability stress screening
Equal allocation, method of, 63
ESD, *see* Electrostatic discharge
ESL, *see* Expected service life
ESS, *see* Environmental stress screening
Exclusive OR gate, 322
Expected service life (ESL), 259
Experimental design, *see* Design of experiments
Exponential distribution model, 20, 92–94, 103
Eyring model, 192

Failure in time (FIT), 89
Failure mode and effects analysis (FMEA), 14, 23, 82, 150, 151–156, 307–308, 311, 317–318, 326
Failure modes and effects criticality analysis (FMECA), 329, 330
   and criticality index, 330–333
Failure rate, 20–21, 53, 60, 89, 98, 162–163, 165, 270
   based on physics of failure, 168
   of a network of components in parallel, 60
   of a redundant system, 60
   of three-element system, 60

Failure truncated category, 98
Failure-truncated test, 98
Fault tree analysis (FTA), 23, 318, 330, 333
   Boolean operation, 333–334
   case study, 319–320
   fault tree diagram symbols, 320–322
   FTA P_1 calculations, 334–336
   process, 322–323
FCA, *see* Field corrective action
FDA, *see* Food and Drug Administration
FEA, *see* Finite element analysis
Field corrective action (FCA), 205
Field failure investigation process, 295
   analyze, 299
      failure analysis and root cause determination, 299–300
      root causes, 300–301
   control, 303
      effectiveness and control, 303
   corrective action and preventive action (CAPA) core team, 304–305
   define, 296
      failure isolation and scope determination, 297–298
      failure observation and determination, 296–297
   improve, 302
      corrective actions and verification, 302–303
   keys to successful investigations, 303–304
   measure, 298
      failure verification, 298–299
Field-related sources of noise, 282–283
   manufacturing quality variations, 290–293
   mismatch of field data and selected reliability model, 285–288
   mixed-failure mechanisms, 293–295
   reliability model and field data misalignment, 283–285
   use profile discrepancies, 288–290
Final reliability model, 84, 270
   Bayes' theorem, 270–271
   reliability model case study, 271–273
Finite element analysis (FEA), 129
Fires burning, 42
FIT, *see* Failure in time
Five-gate Stage-Gate® model, 27
FMEA, *see* Failure mode and effects analysis
FMECA, *see* Failure modes and effects criticality analysis
Food and Drug Administration (FDA), 37, 82
Frequency, defined, 157
FTA, *see* Fault tree analysis

Gamma distribution, 94–95, 270
  chi-square distribution, 97–98
  cumulative damage, application to, 95
  failure-truncated test, 98
  pressure sensor first failure after partial
      physical damages (case study), 95–97
  time-truncated test, 98
Gate symbols, 322
Gaussian distribution, 80–81, 158
Geometric distribution, 106
Grouped data reliability growth model, 236–244

Hall-effect sensor, 129
HALT, *see* Highly accelerated limit testing
Harm, defined, 14, 307, 311
HASS, *see* Highly accelerated stress screening
Hazard, defined, 311
Hazardous situation, 14
  defined, 311
Hazard rate, 89
Highly accelerated limit testing (HALT), 3, 4,
      18, 21–22, 47, 48, 82, 83, 179, 180, 185,
      186–188, 269, 275
Highly accelerated stress screening (HASS), 22,
      84, 266, 267
History of reliability, 2–4
HoQ, *see* House of quality
Hot temperature step stress test, 185
House of quality (HoQ), 44, 45
Humidity aging, 216–217
Hypergeometric distribution, 106–108

IBM, 41, 42
Inherent strength/capabilities of the product, 7
Inhibit gate, 322
Instantaneous failure rate, 233, 237, 242
Integrated system, hardware and software
      reliability growth of, 244–245
Interface documents, 73
Irreversible failures, 148–149

Labeling, 73
LCD, *see* Liquid crystal display
Lead–acid battery remaining capacity (case
      study), 220–222
Lead-free soldering process, 149
Learning Station, 79, 152
Life cycle management, V-model for, 29–31
Life cycle of a product, defined, 27
Life expectancy, 20
  calculations, 156
      for creep–fatigue interactions, 159–162
      design life, reliability, and failure rate, 162

  for pure creep conditions, 159
  for pure fatigue conditions, 156–157
  for random vibration conditions, 158–159
Life metric, 54
Limited sample availability, 209–210
Liquid crystal display (LCD), 181
Lognormal distribution, 20, 99–100, 103–104

Malfunction, 181
Manufacturing defects, accounting for, 255
Manufacturing quality variations, 290–293
MAUDE (Manufacturer and User Device
      Experience), 82
Maximum likelihood estimation (MLE)
      method, 233, 237, 239, 288
MCBF, *see* Mean cycle between failures
MCF, *see* Mean cumulative function
Mean cumulative function (MCF), 54–55,
      259–260
Mean cycle between failures (MCBF), 245
Mean time between failure (MTBF), 4, 20, 52–53,
      76, 90, 94, 97, 98, 112, 166–167, 227,
      242–243, 245, 252–254, 269, 283
Mean time between service (MTBS), 4
Mean time between swaps (MTBS), 4
Mean times between failures (MTBFs), 304
Mean time to failure (MTTF), 4, 53, 89, 94, 196,
      202, 208
Mean time to repair (MTTR), 4, 51, 53, 90
Measurement system analysis (MSA), 317
Migration, 151
MIL-HDBK-217, 168
Miner's index, 156, 171
Mixed-failure mechanisms, 293–295
MLE method, *see* Maximum likelihood
      estimation method
Monte Carlo analysis, 4, 137–139, 286–287
MSA, *see* Measurement system analysis
MTBF, *see* Mean time between failure
MTBS, *see* Mean time between service;
      Mean time between swaps
MTTF, *see* Mean time to failure
MTTR, *see* Mean time to repair

New product development (NPD), 322
Noise factors, 176, 224
Normal distribution model, 91–92
NPD, *see* New product development

OBF, *see* Out-of-box failure
OEM, *see* Original equipment manufacturer
Off-the-shelf/custom-designed components, 181
Operating conditions requirement, 58

Organizational DfR maturity and reliability
    competency, 8
  cross-functional process, 9–13
OR gate, 322
Original equipment manufacturer (OEM), 5, 8,
    196, 248, 279
Out-of-box failure (OBF), 51, 55, 255

Parameter diagram (P-diagram), 177–178, 224
  automation of functions in the test, 227
  developing measurement baselines, 227
  keeping records, 225–227
  testing data generation and outputs, 227
  testing parameter settings, 227
Parts count formulation of reliability, 64
"Parts count" method, 164
Parts database, 24
Part-to-part variations, 141
PCB, *see* Printed circuit board
PCBAs, *see* Printed circuit board assemblies
P-diagram, *see* Parameter diagram
PdM, *see* Predictive maintenance
PFMEA, *see* Process failure modes and effects
    analysis
Physics of failure (PoF), 20, 147, 175
  failure classifications, 147
    chemical failures, 149–150
    failure modes and mechanisms, 150–151
    irreversible failures, 148–149
    progressive failures, 149
    reversible failures, 148
    sudden failures, 149
  failure rates based on, 168–172
Planning reliability testing of a system, 224
PM, *see* Preventive maintenance
PoF, *see* Physics of failure
Poisson distribution, 108–109
POS, *see* Proof of screen
PRD, *see* Product requirements definition;
    Product requirements document
Predictive and analytical tools in design, 117
  design of experiments (DoE), 141
    case study, 142–143
    DoE test design, 143–144
    test runs and output signal strength,
      144–147
  device failure rate prediction, 162
    failure rates based on physics of failure,
      168–172
    using databases, 163–168
  failure modes and effects analysis (FMEA),
    151–156
  life-expectancy calculations, 156

for creep–fatigue interactions, 159–162
design life, reliability, and failure rate, 162
for pure creep conditions, 159
for pure fatigue conditions, 156–157
for random vibration conditions, 158–159
physics of failure (PoF), 147
  chemical failures, 149–150
  failure modes and mechanisms, 150–151
  irreversible failures, 148–149
  progressive failures, 149
  reversible failures, 148
  sudden failures, 149
safety factor or design margin, 126
  engineering analysis and numerical
    simulation, 129–130
  statistical approach, 126–129
  stress derating, 130–132
stress versus strength, 117
  cascading the use profile into component
    specifications, 118–120
  interaction of strength and stress
    distributions, 123–125
  uncertainty in strength and stress, 120–123
tolerance analysis, 132
  functional tolerance concerns, 141
  Monte Carlo analysis, 137–139
  root sum of squares (RSS) method,
    135–137
  unintended consequences, 139–141
  worst-case analysis (WCA), 133–135
Predictive maintenance (PdM), 280
Pressure sensor first failure after partial
    physical damages (case study), 95–97
Preventive maintenance (PM), 6
  of engineering equipment, 275–280
Principle of superposition, 211
Printed circuit board (PCB), 150
Printed circuit board assemblies (PCBAs), 21,
    78, 81, 148, 181, 186, 285
Priority AND gate, 322
Probability density function, 88
Probability distribution function, 128
Probability of failure, 89, 289
Probability ratio sequential testing (PRST), 247
  case study, 248–250
  χ^2 approach, 250–252
Process failure modes and effects analysis
    (PFMEA), 14, 152, 308, 317
Producer's risk, 248
Product development models, 27
  engineering activities, 31–32
  V-model for life cycle management, 29–31
Production reliability stress screening, 179

Production screening testing, 273–275
Product requirements definition (PRD), 35
Product requirements development, 39
    defining the product to be developed, 39–40
    measuring what customers need, 40–41
    product requirements document (PRD),
        42–43
    quality function deployment (QFD), 43–45
        requirements decomposition, 45–47
    translation from needs to requirements, 42
Product requirements document (PRD), 34, 37,
        42–43
Product risk, 13–14
    reliability and its association with, 24–25
Product risk management, 307
    fault tree analysis (FTA), 318
        case study, 319–320
        fault tree diagram symbols, 320–322
        process, 322–323
    risk management tool, 307
        controlling critical or safety-related
            items, 317
        failure modes and effects analysis, 308–311
        hazard, hazardous situation, and harm,
            311–316
Product surveillance team, 79
Product use profile, 18
Progressive failures, 149
Proof of screen (POS), 275
PRST, *see* Probability ratio sequential testing
Pure creep conditions, life expectancy for, 159
Pure fatigue conditions, life expectancy for,
        156–157

QFD, *see* Quality function deployment
Quality, definition of, 7
Quality function deployment (QFD), 43–45
    requirements decomposition, 45–47

RAC, *see* Reliability Analysis Center
RACT, *see* Risk assessment code table
Radio Corporation of America, 3
Random vibration, 217–218
Random vibration conditions, life expectancy
        for, 158–159
Rapid thermal cycling test, 185
Rayleigh model, 244
RBD, *see* Reliability block diagram
RDT, *see* Reliability demonstration testing
Redundant system, 60
Regression coefficient vector, 144
Reliability, defined, 1, 7
Reliability allocation, 23

Reliability Analysis Center (RAC), 330
Reliability at time, *see* Survivor distribution
        function
Reliability block diagram (RBD), 20, 59, 60, 63
Reliability-critical items, 177
Reliability data analysis, 22–23
Reliability demonstration testing (RDT),
        18, 22, 188
    accelerated life testing (ALT), 189
        known stress–life relationships, 190–194
        unknown stress–life relationships,
            194–198
    reliability test duration and sample size,
        198–199
        cumulative damage reliability testing,
            210–218
        limited sample availability, 209–210
        single-use or nonrepairable components,
            199–200
        success-run reliability test duration and
            sample size, 201–209
    test duration with anticipated failures, 218
        test design with anticipated failures
            (example), 218–219
Reliability engineering, 2–3
    defined, 23
Reliability feasibility analysis, 177
Reliability growth index, 230, 242
Reliability growth testing (RGT), 22, 179,
        229, 247
Reliability growth tracking model–discrete
        (RGTMD), 236
Reliability maturity matrix, organization
        design for, 10–12
Reliability metrics, 4, 112–114
Reliability modeling, 20–21, 59, 61–63
Reliability outputs, 265
    baseline model and its application, 267
    design-related analyses, 269
    final reliability model, 270
        Bayes' theorem, 270–271
        reliability model case study, 271–273
    predictive maintenance (PdM), 280
    preventive maintenance (PM) of engineering
        equipment, 275–280
    production screening testing, 273–275
    reliability model case study, 267–268
    reliability testing output, 269
        reliability model case study, 269–270
Reliability planning, 51
Reliability plan process, design for, 73, 76
    baseline reliability model, 83
    business plan, 76

design and development plan (DDP), 76–78
design documents, 83
design for reliability deliverables, 84–85
final reliability model, 84
highly accelerated stress screening
    (HASS), 84
needed inputs, 78
    design for assembly (DfA) and design for
        serviceability (DfS), 82–83
    detailed design documents, 81
    failure modes and effects analysis
        documents, 82
    functional requirements document and
        system architecture, 78–80
    test articles, 82
    use profile document, 80–81
reliability growth, evaluation, and
    demonstration testing, 83–84
Reliability–quality–robustness relationship,
    7–8
Reliability requirement allocation, 63
apportionment table, developing, 65–66
case study, 66
    legacy and field reliability data analysis,
        67–69
    reliability gap analysis, 69–70
    reliability planning strategy, 70–71
    reliability requirements feasibility and gap
        analysis, 66
Reliability requirements, 51
based on use profile, 57
    product mission, 57
    product use distribution, 58
    use and operating conditions, 58
based on voice of business, 56–57
based on voice of customer, 57
development, 52, 71–72
    availability, 54
    failure rate, 53
    mean cumulative function (MCF), 54–55
    mean time between failures (MTBF),
        52–53
    mean time to failure (MTTF), 53
    probability of success rate and out-of-box
        failures, 55
    reliability and life, 54
    reliability metrics, 52
reliability modeling, 59–63
reliability requirement allocation, 63
    developing the apportionment table,
        65–66
    legacy and field reliability data analysis
        (case study), 67–69

reliability gap analysis (case study), 69–70
reliability planning strategy (case study),
    70–71
reliability requirements feasibility and
    gap analysis, 66
Reliability standards and guidelines, 339–340
Reliability Stress Analysis for Electronic
    Equipment, 3
Reliability testing, 21–22, 175
degradation testing, 219–220
    lead–acid battery remaining capacity
        (case study), 220–222
highly accelerated limit testing (HALT),
    180, 186
    and realistic failures, 186–188
robustness versus, 175–177
types of, 178
    production reliability stress
        screening, 179
    reliability demonstration test (RDT), 179
    reliability design margin development
        and characterization, 179
what to test and how to test, 177–178
Repairable systems reliability over service life,
    259–260
Research and development (R&D)
    department, 73
Response surface, 141
Restriction of hazardous substances (RoHS),
    148, 149
Return on investment (ROI), 5
Reversible failures, 148
RGT, *see* Reliability growth testing
RGTMD, *see* Reliability growth tracking
    model–discrete
Risk, 325–326
defined, 307
fault tree analysis, 333
    Boolean operation, 333–334
    FTA P_1 calculations, 334–336
relating reliability to, 326
    failure modes and effects criticality
        analysis (FMECA) and criticality
        index, 330–333
    reliability concerns leading to
        hazards, 329
    risk analysis, 327–328
Risk analysis and risk management, 13
Risk assessment code table (RACT), 315
Risk management tool, 307
controlling critical or safety-related
    items, 317
failure modes and effects analysis, 308–311

hazard, hazardous situation, and harm, 311
    risk assessment code table, 313–316
Risk priority number (RPN), 24, 313
Robinson's index, 159
Robustness, 7
    defined, 176
    versus reliability testing, 175–177
RoHS, *see* Restriction of hazardous substances
ROI, *see* Return on investment
Room-temperature vulcanizing (RTV)
    silicone, 188
Root sum of squares (RSS) method, 135–137
    sensitivity analysis, 136
RPN, *see* Risk priority number
RSS method, *see* Root sum of squares method
RTV silicone, *see* Room-temperature
    vulcanizing silicone

Safety factor or design margin, 126
    engineering analysis and numerical
        simulation, 129
        what-if scenarios, 129–130
    statistical approach, 126–129
    stress derating, 130–132
Sample size calculations for service life system
    demonstration testing, 259
Service life system demonstration testing,
    sample size calculations for, 259
Service replaceable units (SRUs), 285
Severity, defined, 311
Single-use or nonrepairable components, 199–200
Six Sigma tool, 39
SMART requirements, 52
Snap-fit feature, 122, 123–124
Software reliability, 3
SoS, *see* System of systems
SPC, *see* Statistical process control
SRUs, *see* Service replaceable units
Standard normal distribution, 128
Standards on general design practices, 339
Statistical analysis techniques, 87
    basic definitions, 88–90
    continuous distributions and reliability
        analysis, 90
        exponential distribution model,
            92–94, 103
        gamma distribution, 94–99
        lognormal distribution, 99–100, 103–104
        normal distribution model, 91–92
        Weibull distribution, 100–102, 104
    discrete distributions and their applications
        to reliability, 104
        binomial distribution, 104–105

expected service calls (case study), 111
geometric distribution, 106
hypergeometric distribution, 106–108
new product launch (case study), 109–111
Poisson distribution, 108–109
reliability metrics and the bathtub curve,
    112–114
Statistical process control (SPC), 303
Strengths, weaknesses, opportunities, and
    threats (SWOT) analysis, 40
Stress derating, 130–132
Stress versus strength, 117
    cascading the use profile into component
        specifications, 118–120
    interaction of strength and stress
        distributions, 123–125
    uncertainty in strength and stress, 120–123
Subassembly, 37
Subsystem architecture document, 38
Subsystem design document, 38
Success rate, probability of, 55
Success-run reliability testing
    duration and sample size, 201
        design reliability improvement
            demonstration (case study), 206–208
        failure distribution model and test
            duration (case study), 208–209
        field corrective action testing for urgent
            launch (case study), 205–206
        success-run reliability testing, risk of, 203
    risk of, 203
Sudden failures, 149
Survivor distribution function (sdf), 89
Sustaining product reliability, 281
    field failure investigation process, 295
        analyze, 299–301
        control, 303
        corrective action and preventive action
            (CAPA) core team, 304–305
        define, 296–298
        improve, 302–303
        keys to successful investigations, 303–304
        measure, 298–299
    field-related sources of noise, 282–283
        manufacturing quality variations, 290–293
        mismatch of field data and selected
            reliability model, 285–288
        mixed-failure mechanisms, 293–295
        reliability model and field data
            misalignment, 283–285
        use profile discrepancies, 288–290
    warranty and service plan data review,
        305–306

SWOT analysis, *see* Strengths, weaknesses, opportunities, and threats analysis
System, defined, 33
System accelerated life testing, 260–263
System failure rate, calculating, 164
    case study, 165–168
    component hazard rate, 164
    system hazard rate, 165
System of systems (SoS), 33
System reliability testing, 223
    Duane model, 229–231
        Crow–AMSAA NHPP model, 233–236
        Duane model example, 231–233
    failure modes and effects analysis, 228–229
    grouped data reliability growth model, 236–244
    hardware and software reliability growth of integrated system, 244–245
    manufacturing defects, accounting for, 255
    parameter diagram (P-diagram), 224
        automation of functions in the test, 227
        developing measurement baselines, 227
        keeping records, 225–227
        testing data generation and outputs, 227
        testing parameter settings, 227
    probability ratio sequential testing (PRST), 247
        case study, 248–250
        χ^2 approach, 250–252
    repairable systems reliability over service life, 259–260
    sample size and test duration, 255–256
    sample size calculations for service life system demonstration testing, 259
    system accelerated life testing (ALT), 260–263
    system demonstration test plan, 256–258
    time-truncated MTBF demonstration testing, 252–254

Tarum's analysis, 288
Temperature aging, 213–215
Test articles, 82
Test–find–fix–test, 233
Testing the corners, 179
Thermal cycling, 159
Thermal fatigue, 215–216
Time truncated category, 98
Time-truncated MTBF demonstration testing, 247, 252–254
Time-truncated test, 98
Tolerance analysis, 132
    functional tolerance concerns, 141
    Monte Carlo analysis, 137–139

root sum of squares (RSS) method, 135–137
    unintended consequences, 139–141
    worst-case analysis (WCA), 133
        electrical circuits, 134–135
TR-1100, 3
Transfer function, 141, 144

Unintended consequences, 139–141
Unit under test (UUT), 182, 183, 186, 227
US Department of Defense, 2
Use and environmental profiles, damage caused by, 211–212
Use profile
    discrepancies, 288–290
    document, 79, 80–81
    reliability requirements based on, 57
        product mission, 57
        product use distribution, 58
        use and operating conditions, 58
UUT, *see* Unit under test

Vacuum tube, 2
Vibration step stress test, 185
V-model, 47
    for life cycle management, 29–31
    reliability planning and execution in, 47–49
VoC, *see* Voice of customers
Voice of business, reliability requirements based on, 56–57
Voice of customers (VoC), 32, 34–35, 40, 55
    reliability requirements based on, 57
Voice of stakeholders (VoS), 34–35, 37
VoS, *see* Voice of stakeholders
Voting gate, 322

Warranty and service plan data review, 305–306
Waste from electrical and electronic equipment (WEEE), 149
WCA, *see* Worst-case analysis
WEEE, *see* Waste from electrical and electronic equipment
Wei–Bayes equation, 201
Weibull chart, 286
Weibull distribution, 20, 100, 103, 104, 288, 291
    modified Weibull distribution, 101–102
    shape factor, 198, 236, 238, 289
What-if scenarios, 129–130
Whiskering, 151
Worst-case analysis (WCA), 133, 137
    electrical circuits, 134–135

Z score, 91

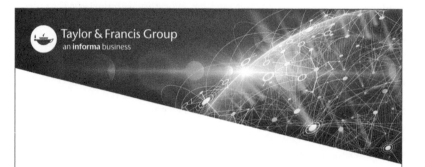